D0640650

HDTV

Advanced Television for the 1990s

HDTV

Advanced Television for the 1990s

K. Blair Benson
Donald G. Fink

Intertext Publications
McGraw-Hill Publishing Company, Inc.

New York St. Louis San Francisco Auckland Bogotá
Hamburg London Madrid Mexico Milan Montreal
New Delhi Panama Paris São Paolo
Singapore Sydney Tokyo Toronto

Shortly after completing his work on *HDTV: Advanced Television for the 1990s*, Blair Benson died. This book is dedicated to his memory.

Library of Congress Catalog Number 90-81045

10 9 8 7 6 5 4 3 2 1

ISBN 0-07-020983-9

Intertext Publications/Multiscience Press, Inc.
One Lincoln Plaza
New York, NY 10023

McGraw-Hill Publishing Company, Inc.
1221 Avenue of the Americas
New York, NY 10020

Composed in Ventura Publisher by Context Publishing Services, San Diego, CA

Contents

Foreword

The clock is ticking on the introduction and widespread use of high-definition television in consumer, industrial, and military markets. Japanese manufacturers of consumer electronics products are preparing to ship HDTV video cassette recorders paired with HDTV receivers to the United States. Cassette duplicators in Japan and the United States are equipping to meet the limited initial demand for VCR program material. The hardware prices will be high, but not excessive for the videophile.

The ramifications of HDTV have expanded beyond technology and marketing into the realm of politics. The political and legislative questions raised by the push for "high-def" promise to be far more difficult to resolve than the technical problems. From Capitol Hill to the Pentagon, HDTV has become the lightning rod for marshalling concern over the future of "good old" American leadership. Fears abound regarding the negative effects that a lack of domestic HDTV expertise could have on the American electronics industry in general, and military electronics in particular.

The technical, economic, and political implications of HDTV have also caused a frenzy of activity in standardization bodies in the United States and around the world. To paraphrase Winston Churchill: "Never have so many talked about so much for so long." Serious interest dates back to 1974 when the Japanese Broadcasting Company (NHK) gave its first public demonstrations of high-definition television. By 1980 interest gained momentum in the United States, and shortly thereafter in Europe. The Europeans' primary motive was to narrow Japan's technological lead.

The international standardization disputes that have accompanied the technical developments are by no means a simple matter of

North America and Japan versus Europe. Of the 500-million monochrome and color receivers now in use worldwide, roughly half would be affected by any new frame rate. Thus, 625-line countries are opting for a 1250/50 standard, rather than the Japanese/American 1125/60.

It is important to recognize the distinction between terrestrial transmission of HDTV and the production of programs. The film and videotape program production industries were among the first to use HDTV technology, albeit to a very limited degree. Their use of the medium will continue to expand independently of the regulatory decisions concerning transmission of HDTV programs. It may be the programs themselves that play the dominant role in the ultimate acceptance of a common worldwide defacto HDTV production standard. The bulk of mass-appeal progamming already available, or currently being shot, comes from the United States. Original material in 1125/60 HDTV is converted to NTSC or PAL and SECAM for distribution. Adoption of a universal HDTV standard using a 60-Hz frame rate is a logical simplification for distribution. An additional benefit gained would be an increase of 5:1 in permissible picture brightness for an equivalent perception of inter-field flicker.

Among the applications considered for HDTV has been the use of video techniques directly for movie production, bypassing the use of film in part or completely. The claims are enhanced production quality and less time for postproduction editing resulting, in turn, in lower costs. In practice, the motion picture industry is in no hurry to abandon a medium that has served it well for the better part of this century and is continuing to improve in quality and flexibility. Film is unquestionably the universal production standard. Nevertheless, HDTV will make inroads in editing and special effects, but the advantages of HDTV for production remains to be proven.

The most immediate applications for HDTV in other than the entertaiment industry are in non-broadcast fields such as medical diagnosis, industrial communications and training, and military imaging. The military has, in fact, been using HDTV systems for some time.

High-definition television has made great strides in the relatively short time it has been under development. However, more work involving primarily the resolution of political and economic questions, rather than technical, remains. This effort is extremely important in its long-range effect on both consumer and professional electronics. It is not without precedence. A similar controversey raged over color television. The stakes were high and the posturing ran rampant. Technology gave way to politics only temporarily. Through it all the television industry survived and prospered. As with color, HDTV will arrive when television viewers are ready to accept it.

Jerry Whitaker
Associate Publisher,
Broadcast Engineering
Video Systems

Preface

On December 17, 1953 the Federal Communications Commission approved color television broadcasting standards, compatible with the existing 525-line monochrome service. The signal system was based upon specifications developed by the National Television Systems Committee (NTSC), a voluntary group of technical experts from research laboratories, manufacturers, and broadcasters. Although limited color broadcasting service to the public in the United States began on January 23, 1953, it was not until 1964 that a significant number of households were equipped with color receivers and a substantial number of network program hours were produced in color. Color broadcasting in Europe, compatible with their 625-line PAL (*P*hase *A*lternation *L*ine-rate) and SECAM (*S*equential *C*ouleur *a*vec *M*emoire) systems, was delayed until the mid-1960s.

In the ensuing years, using NTSC, PAL, and SECAM systems, broadcasters have been able to provide a highly successful color television service to home viewers within the bandwidth of existing VHF and UHF channels originally intended for monochrome signal transmission. At the normal viewing distance of direct-view screens in general use, the maximum resolution capability of 525- and 625-line systems has been adequate. Any increase in resolution would not be discernible and, consequently not justified. Furthermore, the compatibility requirement and the bandwidth restriction of available broadcast channels place limitations on the maximum resolution and image format that can be transmitted using conventional signal encoding and modulation.

However, the trend to larger direct-view screens and improvements in projection systems result in a greater perception of fine detail by viewers and a demand for an increase in resolution. In addition, the greater sense of presence created by a larger screen can be increased further by a wider display similar to that used for present-day motion picture theater presentations.

Historically, in 1968 the Japan Broadcasting Corporation (NHK) initiated a research program to develop a noncompatible high-definition system requiring a wider transmission bandwidth than that available for 525-line terrestrial broadcasting. The work by NHK Laboratories triggered the development and manufacture of HDTV equipment and the inauguration of a satellite HDTV broadcasting service in Japan. Regular HDTV one-hour-a-day program transmissions via satellite using the 1125/60 system, started by NHK in June 1989, are planned to be increased to eight hours a day in 1990.

In the United States, in lieu of FCC Rules and Regulations for HDTV broadcasting, limited program production activity is under way using the 1125/60 system with the recorded videotapes down-converted to 525-line standards for the broadcast and VCR markets. Concurrently, several other systems, both compatible and non-compatible with NTSC, have been proposed as standards for broadcasting.

The proliferation of system proposals has raised questions among the engineers, businessmen, and the general public regarding the current status and future introduction of high-definition television as a home entertainment medium. The choice of a system will have a major impact on broadcasters' capital and operating costs, and on program production and distribution. On the other hand, for the manufacturing industry, HDTV promises to provide a market for solid-state components and integrated circuits that may equal or exceed that of the personal computer.

Meanwhile, standardization efforts in the United States by engineering societies and industry have prompted the Federal Communications Commission to announce that an HDTV broadcast standard will be selected in 1993 after completion of tests and a study by the Advanced Television Systems Committee (ATSC). Thus, hardly a day passes without a news item in the daily press referring to technical developments or legislative actions relative to HDTV.

In the first two chapters *HDTV: Advanced Television for the 1990s* provides an overview of developments which have led to the present status. The following three chapters are devoted to the fundamentals underlying changes from conventional video technology to high-definition imaging: fuller realization of the demands of natural vision, digital processing of video signals, and the spatio-temporal analysis of signal content that permits the insertion of additional information into conventional channels.

Following are chapters on the implementation of HDTV service, including the impact of noncompatibility and the requirements for compatibility. One chapter is devoted entirely to NHK 1125/60 Hi-Vision, the only system in use at the time of publication. This is followed by two chapters on single and multi-channel allocations, their implementation in the several distribution media, and in studios, transmitters, and receivers.

A chapter on picture signal generation by photoconductive tubes and solid-state sensors and signal processing is followed by coverage of HDTV signal distribution and transmission in studios and the all-important requirements for conversion to HDTV capability, including the need and means for use of fiber optics for transmission.

Another chapter is devoted to display technology, not only for home viewing but in the many other entertainment and business applications.

Motion picture production using film and videotape and the means for duplication of the latter on film by electron-beam recording are dealt with in Chapter 13.

The authors have been well aware of the rapid pace at which every aspect of the HDTV profession and industry have moved during the writing of this book. Chapter 14, written just prior to release of the manuscript to the publisher, after a brief review of HDTV system development, provides a last-minute summary of events and progress on the technical standardization and legislative fronts and the requirements for implementation of simulcast broadcasting systems by stations, cable systems, and networks. Lastly, new markets for HDTV technology are suggested.

K. Blair Benson
Donald G. Fink

Chapter

1

High-Definition Television—Its Status and Prospects

1.1 HDTV Defined

High-definition television is defined by the image it presents to the viewer: its detail, its aspect ratio, and the distance at which it is to be viewed.

1.1.1 Detail

The HDTV image has approximately twice as much luminance definition horizontally and vertically as do the 525-line NTSC system and 625-line PAL and SECAM systems (the "conventional systems"). The total number of luminance picture elements (pixels) in the image is therefore four times as great, and the wider screen adds one quarter more.

The increased vertical definition is achieved by employing more than 1000 lines in the scanning patterns; values currently in use or proposed are 1050, 1125, and 1250 lines. The increased luminance detail in the image is achieved by employing a video bandwidth approximately five times that used in the conventional systems. Additional bandwidth is used to transmit the color values (chrominance) separately, so the total bandwidth is from six to eight times that used in the existing color television services.

1.1.2 Aspect Ratio

The HDTV image is 25 percent wider than the conventional image. The ratio of the image's width to its height in existing and most proposed HDTV systems is 16/9 = 1.777, whereas the conventional standard is 4/3 = 1.333. The change follows the trend of motion pictures from a ratio of 4/3 prior to 1953 to a wide-screen format. The widths of the majority of motion-picture screens are now 85 percent wider than they are high, and some are 135 percent wider.

1.1.3 Viewing Distance

Since the eye's ability to resolve detail is limited, the more detailed HDTV image should be viewed more closely than is customary in the conventional systems in order to realize the benefits of the higher definition image.

Full visual resolution of the detail of conventional television is available when it is viewed at a distance equal to about six times the height of the display. The HDTV image should be viewed from a distance of about three times the picture height for the full detail to be resolved. If viewed at six times the height of the picture, the extra cost of the HDTV receiver is wasted so far as pictorial detail is concerned.

It follows that the HDTV image must be larger than the conventional one. When viewed at six feet, the HDTV image must be not less than two feet high, and its diagonal is then not less than four feet, larger than any conventional cathode-ray tube display. Displays capable of presenting such large images represent a major fraction of the HDTV receiver cost.

1.2 The Market for HDTV

An important impetus supporting the development of HDTV is the maturity of the market for existing services. Color television service was first authorized in 1954, became an active industry in the 1960s, and has grown in 30 years to be the primary component of the consumer electronics industry. The 1989 statistics show one television set in use for every nine people on earth and that 54 percent of the 605,000,000 sets in use were color receivers. In most of the developed nations there is hardly a household that is not served; in the United States in 1989 there were 655 sets in use for every 1000 people, an average of nearly two (actually 1.8) for every home served by television broadcast or cable service. Table 1.1 gives country-by-country details of this penetration of the consumer television market.

Extrapolation of these figures to estimate the market for HDTV receivers is difficult. In the first place, there is scant evidence that a set owner would be willing to buy an expensive HDTV receiver (estimated to start at about $2000) to obtain performance superior to what he now enjoys. There is, in fact, some evidence[1] to the contrary gathered at MIT. Second, there is considerable concern that the additional channel space required for HDTV signals cannot be found in the populous areas of the terrestrial broadcast service. This limitation would not, however, severely constrain the cable television service, which now serves more than half the homes in the United States. Third, steady improvements in the quality of the images provided by the conventional services have set a high standard against which HDTV service must compete.

Despite these uncertainties, the potential market for HDTV components and receivers is very strong. One area having great potential is the market for semiconductor products. The development of HDTV receiver technology could not have proceeded without the availability of large-scale, inexpensive integrated circuits.

If only two percent of the 324,000,000 color television receivers in use in 1989 were replaced annually by HDTV receivers, their number would exceed the 3.5 million annual production of personal computers for the home market. If a 10 percent annual sales rate for HDTV receivers were achieved, their use of integrated circuits might well represent the major market of the semiconductor industry. This possibility has not been ignored by those concerned with such markets. A consortium of the largest semiconductor manufacturers in the United States was formed in 1988 to plan for just such an eventuality.

1.3 HDTV and the Movie Industry

Motion pictures have offered both the inspiration for and the first application of HDTV technology. The wide screen initially was adopted by movie producers to offer a broader scope of action. But it was soon found that theater patrons seated in the center front seats

Table 1.1 Worldwide Television Markets (populations of 10 million or more—1989 estimates)

Geographic area[a]	Population in millions	TV receivers in use (millions)[b]			Receivers per thousand inhabitants	Television systems[c]
		Color	Mono	Total		
North America						
Canada	25.3	8.8	2.9	11.7	462	NTSC
United States	247.5	100.0	62.0	162.0	655	NTSC
Mexico	88.1	6.9	8.8	15.7	178	NTSC
Central America	23.5	0.57	0.54	1.11	47	NTSC
Caribbean						
Cuba	10.6		0.7	0.7	66	NTSC
Others	20.9	1.24	3.84	5.08	243	NTSC SECAM
South America						
Argentina	32.6	0.5	5.5	6.0	184	PAL
Brazil	154.0	14.0	10.0	24.0	156	PAL
Chile	12.9	0.67	1.67	2.34	181	NTSC
Columbia	31.8	0.5	2.5	3.0	94	NTSC
Ecuador	10.0	1.0	1.0	2.0	200	NTSC
Peru	21.8		2.4	2.4	110	NTSC
Venezuela	19.2	0.8	1.85	2.65	146	NTSC
Others	15.2	0.29	0.77	1.06	74	PAL NTSC SECAM
Western Europe						
Belgium	9.9	2.23	0.75	2.98	301	PAL
France	55.8	20.24	6.33	26.57	476	SECAM
West Germany	60.2	20.52	4.81	25.33	421	PAL
Greece	10.0	0.4	1.4	1.8	180	SECAM
Italy	54.7	7.82	6.71	14.53	266	PAL
Netherlands	14.7		4.73	4.73	322	PAL
Portugal	10.2	0.22	1.41	1.63	160	PAL
Scandinavia	22.6	7.81	1.50	9.31	412	PAL
Spain	39.0	7.2	5.4	12.6	323	PAL
United Kingdom	56.8	16.00	2.68	18.68	329	PAL
Others	17.1	5.09	0.84	5.93	347	PAL
Eastern Europe						
Bulgaria	9.0	0.25	1.65	1.90	211	SECAM
Czechoslovakia	15.7	0.87	3.49	4.36	278	SECAM

(Continued)

Table 1.1 ***(Continued)*** **Worldwide Television Markets (populations of 10 million or more—1989 estimates)**

Geographic area[a]	Population in millions	TV receivers in use (millions)[b]			Receivers per thousand inhabitants	Television systems[c]
		Color	Mono	Total		
East Germany	16.7	2.23	3.90	6.13	367	SECAM
Hungary	10.6	0.50	2.50	3.00	283	SECAM
Poland	38.4		9.89	9.89	256	SECAM
Romania	23.2		3.91	3.91	169	SECAM
U.S.S.R.	287.0	29.20	58.10	97.30	339	SECAM
Yugoslavia	23.8	.50	3.62	4.12	173	PAL
Asia						
Bangladesh	112.8	0.07	0.30	0.37	3	PAL
Burma	39.9	0.06	0.01	0.07	2	NTSC
China, Mainland	1070.0	1.01	8.17	9.18	9	PAL
China, Taiwan	20.3	5.00	0.78	5.78	285	NTSC
India	833.4	1.80	7.90	9.70	12	PAL
Indonesia	187.7	1.69	4.81	6.50	35	PAL
Japan	123.2	30.77	1.69	32.46	263	NTSC
Korea, South	45.2	4.36	2.95	7.31	162	NTSC
Malaysia	16.9	1.23		1.23	73	PAL
Nepal	18.8	0.20	0.01	0.21	11	PAL
Pakistan	110.4		1.51	1.51	14	PAL
Sri Lanka	17.5	0.19	0.31	0.50	29	PAL
Thailand	55.0	1.84	2.29	4.13	75	PAL
Vietnam	66.7		0.50	0.50	75	MONO
Africa						
Algeria	25.1		1.140	1.140	46	PAL
Egypt	54.8	0.350	3.850	4.200	77	SECAM
Ghana	14.8	0.035	0.135	0.170	11	PAL
Kenya	23.7		3.833	3.833	162	PAL
Madagascar	11.1	0.010	0.040	0.050	5	SECAM
Morocco	25.4	0.132	0.989	1.121	44	SECAM
Mozambique	15.3	0.004	0.001	0.005	<1	PAL
Nigeria	115.1	0.800	3.400	4.200	37	PAL
South Africa	35.6		2.000	2.000	56	PAL
Sudan	25.0		0.090	0.090	4	MONO
Tanzania	24.7	0.015	0.003	0.018	<1	PAL
Uganda	16.8		0.070	0.070	4	PAL
Zaire	34.0	2.000	1.000	3.000	88	SECAM

(Continued)

Table 1.1 *(Continued)* **Worldwide Television Markets (populations of 10 million or more—1989 estimates)**

Geographic area[a]	Population in millions	TV receivers in use (millions)[b]			Receivers per thousand inhabitants	Television systems[c]
		Color	Mono	Total		
Others	80.1	0.572	1.687	2.259	28	PAL SECAM MONO
Mideast States						
Iran	51.0		2.000	2.000	39	SECAM
Iraq	17.6		0.500	0.500	28	SECAM
Saudi Arabia	12.7	1.500	0.050	1.550	122	SECAM PAL
Syria	12.2	0.400	1.000	1.400	115	SECAM
Turkey	55.4	3.681	5.857	9.538	173	PAL
Others	19.2	2.200	1.750	3.950	206	PAL SECAM MONO
Other Countries						
Australia	16.1	5.011	0.098	5.109	317	PAL
Philippines	62.0	0.488	3.510	3.998	64	NTSC
New Zealand	3.4	0.863	0.072	0.935	275	PAL

[a] Caribbean, others: El Salvador, Jamaica, Dominican Republic, Antigua, Bahamas, Antilles, Virgin Islands (all NTSC), Haiti (SECAM).
Central America: Costa Rica, Guatemala, Honduras, Nicaragua, Panama (all NTSC).
South America, others: Bolivia (NTSC), French Guinea (SECAM), Paraguay (PAL), Surinam (NTSC), Uruguay (PAL).
Western Europe: Scandinavia—Denmark, Finland, Norway, Sweden (all PAL). Others — Austria, Ireland, Switzerland (all PAL).
Mideast States, others: Jordan (PAL), Israel (PAL), Lebanon (SECAM), Oman (PAL), Yemen (MONO).
Africa, others: Bahrain (PAL), Benin (SECAM), Central African Republic (SECAM), Congo (MONO), Djibouti (SECAM), Gabon (SECAM), Guinea (SECAM), Liberia (PAL), Libya (SECAM), Niger (SECAM), Qatar (PAL), Seychelles (PAL), Senegal (SECAM), Sierra Leone (PAL), Swaziland (PAL), Togo (SECAM), Tunisia (PAL/SECAM), Zambia (PAL), Zimbabwe (PAL).

[b] *Source:* Television and Cable Factbook No. 57, Warren Publishing Co., Washington, DC, 1989.

[c] *Source:* Montreux Television Symposium Committee (Reference 5).

enjoyed a sense of presence and of participation in the performance not offered by the narrow screen. Psychophysicists attributed this to the fact that the wide screen then occupied a substantially larger part of the visual field of view. This understanding has carried over to the development of HDTV, in which a wider screen, viewed closely, has been a principal design objective.

The application of television to motion picture production has, until the advent of HDTV, been limited to the use of conventional systems. A television camera and videotape recorder have been used, alongside the film camera, to permit prompt review of the "takes" prior to film development.

Beginning in the early 1980s, film producers were offered a high-definition television system capable of producing images having the full detail of 35-mm film. The scenes could be recorded on videotape, reviewed at once, then edited and otherwise put into final form, and finally transferred to film for exhibition in theaters. Since the intermediate delays for film processing were eliminated, much time was saved, and estimates of production cost savings as high as 35 percent were claimed. There are, however, many limitations to the use of HDTV in the film industry. A review of its status has been prepared by Kline.[2]

The HDTV system used for film production, including its cameras, monitors, recorders, editing consoles, and tape-to-film transfer devices, was developed in Japan beginning in 1978. The leading contributors were the Japan Broadcasting Corporation (NHK) and Sony, and the system became known as NHK Hi-Vision. The system is described in detail in Chapter 7. In 1989 it was the only HDTV system in regular use albeit experimental. The specifications of 1125 lines, 60 fields per second, 2:1 interlacing, 20-MHz luminance bandwidth, and 10-MHz chrominance bandwidth, were proposed as a standard for television studio production, but the Committee Consultive International Radio (CCIR), the international body responsible for such standards, has deferred action until 1991 pending study of alternative proposals.

1.4 The Prospects for HDTV Broadcasting

Until 1989 there was no HDTV service to the public. The first version of the NHK system known as MUSE (see Section 2.7) began operation in Japan in late 1989. MUSE is a direct-broadcast-from-satellite (DBS) system that employs the 1125-line image at its input, but converts it to the narrower channel required by the satellite transponders. This conversion retains the full detail of the 1125-line image, but only when the scene is stationary. When motion occurs, the definition is reduced by about 50 percent. If a moving object appears against a stationary background, this loss is often not too evident. But if the whole scene is in motion, as when the camera moves ("pans") to follow action, the whole image suffers the 50 percent loss of detail. Despite this loss, tests of the MUSE system have attracted a large following in Japan, and many owners of conventional receivers have purchased the converter necessary to adapt the MUSE transmissions to them, well in advance of the scheduled inauguration of the service.

The inauguration of HDTV service in other nations is not expected to start until 1992 at the earliest. Meanwhile, dozens of proposals have been made for systems that could provide HDTV service or advanced versions of the conventional services. The major proposals are treated in Chapters 8 and 9.

The planning for HDTV service to the viewing public has been complicated by the availability of four principal methods of distributing programs: terrestrial ("over-the-air") broadcasting, cable service by coaxial or fiber-optic cables, direct broadcasting from satellites, and video recorders using magnetic tape or optical disks. Of these, the oldest and most widespread medium, terrestrial broadcasting, faces the greatest difficulty in finding the additional channel space required by the HDTV service. Sections 2.9 through 2.13 are devoted to the channel-space requirements of the four basic media and the limits faced by the terrestrial service.

1.5 HDTV Channel Requirements

The requirement (Section 1.1) for a video bandwidth six to eight times as wide as the conventional system poses the central problem of HDTV design: how to obtain the requisite channel space in each of the basic program distribution media. In the NTSC system, for example, the video baseband is 4.2 MHz wide; six times this is 25.2 MHz, eight times is 33.6 MHz. Taking these figures at face value, without applying any technique of bandwidth reduction, we find that additional channel space of from 21.0 to 29.4 MHz would be required, equal to the space occupied by four to five additional 6-MHz channels.

Meeting such a demand would, even if possible in practice, be extremely wasteful of spectrum resources, regardless of the medium. So means have been sought to reduce the channel space requirement to, at best, no extra channel or, at worst, one extra channel. Such systems have become known as "single-channel" and "dual-channel," respectively; examples are described in Chapters 8 and 9.

Such compression of channels involves a number of compromises. In one case, the trade-off is between higher definition and precise rendition of moving objects. On the one hand, it is feasible to defer the transmission of a portion of the detail, spreading its signal over a longer time and thus reducing the bandwidth it occupies. But if motion is present in the scene over this longer interval, the deferred detail does not occupy its proper place, and smearing, ragged edges, and even gross distortion of shape can occur.

The opposite compromise is taken when additional detail is forced into a fixed time of transmission. This was the compromise, very successful as it turned out, adopted in the development of compatible color television. The time of transmitting one frame in monochrome television is by no means fully occupied with signal information, there being more unoccupied time than that occupied. By arranging the additional chromatic information needed to transform the image from monochrome to color so that it fills the unoccupied frequencies, the monochrome signal remains essentially unchanged and continues to be available for monochrome receivers. Receivers designed for color accept both sets of information and combine them to produce the color display.

This process is not free of error. Great difficulty is experienced in color receivers in keeping the two sets of information separate until they are ready for the process that combines them. A major objective of all advanced and high-definition system development is to preserve this signal separation, even if it requires additional bandwidth to do so.

The color television channel, after the addition of chromatic to monochromatic information, is still not fully occupied. It is often possible to add additional information to the conventional channel (Sections 2.11 and 5.7) without gross contamination of the color service to which the channel is assigned. This is the primary approach to the single-channel system, although there are many other unoccupied times (between scanning lines and fields) that are prospective recipients of additional information.

In the dual-channel approach to HDTV system design, all or part of an additional channel is required to accept additional information, received only by the HDTV receiver, to increase the definition and the width of its display. There are several variations in this technique, depending on the functional relationship of the extra channel to the conventional channel. Thus arises the issue of compatibility—the extent to which the existing service is protected from the encroachments of the new.

1.6 The Compatibility Issue

Those designing advanced and HDTV systems must contend with the impact of their work on the existing services. The majority have elected to develop systems that, while offering a superior service to new receivers, continues to serve the existing audience. There are powerful economic and political reasons for *compatibility* between the new services and the old. The investment by the viewing audience in the 605 million receivers in use in 1989, at a conservative dollar value of $200 each, reached the imposing total of $121 billion. Moreover, this investment was spread widely among the inhabitants of most countries. Table 1.1 lists nearly 40 nations with populations of 10,000,000 or more where the number of receivers per thousand population ranged from 110 to 665 in 1989. It is not surpising that, without exception, all new television services, notably the National Television System Committee (NTSC), Phase Alteration Line (PAL), and Sequential avec Memoiré (SECAM) color systems, have been introduced in a manner that preserved the existing service. PAL and SECAM were, in fact, incompatible with the 405-line and 819-line monochrome services in use when they were introduced in 1967. So new channels had to be found for the color services, and the monochrome services continued uninterrupted until their audiences had dwindled to the point that it was politically feasible to discontinue them. The 405-line service in Great Britain had a life span of 50 years, 19 of which ran parallel to the PAL color service.

This history has weighed heavily on the deliberations of the Federal Communications Commission concerning standards for HDTV. In August 1987 that body issued a document[3] that, at least tentatively, established compatibility as a requirement for the system it would ultimately choose for the service. Independently, many system developers have concluded that a noncompatible service would meet with stiff resistance in the manufacturing industry and the marketplace. The result is that a substantial majority of the systems proposed to the Commission employ the NTSC field-scanning rate of 59.94 Hz, rather than the 60-Hz rate of the 1125-line NHK system. The difficulties of transposing from one of these frequencies to the other (Section 6.6) preclude their joint use in receivers; one or the other may be used, but not both (short of complete duplication of circuit functions).

The impact of compatibility on advanced and HDTV system design is so great and the technical issues so complex that a separate chapter (Chapter 6) is devoted to them.

1.7 Outlook for Terrestrial Broadcasting

Plans for terrestrial HDTV broadcasting are underway in Japan, the United States, and Europe. In Japan, the initial service will be offered by the NHK-MUSE system of direct broadcast from satellites (Section 2.7). In addition, satellite transmissions of the NHK Hi-Vision system (not the modified MUSE system) were planned for display in public places and elsewhere outside the home audience.

In the United States, jurisdiction over terrestrial broadcasting is exercised by the Federal Communications Commission (FCC). In 1987 the FCC issued a tentative ruling[3] to the effect that the channel assignments for the HDTV system to be standardized would be limited to the then existing VHF and UHF channels. The Commission stated that it would issue the standards for the HDTV service after testing of the many alternative proposals then before it, based on the recommendation of its Advisory Committee on Advanced Television Systems. An industry advisory organization, the United States Advanced Television System Committee (ATSC), was organized in 1985 and set up an industry-supported laboratory, the Advanced Television Test Center (ATTC), for testing systems and equipment. Field tests of proposed terrestrial HDTV broadcast systems, transmitted over two UHF stations near Washington, DC, began in December 1988 and are continuing.

The tentative plan of the FCC to confine channel allocations for terrestrial HDTV broadcasting to the VHF and UHF channels has raised doubts that dual-channel HDTV service can be accommodated. In the populous areas where market support is available, the existing allocations to VHF stations leave no room for extra channel space for HDTV. Attention has therefore been focused on the UHF spectrum. But even in this less populated region, protection of the existing service against interference from HDTV transmissions may not be possible in the larger market areas. Details of this problem are treated in Section 2.12.

In Europe, plans for HDTV terrestrial broadcasting have been undertaken by the European Economic Community, under the byword

Eureka. The basis is an extension[4] of the MAC system of satellite distribution, which would deliver an HDTV service to terrestrial stations. Direct broadcast of the HDTV service would also be be available to the home audience.

1.8 Outlook for Cable Television

The cable television industry in the United States enjoys two advantages in planning for HDTV—regulatory and technical. In 1984, Congress removed control of the industry's price structure from the localities granting its franchises. The cable proprietors thus have the freedom to set their rates for service at whatever levels the market will bear and to increase those rates substantially when the HDTV service is offered. Moreover, the FCC has no jurisdiction whatever over the standards and practices adopted by the cable service. Thus, the cable industry is free to adopt a system of HDTV that meets its needs, without reference to the standards adopted by the terrestrial broadcasters, who remain under FCC jurisdiction. Cable industry spokesmen have insisted that, if the constraints faced by the terrestrial broadcasters lead the FCC to adopt a "compromise standard" short of HDTV's full potential, the cable industry will not be bound by that standard. Thus, two standards for HDTV in the United States are a definite possibility. Much attention is being paid to the ways in which the two standards might differ, thus allowing a superior service to be offered by the cable systems, and the ways in which they should not differ, to avoid substantial increases in the cost of HDTV receivers.

Faced with this outlook, the cable industry has set up its own advisory committee and has jointly financed a test center, in parallel to the ATTC set up by the noncable forces. Cooperation exists between the two centers, and it appears that the technical and economic questions facing both camps will be thoroughly and jointly aired.

The technical advantage possessed by cable television resides in the fact that its systems are self-contained and shielded against the effects of outside interference. Moreover, the number of channels available is limited only by the capacity of the circuits and the associated equipment. The existing coaxial cable systems in the United States have access to the continuous frequency range from 54 to 650 MHz, within which 94 interference-free channels can be offered to those equipped to receive them. This compares with 68 interference-prone channels available in the American terrestrial broadcast spectrum.

It thus appears that if the dual-channel approach to HDTV proves necessary, cable television can readily encompass the spectrum requirements. Moreover, there is even more channel space available to cable as it adopts fiber-optic technology. Many cable systems are now transforming their distribution circuits (those carried on poles along streets) from coaxial to fiber, and experiments using fiber in the drop circuits to individual homes are under way. The ultimate plan is to use optical fiber throughout all systems whose traffic can bear the cost of its installation. One incidental question is whether the regional telephone companies, whose use of fiber optics is far advanced, will gain the right to offer television cable service to subscribers in their local circuit areas and, in time, to become full proprietors of the service, in direct competition with the cable industry. Bills to permit this are under discussion in the U.S. Congress. Whoever is the eventual proprietor, the availability of HDTV service through wire or fiber connections to the home is technically above question, and subject only to economic forces and legislative decisions.

1.9 Outlook for Direct Broadcast Satellite Service

The DBS service currently proposed for HDTV occupies frequencies in the 12- and 22-GHz bands, and the bandwidths of each channel have been set at 24 MHz (in the Americas) and 27 MHz (in Europe and Asia). Frequency modulation is used with approximately a 3-to-1 ratio between the baseband modulation and the channel width. The maximum video bandwidth thus falls between 8 and 9 MHz. The Japanese MUSE HDTV system (Section 2.7) operates with an 8.1-MHz video band, the European MAC system (Section 2.6) with 8.5 MHz.

These bandwidths are barely sufficient for a multichannel HDTV system, e.g., one using the 6-MHz NTSC channel plus an auxiliary channel of 2–3 MHz. If the full channel requirement for the NHK Hi-Vision system (up to 30 MHz) is to be accommodated, the signal must be compressed in frequency. In the MUSE DBS system this is accomplished by transmitting the signals representing every third picture element, thus delaying the rendition of the full detail for three times the normal period, and reducing the bandwidth by a factor of three. This delay causes a reduction in the rendition of detail whenever any part of the scene is in motion. Such a compromise between the definition of stationary and moving objects is inevitable in any HDTV system using time-delay bandwidth compression. The wide approval of the MUSE system by the Japanese during its field tests has shown that such compromises are acceptable.

Other DBS HDTV systems under development also employ some form of bandwidth compression. These include, in Europe, the Eureka EU-95 project, a 1250-line interlaced version[5] of the conventional MAC system. In the United States, the Philips HDS-NA (High-Definition System—North America) system offers a 1050-line, 59.94-field interlaced DBS service in which no time-delay of the detailed information is involved, so no deterioration of the display of moving objects occurs. Although all these DBS systems have, as their primary audiences, homes equipped with satellite-reception antennas, they will also serve to distribute HDTV service to the terrestrial broadcasters and cable systems for retransmission to their audiences.

1.10 Outlook for Videotapes and Disks

Videotape and disk recorders capable of covering the wide bandwidths of HDTV service are available for use in studios. They are complex and expensive devices, and their adaptation to domestic use can be expected to be delayed for many years, as was the case in tape recorders for the conventional systems. But activity in this field is not wholly dependent on the broadcast industry, since prerecorded tapes and disks are a major part of the market. Now that two-thirds of the homes enjoying television in the United States have tape recorders, the threat of an independent development of HDTV for that audience is of concern to the broadcasters. One such independent development is the "Super-VHS" tape recorder (Section 2.2) which combines component processing with improved tape to obtain 400 lines of horizontal resolution, compared with the upper limit of 330 lines in the NTSC broadcast system.

The first introduction of tape and disk recording into the HDTV field is expected to come with the MUSE system (Section 2.8). This is a variant of the NHK 1125-line HDTV system designed for the direct broadcast by satellite (DBS) service in Japan. Time-compression of the 30-MHz NHK signal permits the MUSE system to operate with a video bandwidth of only 8.1 MHz, well within the scope of domestic tape recording.

The use of MUSE signals on videotape and disks for domestic service is tempered by the fact that the MUSE HDTV format is incompatible with the conventional systems of terrestrial broadcast. By 1990, Japan will have its MUSE DBS system in operation and will be offering HDTV receivers to the Japanese market. Tape and disk recorders for the MUSE system are also planned for the domestic market. American and European markets will be targets of the Japanese exporters. Thus, MUSE receivers and recorders could be offered to these markets, without reference to the broadcast services. There is even the possibility that Japanese MUSE DBS transmissions could be aimed to cover the western regions of the American and Canadian audiences.

Whether any of these steps will actually be taken by the Japanese industry, only time will tell. But the mere possibility of such an independent introduction of HTDV service, using the MUSE technology, has aroused concern among the industries and governments of Europe and North America.

1.11 HDTV Production and Broadcast Plant Design and Operations

Essentially all the equipment in the broadcast plant, from cameras and lighting the studio to the transmitter and its antenna, must be reexamined, and in most cases replaced, to meet the wideband demands of HDTV production and distribution. This is true not only in all live-camera work, inside and outside the studio, but also in telecine operation, and in recording and editing operations. Cable systems must also be redesigned if and when they use dual-channel operation, since the two channels must possess precisely matched amplitude and time-delay characteristics. Ghost-causing reflected signals must also be kept far below the levels now acceptable in the cable service.

That these requirements can be met in the studio, in outside broadcast, and in recording and editing has been amply demonstrated by the use of the NHK Hi-Vision system in motion picture production. But these uses are comparatively straightforward compared with those imposed by the time constraints and scheduling requirements of the broadcast industry. In particular, the need for real-time switching and routing in broadcast operations has drawn attention to the need for standardization of the signal format to be used within studio operations.

This need was recognized by the International Radio Consultative Committee (CCIR) in the early 1980s. A Study Group of this body voted unanimously in 1985 to recommend, for studio operations, standards based on the NHK Hi-Vision specifications: 1125-line scanning, interlaced 2-to-1 at 60 fields per second, with specific characteristics for color primaries, white balance, and gamma correction. It was understood that these standards applied only to the internal operations of the broadcasters, who would convert them by digital means to whatever terrestrial HDTV broadcast standard applied in their respective service areas.

When this recommendation came before the CCIR in 1986, several member countries that employed the PAL and SECAM standards, neglecting the question of interfield flicker, pointed to the loss of quality associated with conversion from the 60-Hz proposed HDTV standard to the 50-Hz standard of their conventional systems. This objection caused the CCIR to take no action on the recommendation of its Study Group, voting to delay a decision until 1990, pending study of alternative HDTV production standards.

There was strong support of the Study Group recommendations in the United States. The Society of Motion Picture and Television Engineers (SMPTE) drew up a version[6] of the 1125-line standard in 1988 and proposed its adoption by the American National Standards Institute (ANSI). Others objected that the proposed standard would be difficult to convert to the NTSC system, particularly if the HDTV version of that system was to be compatible with the existing service. The National Broadcasting Company recommended that the HDTV studio standard be set at 1050 lines (525 × 2) and that the field rate of 59.94 Hz be retained, rather than the 60 Hz of the ANSI standard. The Philips Laboratories and the David Sarnoff Laboratories concurred and based their HDTV systems on the 1050-line, 59.94-field specifications. The SMPTE undertook to reconsider their action in 1989. The ultimate resolution of this argument on HDTV studio standardization in the United States is reported in Chapter 14.

One important aspect of the many approaches to standardization of HDTV studio operations is its impact on the manufacturers of equipment. One approach adopted in camera design is to employ removable integrated circuits within the camera assembly. By replacing the chips suitable for one standard by others, one such camera[7] can be adapted on any of six different standards: NTSC, PAL, SECAM (each in interlaced or progressive scan), and three HDTV scans at 1050, 1125, or 1250 lines, at field rates of 50, 59.94, or 60 Hz, 16/9 aspect ratio. One characteristic of HDTV cameras is their need for high illumination of the scene. This camera is rated at 1000 lux at a lens opening of *f*/2.8 and a low level of 100 lux at *f*/1.5. These figures compare with the 7 lux at *f*/2.1 available in conventional color cameras for home use that produce acceptable images, although far below broadcast quality. The multistandard camera, in early production, was reported to cost $400,000.

1.12 System Terminology

Television systems are identified in five principal categories:

Conventional systems The NTSC, PAL, and SECAM systems as standardized prior to the development of advanced systems.

Improved definition systems (IDTV) Conventional systems modified to offer improved vertical and/or horizontal definition (Sections 2.2 and 2.3).

Enhanced definition systems (EDTV) Same as IDTV.

Advanced systems In the broad sense, all systems other than conventional systems. In the narrow sense, all systems other than conventional and HDTV.

High-definition systems (HDTV) Systems having vertical and horizontal resolutions approximately twice those of the conventional systems.

Simulcast systems Transmission of conventional NTSC, PAL, or SECAM on existing channels and HDTV transmission of the same program on one or more additional channels.

Production systems Systems intended for use in the production of programs, but not necessarily in their distribution.

Distribution systems Terrestrial broadcast, cable, satellite, video cassette, and video disk methods of bringing programs to the viewing audience.

References

1. Neuman, R. MIT Media Laboratory Testimony before House Science Committee, United States Congress, March 1989. Reported in *TV Digest* 29: No. 23, p. 12 (March 27, 1989).
2. Kline, D. D. Can Hollywood and HDTV Be Friends? *IEEE Trans. Consumer Electronics* 34: 48–53 (February 1988).
3. Federal Communications Commission. In the Matter of Advanced Television Systems and Their Impact on the Existing Television Service. Docket No. 87-268, August 20, 1987. Tenta-

tive Decision and Further Notice of Inquiry. Docket 88-288, September 1, 1988.

4. Raven, J. G. High Definition MAC: The Compatible Route to HDTV. *IEEE Trans. Consumer Electronics* 34: 61–63 (February 1988).
5. Jurgen, R. K. High-Definition Television Update. *IEEE Spectrum* 25: 56–62 (1988).
6. SMPTE Standard 240M, Television-Signal Parameters-1125/60 High Definition Production System. Society of Motion Picture and Television Engineers, White Plains, NY, 1988.
7. Multistandard HDTV-Camera, Model KCH 1000, BTS Broadcast Television Systems GMBH, a joint company of Bosch and Philips, Darmstadt, Federal Republic of Germany, and Mahwah, NJ 07430.

Chapter

2

Techniques of Advanced Television Systems

2.1 Developments Leading to Advanced Television Systems

The development of advanced television service has involved many new techniques. Among them are:

Improved performance of conventional systems:

Reception: higher horizontal and vertical resolution, digital processing, large displays.

Transmission: new cameras, image enhancement, adaptive prefilter encoding, digital recording, and signal distribution.

Signal compression for direct satellite broadcasting.

Auxiliary signals within conventional channels.

Allocation and occupancy of augmentation channels.

Concurrently, an extensive study has been undertaken concerning the different characteristics of the major systems of program distribution: terrestrial broadcasting, cable distribution by wire or fiber optics, satellite broadcasting, magnetic and optical recorders. A major purpose of this study is to determine how the wide video basebands of HDTV can be accommodated in each system, and whether a single HDTV standard can embrace the needs of all of them.

This work has not only provided many of the prerequisites of HDTV but, by advancing the state of the conventional art, it has established a higher standard against which the HDTV industry must compete. The general principles of the techniques most directly affecting the development of HDTV are treated in this chapter. The specifics of their implementation appear in later chapters.

2.2 Improvement in Horizontal Resolution

The composite transmission of luminance and chrominance in a single channel is achieved in the NTSC and PAL systems by choosing the chrominance subcarrier to be an odd multiple of one half the line-scanning frequency. This causes the component frequencies of chrominance to be interleaved with those of luminance. The intent of this arrangement is to make it possible to separate the two sets of components in the receiver, thus avoiding interference between them prior to the recovery of the primary color signals for the display.

In practice this process has been fraught with so much difficulty that it has imposed a substantial limitation on the horizontal resolution available in receivers. Signal intermodulation arising in the bands occupied by the chrominance subcarrier signal have produced degradations in the image known as *cross-color* and *cross-luminance*. The first causes a display of false colors to be superimposed on repetitive patterns in the luminance image, while the second causes a crawling dot pattern, primarily visible around colored edges.

These effects have been sufficiently prominent that manufacturers of NTSC receivers have tended to limit the luminance bandwidth to less than 3 MHz below the 3.58-MHz subcarrier frequency and far short of the 4.2-MHz maximum of the broadcast signal. This causes the horizontal resolution in such receivers to be confined to about 250 lines, i.e., three-quarters of that offered by the broadcast signal. The filtering employed to remove chrominance from luminance is a simple *notch filter* tuned to the subcarrier frequency.

It is clear that an important improvement in the definition of conventional receivers would result if the signal mixture between luminance and chrominance could be substantially reduced, if not entirely eliminated. This has become possible with the advent of the *comb filter*.

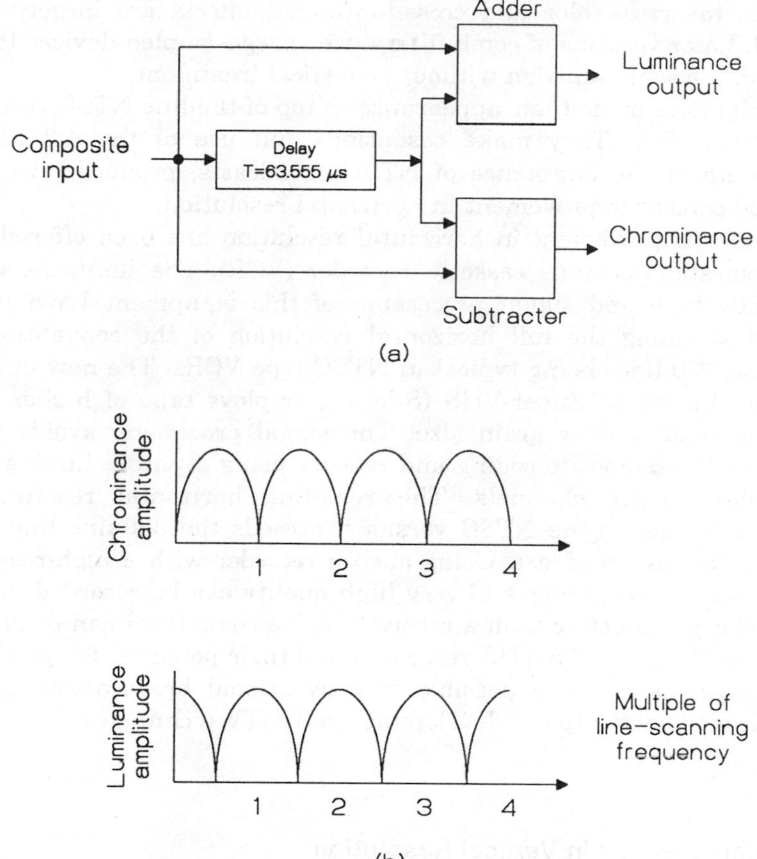

Figure 2.1 Comb filter introducing one line-scan delay.

A common version of the comb filter consists of a glass fiber connected between two transducers: video-to-acoustic and acoustic-to-video. The composite signal is fed to one transducer, producing an acoustic analog version, which reaches the far end of the fiber after an acoustic delay equal to one line-scan interval (64 μs for PAL and SECAM, 63.555 μs for NTSC). There the delayed electrical output version is recovered.

When the delayed output signal is added to the input, the sum (Figure 2.1) represents luminance nearly devoid of chrominance content. Conversely, when the delayed output is subtracted from the input, the sum represents chrominance similarly devoid of luminance. When these signals are used to recover the primary-color signals, the cross-color and cross-luminance effects are largely removed. Later versions of comb filters use charge-coupled devices that perform the same function without acoustical treatment.

Comb filters made their appearance in top-of-the-line NTSC receivers about 1981. They make essentially full use of the 4.2 MHz bandwidth of the luminance of NTSC broadcasts, producing better than 30 percent improvement in horizontal resolution.

Another improvement in horizontal resolution has been offered in the domestic videotape cassette recorder (VCR). The limitations of magnetic tape and signal processing of this equipment have prevented attaining the full horizontal resolution of the conventional systems, 250 lines being typical in NTSC-type VCRs. The new development, known as Super-VHS (S-VHS), employs tape of higher retentivity and smaller grain size. The signal processing avoids the limits of the composite coding and records using separate luminance and chrominance channels. The resulting horizontal resolution, about 400 lines in the NTSC versions, exceeds the 330-line limit of that system as broadcast. Using such a recorder with a high-resolution video camera, images of very high quality can be recorded, having a sharpness better than a receiver with a comb filter can display.

The appearance of S-VHS recorders and their potential for producing images better than possible in conventional broadcasting gave significant impetus to the development of HDTV techniques.

2.3 Improvement in Vertical Resolution

At a later date an increase in the vertical resolution of displays was offered by introducing progressive scanning in receivers designed for the NTSC, PAL, and SECAM services. In late 1988 an American subsidiary of a Dutch company introduced progressively scanned NTSC-type receivers[1] employing the elaborate digital processing required to convert the interlaced signal to progressive scanning and to provide other improvements in the reception process.

In the progressively scanned mode intended for the NTSC service there are 525 lines in each field versus the 265.5 lines per field as broadcast. An extra "blank" line is present in the display between each pair of "active" lines. The blank line may be filled in by interpolation of the video information from the active lines above and below it, plus that from the corresponding line of the previous field.

To perform this task, the composite input signal is analog-to-digitally converted to 7-bit words at 13.5 MHz. After further processing, the video signal from one field is stored digitally in a field-store containing capacity for 2,560,000 bits (ten 64K × 4 RAM chips). One field later, bits representing each picture element in the then-present lines preceding and following the blank line are retrieved from line-stores while the congruent line from the previous field is read out from the field-store. These bit streams, brought into precise synchronism by delay circuits, are compared to select the digital form of the blank-line signal, picture element by picture element. This signal is then converted to analog form, representing the *Y*, *R-Y*, and *B-Y* content of the blank line during each line scan. These are matrixed to *R-G-B* in the conventional manner.

There are five optional methods of filling in the blank lines, three of which are available to the televiewer. In the first, the blank lines are left blank, i.e., only the active lines are visible. They constitute the standard interlaced scan, not often used by the viewer, but of interest in demonstrating the improvement afforded by progressive scanning by the dealer at the time of sale. In the second option, the content of the preceding line is simply repeated to fill the blank line. This mode is used when displaying freeze-frame images. In the third mode, used during normal operation, the median value (not average value) of each of three picture elements is selected, that is, from the preceding and following lines and from the congruent line of the previous field. The use of the median value minimizes the smearing of objects in motion that would otherwise occur due to time delay from one field to the next. It also preserves the full degree of the vertical resolution as transmitted, i.e., the inherent 40 percent improvement offered by progressive scan. Progressive scan also avoids other defects associated with interlaced scanning (Section 3.17).

Other possible methods of filling in the blank lines are not used. Averaging the content of the preceding and following lines is not used because it would reduce the vertical resolution. Merely inserting the content of the congruent line from the previous field is not used because of the smearing created in the display of objects or scenes in motion. These defects are avoided by selecting picture elements separately in the normal operating mode.

This receiver employs comb filtering to utilize the full luminance bandwidth of the system and uses an elaborate digital memory to reduce noise and to create a variety of picture-in-picture effects. Similar receivers have been introduced in Japan and Europe.

2.4 Display Dimensions

For the HDTV display[2,3] to be fully effective it must be wider and larger than that suitable for conventional service. At the typical viewing distance of six feet, and at the optimal viewing ratio for HDTV of three times the picture height, the corresponding picture height is 24 inches. Its width at the 16/9 aspect ratio is 43 inches, and the diagonal is 49 inches. Such an image can be produced only by a large and expensive display device, one that can be expected to represent the major item in the HDTV receiver cost.

Large, direct-view, cathode-ray type displays have been produced for the conventional services, but their maximum diagonal has been reached at approximately 40 inches. Larger tubes would require so large a cabinet that the receiver could not be maneuvered around corners during delivery to the home.

The alternative, the projection display, is more manageable.[4] Such displays have achieved high levels of performance in detail, brightness, contrast, and color registration at dimensions adequate for high-definition images. But professional displays, adequate for HDTV images having 500,000 or more picture elements, are priced in excess of $100,000. Lower-priced projection displays, which have an important place in the domestic market for the conventional systems, do not have the detail requisite for HDTV use. Attention to this problem has occupied the planning bodies of both the FCC and the trade associations.

While the question of display size is receiving attention, there remains the difference in the shape of the display for conventional versus HDTV images. Figure 2.2 shows the alternative losses of image areas. If a conventional 4:3 display receives an HTDV 16:9 transmission with the screen height filled (Figure 2.2a), the left and right portions of the wide image (25 percent of the image area) are hidden. If the 4:3 screen width is filled by the 16:9 image, the top and bottom of the display (25 percent of the image area) are blank (Figure 2.2b). Conversely, if an HDTV receiver displays a conventional 4:3 image at full display height (Figure 2.2c) the right and left edges (25 percent of the display area) are blank.

In projection displays these changes in the shape of the image may not be obvious if the screen size is adequate for the dimensions of whatever image is being displayed. But in the cathode-ray displays of conventional receivers, the blank areas are clearly outlined by the

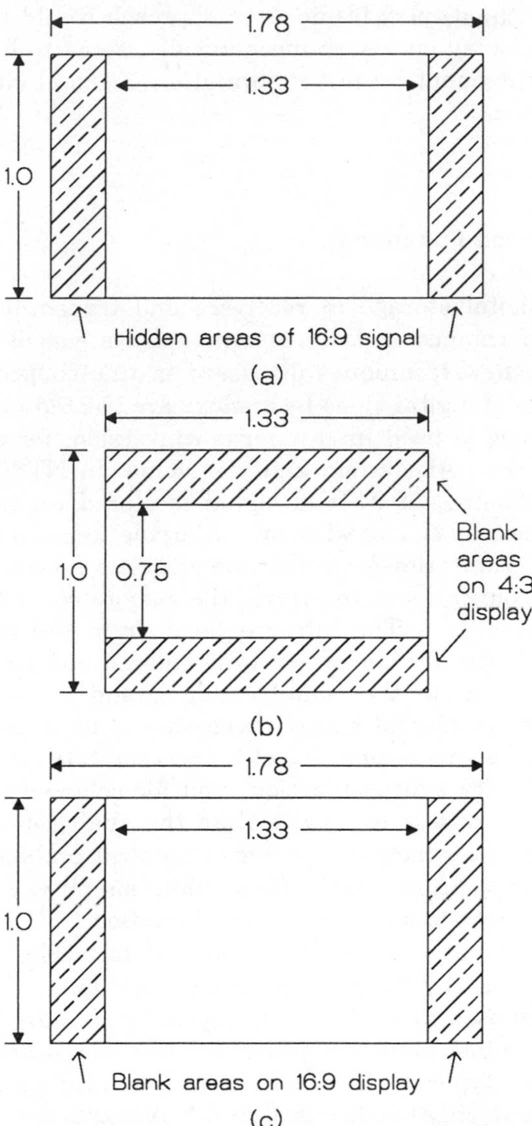

Figure 2.2 Unused areas of conventional and HDTV displays: (a) HDTV image displayed on a conventional receiver at full screen height; (b) HDTV image on conventional receiver at full screen width; (c) conventional display on an HDTV receiver at full screen height.

fixed edges of the display frame. One approach would be to shift the edges of the display enclosure mechanically, so as to hide the blank areas. The adjustment could be manual or, if signal clues were provided, automatic.

2.5 Digital Video Functions

The use of digital storage in receivers and transmitters to obtain improved performance in the conventional systems is a prominent feature of the new techniques discussed in this chapter; the fundamental aspects of digital video technology are treated in Chapter 4.

Digital storage is used in standards conversion, for example from the NTSC to the PAL system and vice versa. In NTSC-to-PAL conversion the 525-line, 59.94-field signal is stored digitally and read out at 625 lines, 50 fields, with line interpolation and time delay to achieve appropriate transfer of the line and field contents.

In current conventional receivers, the remote control function relies on digital control. The infrared beam from the control unit is digitally encoded to store channel assignments and to select among them, adjust or mute the sound channel, and to turn the power source on and off. Digital signal processing is used also in receiver circuits following the second detector (Section 4.9) for separation of scanning and color synchronization, and for color signal matrixing. Digital circuits are less expensive than the analog circuits they replace, and they are substantially free of the drift problems associated with time, temperature, and voltage, thus simplifying the maintenance of equipment to a go, no-go, determination.

Digital techniques also are an important factor in the quality of the received image by their use in broadcast plants. The analog signals commonly used in studio recording and signal distribution must be protected against noise and other degradations over a wide range of signal levels, typically 50 dB. In contrast, video signals in binary digital form have only two levels: 0 and 1. As explained in Chapter 4, this difference permits almost perfect transmission, recording, and rerecording of video information in digital form. Whereas the second or third generation of rerecording of analog videotape shows evident deterioration, as many as ten or more generations of digitally recorded tape retain broadcast quality. The digital recorder is not a simple device: to achieve its protection against noise it must employ a very wide bandwidth, typically 125 MHz. Nevertheless, it has been reduced recently to practical broadcast application.

The same protection against noise is equally valuable in distributing video signals within the broadcast plant up to the transmitter. The bandwidth of digital signals cannot be handled by a single coaxial circuit, so as many as eight parallel coaxial circuits are used.

The digital distribution of video signals within the broadcast plant is a topic of such universal importance that it has achieved worldwide standardization. CCIR Recommendation 601, adopted by the NTSC, PAL, and SECAM countries, specifies that the video signal is to be encoded in digital sample words of 8 bits (representing 256 signal levels). The luminance signal is sampled at 13.5 megasamples per second (Ms/s), the two chrominance signals at 6.75 Ms/s, and the signals are transmitted in time sequence, not compositely. There are thus 13.5 + 6.75 + 6.75 = 27 million samples to be taken each second, and each is composed of 8 bits. The total bit rate then amounts to 27 × 8 = 216 megabits per second (Mb/s), equivalant to a base bandwidth of 108 MHz for the video information alone. When guardbands and other control functions are added, the baseband actually used is of the order of 115 MHz. Clearly, in the absence of fiber optics, the digitized signal cannot be distributed over long distances unless it is reduced to the form of digital videotape.

2.6 Luminance-Chrominance Filtering in Transmission

The use of comb filters in receivers (Section 2.2) to separate the luminance and the chrominance components suggests that a similar function can be performed at the transmitter during the luminance-chrominance encoding process. Such a system was proposed by Strolle[2] in 1986. Later work has been reported by Faroudja,[5] who has demonstrated a system employing comb filters of advanced design with provisions for adapting their characteristics to changes in image content. By these means, the luminance-chrominance component separation is greatly increased. Some of this improvement may be seen in receivers using only notch filters; the improvement is greater when the receiver has a comb filter, and still greater improvement can be obtained if the receiver has a field- or frame-store.

In principle, full attainment of the greater separation of the luminance and chrominance components requires that the comb filter in the receiver have complementary characteristics to that used at the transmitter. The filter commonly used in receivers (Section 2.2) delays the composite signal by the duration of one line scan only before separation occurs.

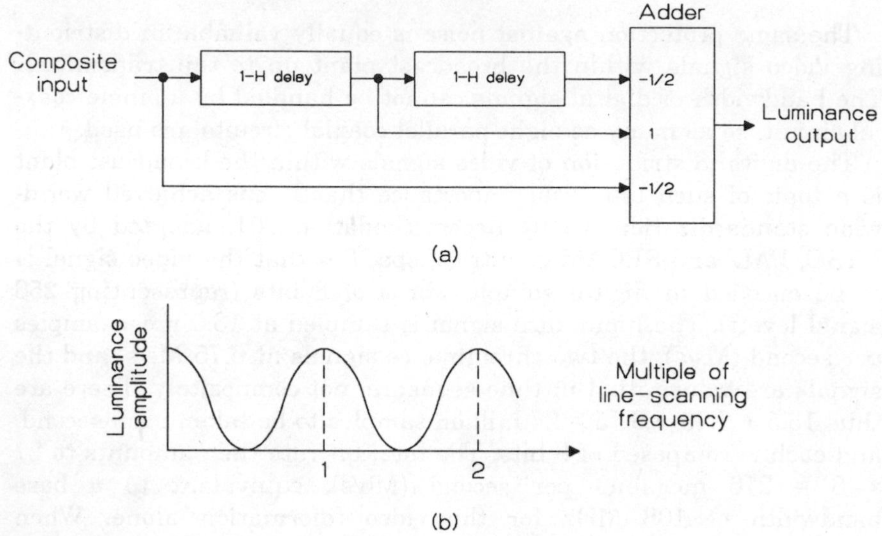

Figure 2.3 Comb filter introducing two line-scan delays, showing luminance passband for comparison with that of a single line-scan delay filter (Figure 2.1b).

Filters using two or more line delays have been designed but have been deemed uneconomical for use in receivers. The greater the number of line delays, the sharper is the cutoff of the filter passband, and the more complete is the separation. Faroudja has employed "2-H" comb filters (delays of two line scans) as a suitable compromise between performance and complexity. Figure 2.3 shows a simplified diagram of a 2-H comb filter and its luminance passband. See Figure 2.1b for comparison with the 1-H performance.

A typical NTSC encoder is shown in Figure 2.4. It performs the familiar operations of producing the *I* and *Q* signals, limiting their passbands, using them to modulate the subcarrier in c quadrature and adding the modulated subcarrier to the luminance, blanking, and synchronizing signal waveform. Figure 2.5 shows the prefiltered encoder demonstrated by Faroudja. The *Y* signal input is applied to two complementary filters, which separate the low video frequencies up to 2.3 MHz from the higher range of 2.3–4.2 MHz. The latter frequencies are passed through a 2-H comb filter, confining their spectral components. This signal is then added back to the lower frequencies to produce the comb-filtered luminance signal. The subcarrier signal, modulated by *I* and *Q*, is then passed through a separate 2-H comb filter which similarly confines its spectral components.

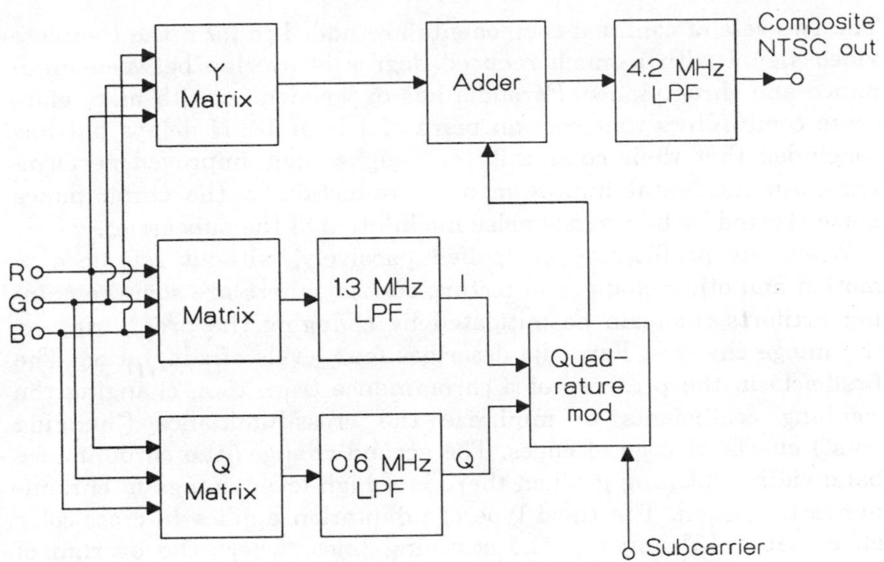

Figure 2.4 Luminance-chrominance encoding operations of the standard NTSC system.

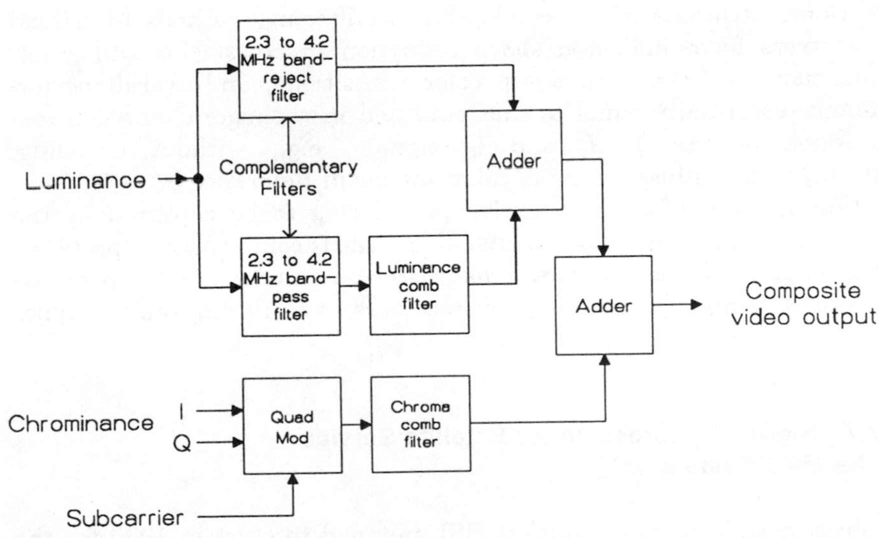

Figure 2.5 Prefilter encoder based on Faroudja's design, using two 2-H comb filters in the chrominance sideband region from 2.3 to 4.2 MHz. (From Reference 5.)

The two sets of confined components are added to form the complete video signal with a much reduced degree of overlap between luminance and chrominance. Faroudja has experimented with more elaborate comb filters that contain pairs of 1-H and 2-H delays but has concluded that their complexity outweighs their improved performance. An incidental improvement is reduction in the chrominance noise created by luminance noise modulation of the subcarrier.

When the prefiltering is applied passively, without reference to motion and other changes in picture content, there are some remaining artifacts that can be mitigated by changing the prefiltering as the image changes. Faroudja describes four levels of adaptation. The first acts in the presence of a chrominance transition, changing the combing coefficients to minimize the cross-luminance ("hanging dots") effects at colored edges. The second changes the chrominance bandwidth, widening it when there is a high-level change in chrominance transition. The third type of adaptation applies to cross-color effects at a 45° angle to the scanning lines, where the overlap of chrominance and luminance is greatest. In this third process, the average of adjacent chrominance picture-element signals is taken at a luminance transition. The fourth process uses a sharp luminance transition as a reference and synthesizes the corresponding chrominance transition over a frequency band extending above the 1.3 MHz limit dictated in the NTSC standards.

Demonstrations of these adaptive prefiltering methods to critical observers have indicated sharp reduction in cross-color and cross-luminance effects, with sharp color transitions, and overall performance essentially equal to that obtained in separate channel transmission of the *Y*, *I*, and *Q* signals, e.g., without encoding. (Comparative illustrations in color appear in Reference 3.)

For the full effect of adaptive prefiltering to be captured by the receiver, it must employ a double-delay (2-H) comb filter. If prefiltering is generally adopted by broadcasters, manufacturers can be expected to employ the 2-H type in receivers now offering the 1-H type.

2.7 Signal Compression for Satellite Service—The MAC System

Direct broadcast by satellite (DBS), planned to start in 1990 for the PAL and SECAM audiences in Europe, is to use channels in the 12-GHz and 22-GHz bands. The established channel widths for the

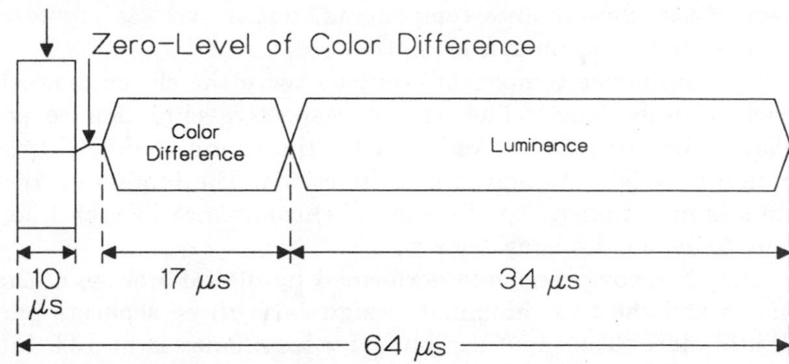

Figure 2.6 Signals occupying the line time of the D2-MAC system of direct satellite broadcast to the PAL and SECAM audiences in Europe: sound/data, color-difference, and luminance. (From Reference 7.)

satellite service have been set at 24 MHz in the Americas and 27 MHz in Asia and Europe, using frequency modulation.

In order to obtain adequate picture-signal performance of enhanced 625-line, 50-fields/s images, the European DBS service on FM channels of this bandwidth, the video baseband must be limited to about 8 MHz. This is accomplished in the European service by means of time-compressed signals using a system specially designed for DBS known as *multiplexed analogue components* (MAC). This system (Figure 2.6) transmits luminance and chrominance in time sequence, thus avoiding the signal mixture effects described in Sections 2.2 and 2.5. The sum of the bandwidths assigned to these signal components (with additional space for several high-quality sound channels, data, and control functions) exceeds the 8-MHz satellite baseband limit.

The MAC compression scheme readily adapts the 625-line, 50-Hz signals to the requirements of the satellite transponders. An early version (B-MAC)[6] was adopted by Australia for DBS service in 1985. A later version (D2-MAC)[7] is to be used in the European service.

The MAC method of time compression fits the luminance and chrominance components sequentially in the line scanning time of 64 microseconds (μs) of the 625-line, 50-Hz patterns. The sequence of the components is as follows: the second segment of each line scan is reserved for chrominance, but with only one color-difference signal per line, i.e., the *R-Y* component is sent on one line of the field scan, the *B-Y* component on the next line. This reduces the vertical resolution of the chrominance components, but by an acceptable amount relative to the luminance resolution.

The luminance component occupies twice the chrominance time in each scanned line, following the respective chrominance segment. The initial scan time is reserved for the sound channels, synchronization signals, data and control functions. The luminance resolution thus is maintained at twice that of chrominance in each line and at four times for the image as a whole.

The time compression is performed by digital storage of the luminance and the two chrominance signals in three separate parts of a digital store. The input sampling for luminance is at 13.5 MHz and 6.75 MHz for the chrominance signals. After 3/2 and 3/1 time compression, respectively, these signals have the same sampling rate, 20.25 MHz, and they are stored separately at this rate.

The contents of the luminance store corresponding to each scanning line are read out in a pulse train, converted to the analog form and inserted in the scanning line, with the digital readout so timed that it occupies the line time following that reserved for the chrominance interval. This analog signal constitutes the time-compressed luminance content of that scanning line.

Similarly the two chrominance stores are sequentially read out to produce pulse trains and converted to analog signals. These represent successively the *R-Y* and *B-Y* signals and are so timed, as retrieved from digital store, that their analog forms are inserted in the second segments of successive scanning lines. The digital store is refreshed during each field scan. The D2-MAC system includes four digitally encoded sound channels sampled at 32 kHz with 14 bits per sample.

At the receiver the inverse coding processes take place. The MAC analog baseband video signal is digitally encoded and stored. The pulses are then read out of the store at slower rates to restore luminance and chrominance signals to their initial durations. The analog versions are digitally matrixed and separately recovered by digital-to-analog conversion to form the primary color signals.

As in the case of the MUSE system, described below, the reception of the MAC signal by conventional receivers requires a converter to provide the inverse encode-decode digital operations described above. Fortunately the cost of the digital components has become so low that it constitutes a small part of the cost of a MAC-DBS installation.

2.8 Signal Compression for Satellite Service— The MUSE System

The MUSE (*mu*ltiple-sub-Nyquist-*s*ampling-*e*ncoding) system is an adaptation[8] of the NHK HDTV system (Chapter 7) for the DBS service in the 12-GHz band.

As previously noted in the case of the MAC system, the wide base bandwidth of the 1125-line NHK system (more than 20 MHz) cannot be accommodated by the satellite transponders unless the signal is compressed. The MUSE system reduces the total video baseband requirement to 8.15 MHz, suitable for the DBS service.

The 1125-line signal is initially digitally encoded at 48.6 Ms/s (Figure 2.7a). After further processing, the signal controls two filters, one responsive to stationary portions of the image, the other to the moving portions. The latter filter is controlled by two motion detectors, one following the outline of the moving area, and the other its direction of motion. The outputs of the filters are combined and the combined signal is again sampled at the sub-Nyquist rate of 16.2 MHz. The resulting pulse train is converted to the analog form with a base bandwidth of 8.1 MHz. This analog signal frequency modulates the transmitter to the satellite.

The subsampling at one-third the initial sampling rate causes the successive transmission of signals representing every third picture element. Three adjacent picture elements thus appear on three successive scans of the same line. Since the stationary elements do not move during this three-field interval, they appear in their correct positions on the display. Thus, the stationary parts of the image are reproduced with the horizontal resolution (600 lines per picture height) of the 1125-line system.

The elements of moving parts of the image do not recur in their original positions. Their lateral displacement during the interval causes a smearing effect that reduces the resolution by about 50 percent. Since the camera and the viewer have lower acuity for the details of moving objects, the overall image appears to be highly resolved. However, when the camera moves laterally ("pans"), the horizontal resolution of the image as a whole is reduced.

At the receiver, the inverse coding is performed (Figure 2.7b). The demodulated analog signal from the satellite is first sampled at 16.2 MHz. The inverse of the motion-versus-stationary filtering is performed, and the 16.2-MHz samples are up-converted to the 48.6-MHz initial sampling rate. This digital signal is then reconverted to the

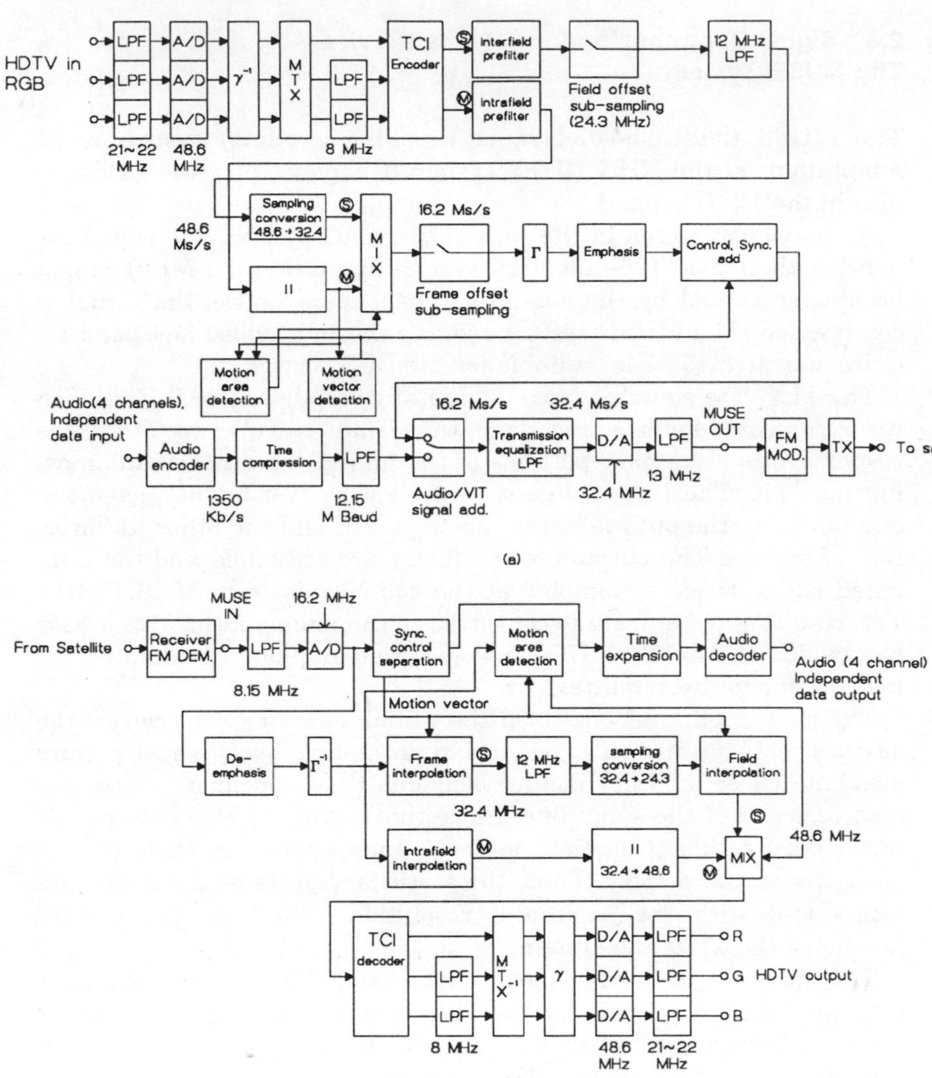

Figure 2.7 Signal processing in the MUSE version of the NHK Hi-Vision system: (a) encoding at the transmitter; (b) decoding at the receiver. The need for complex integrated circuits is evident. (From Reference 8.)

analog components from which the R, G, and B primary signals are reproduced in the video frequency range up to 22 MHz.

These receiver conversions are built into receivers designed for the MUSE system. For reception on conventional receivers, the televiewer must use an external converter to translate the frequency-modulated satellite signal into amplitude modulation for the conventional receiver input. When its cost and that of the dish antenna and mount are considered, the additional cost of the MUSE conversion is not excessive. In 1988, more than 100,000 MUSE-satellite converters were purchased in Japan by owners of conventional (NTSC) receivers in advance of the planned inauguration of the MUSE service in 1989.

The aspect ratio of the MUSE image, as broadcast, is 16/9. Only the center portion is visible on conventional receivers (Figure 2.2a). Conventional receivers are also available with optional wide-scan display. When this is selected (by the viewer or by signal clues), the full width of the MUSE image is shown with blank space above and below (Figure 2.2b).

The sound signal of MUSE is time-multiplexed between the video line scans. It provides a minimum of two high-quality stereo channels, plus provision for additional data and control functions.

2.9 Bandwidth Requirements for HDTV Distribution Systems

The video base bandwidths required for luminance and chrominance in the HDTV service are at least twice those of the NTSC, PAL, and SECAM services. Moreover, in the separate component method of transmission the bandwidths occupied by the three signals (R, G, B, or *Y*, *R-Y*, and *B-Y*) are additive, so the total bandwidth prior to signal processing is of the order of 20–30 MHz.

Compression techniques exist to lower the upper limit of this range substantially, but none has been found that satisfactorily imposes the compressed signal on the conventional channel without impairing the conventional service. In fact, the compatible protection of the NTSC, PAL, and SECAM services has allowed within their channels only limited additions of information (Section 2.10).

It follows that in the development of the HDTV service, it has been decided by most system designers, that additional channel space must be found *outside* the conventional channels of terrestrial broadcasting. Moreover, it is believed that this need applies equally to all present and future proposed distribution methods for HDTV service. However, the *capability* of providing the additional spectrum space differs markedly among the several distribution systems available for HDTV.

The greatest difficulty appears in terrestrial broadcasting, because the available channels are limited strictly by allocations to other users of the spectrum whose services cannot be displaced. The cable-television distribution systems face no such competition, since they are self-contained and protected from outside interference. The range of channels available to them is limited only by the available technology and its cost, and by the channel capacity of receivers. The satellite relay systems can utilize the available frequencies to reach many spaceborne transponders by using narrow beams.

Still another potential means of distribution is the local loop optical-fiber telephone connection to the home. The use of the extremely short wavelengths of infrared light offers a frequency range so extensive that bandwidth, per se, is not a limiting factor. Instead, the major factors are the cost and availability of the service and policy matters governing relations between the common carrier of the fiber connections and program organizations.

Finally, there are the avenues of HDTV service to the home that do not involve broadcast, cable, or satellite distribution means: the video recorder/reproducers using either magnetic tape or optical disks. Magnetic recorders suitable for production of HDTV programs have been available since 1985. It follows that lower-cost consumer products will become available relatively soon.

Since the problem of finding additional channel space has been most pressing in terrestrial broadcasting, factors affecting this problem are the primary concerns in this chapter.

2.10 Bandwidth Requirements for Terrestrial HDTV Service

There have been several approaches to providing better image quality in terrestrial broadcasting, beyond the enhancements to the conventional service previously discussed (Sections 2.2, 2.3, and 2.4). They have in common the attainment of additional channel space for wider basebands, either by finding a place for it within the existing channel or, if that fails, by using space outside the existing channel without serious harm to those broadcasters or others licensed to occupy the space thus preempted. One form of extra space is known as an *augmentation channel*; its purpose is to augment the information carried by the conventional service so as to increase its definition and/or its aspect ratio.

This external channel space has been very difficult to find, particularly in heavily populated sections of the service areas where nearly all of the available channels are occupied. Each such occupied channel must be protected from interference by stations on the same and adjacent channels. This protection is provided by the allocation authorities in each country, who have set up the minimum allowable distances between stations and other restrictions to prevent interference.

2.11 Signal Occupancy within Conventional Channels

The conventional television channels are not fully occupied, as was shown when the chrominance subcarrier was added to the luminance signal to produce the compatible color service. In the NTSC system the subcarrier frequency was chosen to be an odd multiple, 455, of half the line-scanning frequency. The choice was made to assure that the energy associated with the subcarrier would have low visibility on monochrome receivers.

At this frequency the phase of the subcarrier reverses on successive scans, so the subcarrier cycles in each line cancel one another, averaging to zero brightness (although the visual cancellation is often incomplete). A similar cancelling of subcarrier interference occurs when any other odd multiple is chosen. One possibility of adding information to the channel is to select a second subcarrier and to locate it in such a position that it causes little interference to the conventional service.

This approach to additional channel occupancy was analyzed by Fukinuki[9] and co-workers and reported in 1984. In 1988, a proposed system was described by Isnardi[10] that had undergone computer simulation at the David Sarnoff Research Center. The system is named ACTV-I, for "Advanced Compatible Television, First System."

In the ACTV-I system (Figure 2.8), the second subcarrier frequency is the 395th multiple of half the line-scan frequency, i.e., 3.1075171 MHz, below the chrominance subcarrier by approximately 0.5 MHz. This subcarrier is quadrature modulated (in the same fashion as is chrominance) by two auxiliary signals that carry additional information not intended to be visible on conventional displays, but to be recoverable on receivers designed for the ACTV-I system. The modulated subcarrier is attenuated 12 dB and added to the conventional NTSC video waveform. At the simulated receiver, the second subcarrier and its sidebands are passed through a filter, amplified 12 dB, and quadrature demodulated to recover the auxiliary information.

Isnardi's analysis shows, and computer analysis confirms, that certain forms of interference are visible, on close examination, on

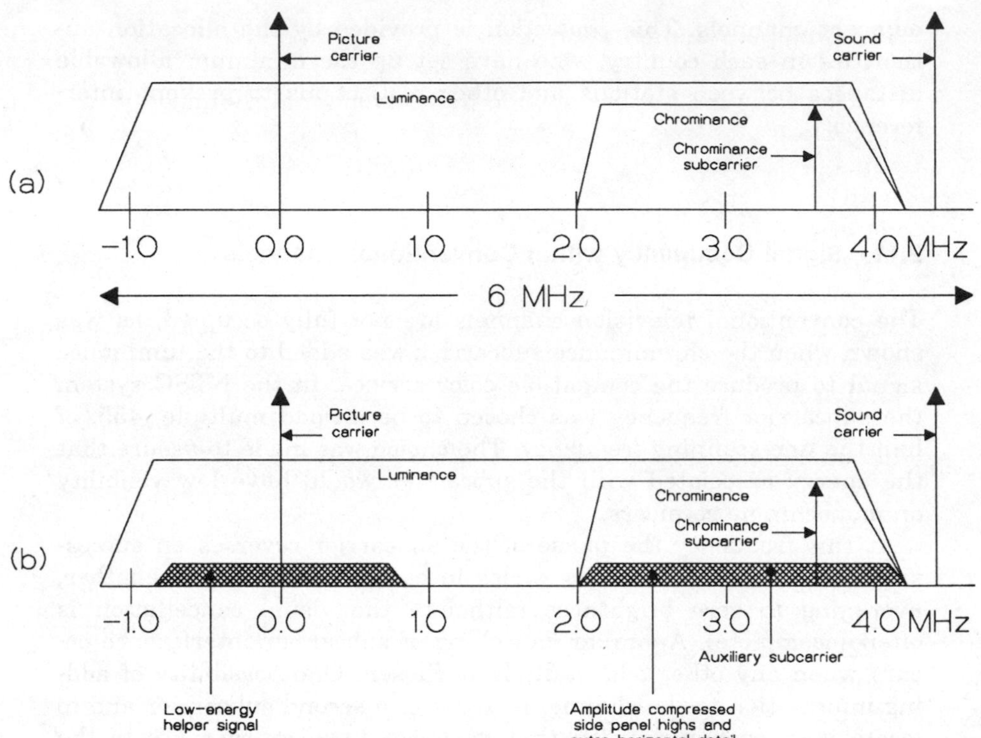

Figure 2.8 Channel occupancy of the ACTV-I advanced television system (b) compared with the NTSC channel (a).

conventional receiver displays when the second subcarrier is present. One such form takes the form of static or flickering vertical color stripes, caused by beats between the horizontal elements of the auxiliary modulation and the chrominance subcarrier.

This interference can be minimized by adjustment of the receiver's chroma and sharpness controls, and their effect is also lessened by the restricted luminance and chrominance bandwidths of most conventional receivers. However, when the receiver uses a comb filter to cover the full 4.2-MHz luminance spectrum, the auxiliary signal components are treated as chrominance and the interference is then clearly evident. The ACTV-I system, originally scheduled for implementation in hardware form by late 1988 or early 1989, has been delayed by the industry investigation of other system proposals and laboratory evaluations.

2.12 Channel Augmentation

The use of spectrum space outside the conventional channels has received close attention of the New York Institute of Technology, the Philips Laboratories, the David Sarnoff Research Center, and the Zenith Corporation. Specifics of their work are treated in later chapters.

This work on channel augmentation has drawn attention to a fundamental question concerning alternative means of program transmission. The availability of additional channel space in the self-contained cable and fiber-optic networks is not in question, but its availability in terrestrial broadcasting is limited and, in the opinion of many cognizant engineers, not universally attainable. In the United States, the question has been brought into sharp focus by the ruling issued in 1988 by the Federal Communications Commission stating, first, that the HDTV standards to be issued by that body must be compatible with the existing NTSC service and, second, that all allocations for the HDTV service would be confined to the existing channels in the VHF and UHF services (channels 2–69; i.e., 55–88, 174–216, and 470–806 MHz).

Since the FCC does not have jurisdiction over the channel allocations used in the cable networks, the implication was that a choice was being offered. In other words, cable systems could adopt channel augmentation, preserving the conventional service and allocating all or part of a second channel for use by HDTV receivers. But terrestrial broadcasters might not have that option. They might be required to keep all the HDTV signal content within the existing channels. The David Sarnoff Research Center, having in mind these alternatives, is developing two systems: the ACTV-I system described above (Section 2.11) and the ACTV-II system (Chapter 9) with an augmentation channel. The prospect that there might be two sets of HDTV standards, one adopted by the cable industry, the other by the terrestrial broadcasters, poses a serious question of public policy. While receivers could be designed to embrace the two standards, they would be more complex and expensive than receivers designed for a single standard.

2.13 Augmentation in Terrestrial Service

The limitations to augmentation in the terrestrial service are under detailed investigation. The basic requirement is protection of existing service. Under the FCC ruling mentioned above, only VHF and UHF television channels are available for augmentation purposes. The VHF channels are so fully occupied in the major market areas that only UHF channels (14–69 and 470–806 MHz, inclusive) are available in these centers. In the majority of cases this would mean matching a VHF existing channel with UHF augmentation. The major differences in direct and multipath propagation, time delay coverage, noise, and other types of interference between the VHF and UHF channels is the first of many problems encountered in this effort. Despite these impending difficulties, major attention is being paid to the UHF channels for augmentation, and what follows here is limited to them.

The FCC has set up specific regulations governing the allocation of UHF channels and their protection from interference. An excellent summary of the FCC rules has been prepared by Pritchard[11] to which reference should be made for further detail. Interference with UHF television reception arises from seven entry points at which typical receivers are sensitive: the channel tuned to, the two adjacent channels, images of the sound and picture carriers, beats with the local oscillator, with the intermediate frequencies, and by intermodulation between station carriers. Table 2.1 lists these sources of interference and the minimum mileage separations set up by the FCC to protect against interference from these sources. It will be noted that interference can be caused by at least one of these sources at channels located at the desired channel, and by channels located above and below it by 1, 2, 3, 4, 5, 7, 8, 14, and 15 channel numbers, and that the minimum separation varies with the type of interference from 20 to 205 miles.

It is apparent, that in choosing a channel for augmentation, close attention must be paid to the distances not only to other stations on the same and adjacent channels, but also to those on widely separated channels to which interference might be caused by the other mechanisms listed in Table 2.1. The distances will be affected by the effective radiated power of the augmentation transmitter, relative to that on which the FCC mileage separations are based. Many other factors, such as local terrain, must be taken into account in this assessment.

It would appear that there are severe restrictions, known as "taboos," on channel selection imposed by these FCC regulations. The question has been raised whether the FCC was unduly conservative in drawing up the geographical separation limits. This has been under close study by FCC and industry advisory groups. It is abundantly clear, whatever the outcome of this examination, that the true extent of interference caused by augmentation can be determined

Table 2.1 Channel Separations for Channels 14–69 (UHF)

Interference type	Channel numbers and separation	Required separation, mi (km)
Co-channel	0	Zone I, 155 (249) Zone II, 175 (282) Zone III, 205 (330)
Adjacent channel	± (6 MHz)	55 (89)
Sound image	± 14 (84 MHz)	60 (97)
Picture image	± 15 (90 MHz)	75 (121)
Local oscillator	± 7 (42MHz)	20 (32)
IF beat	± 8 (48 MHz)	20 (32)
Intermodulation	± 2 through ± 5	20

Approximate zone boundaries: Zone I—Illinois, Indiana, Ohio, Pennsylvania, Maryland, Delaware, New Jersey, Connecticut, Rhode Island, Massachusetts; parts of Wisconsin, Michigan, New York, West Virginia, Virginia, New Hampshire, Vermont, and Maine. Zone II—United States not included in Zones I and III, Puerto Rico, Alaska, and Virgin Islands. Zone III—Florida and Gulf Coast states to the Mexican border.

Source: 73.698 FCC Rules and Regulations, Part 73.610, Table IV.

only after extensive tests in the field. While laboratory testing of the relative performance of the HDTV proposals has been undertaken, the extent of field testing has been limited. The experience of the NTSC in setting up the compatible color standards in the early 1950s is ample evidence that the conclusions of highly experienced engineers may differ significantly from field-test data.

References

1. Naimpally, S., L. Johnson, T. Darby, R. Meyer, L. Phillips, and J. Vantrease. Integrated Digital IDTV Receiver with Features. *IEEE Trans. Consumer Electronics* 34: 410–419 (August 1988). The specific details refer to Models 27J245SB and 31J460SA.
2. Strolle, C. H. Cooperative Processing for Improved NTSC Chrominance/Luminance Separation. *SMPTE Journal* 95: 782–789 (August 1986).

3. Ashizaki, S., Y. Suzuki, K. Mitsuda, and H. Omae. Direct-View and Projection CRTs for HDTV. *IEEE Trans. Consumer Electronics* 34: 91–99 (February 1988).
4. Kawashima, M., K. Yamamoto, and K. Kawashima. Display and Projection Devices for HDTV. *IEEE Trans. Consumer Electronics* 34: 100–110 (February 1988).
5. Faroudja, Y. and J. Roizen. Improving NTSC to Achieve Near-RGB Performance. *SMPTE Journal* 96: 750–761 (August 1987).
6. Lucas, K. B-MAC: A Transmission Standard for Pay DBS. *SMPTE Journal* 95: 1166–1172 (November 1985).
7. Sabatier, J., D. Pommier, and M. Mathieu. The D2-MAC-Packet System for All Transmission Channnels. *SMPTE Journal* 95: 1173–1179 (November 1985).
8. Ninomiya, Y. Ohutsuka, Y. Izumi, C. Gohshi, and Y. Iwadate. An HDTV Broadcasting System Utilizing a Bandwidth Compression Technique—MUSE. *IEEE Trans. on Broadcasting* BC33: 130–160 (December 1987).
9. Fukinuki, T., Y. Hirano, and H. Yoshigi. Experiments on Proposed Extended-Definition TV with Full NTSC Compatibility. *SMPTE Journal* 93: 923–929 (October 1984).
10. Isnardi, M. A. Exploring and Exploiting Subchannels in the NTSC Spectrum. *SMPTE Journal* 97: 526–532 (July 1988).
11. Pritchard, D. H. In *Television Engineering Handbook*. K. Blair Benson (Ed.), McGraw-Hill, New York, 1986, pp. 21.4–21.57.

Chapter

3

Visual Aspects of High-Definition Images

3.1 Objectives of HDTV Service

High-definition television has improved on earlier techniques primarily by calling more fully on the resources of natural vision. This chapter is concerned with those aspects of vision that are basic to the design of the television service. Particular attention is devoted to the techniques required for HDTV service, compared with the methods that are used in conventional NTSC, PAL, and SECAM services.

The primary objective in HDTV has been to enlarge the visual field occupied by the television image. This has called for larger, wider pictures that are intended to be viewed closely. To satisfy the viewer at this closer inspection, the HDTV image must possess proportionately finer detail and sharpness of outlines.

In other respects, the HDTV service has had to embrace a wide range of earlier techniques. Its transmissions should, if possible, be suitable reception on conventional receivers, that is, they should be compatible with the conventional services. These use a signal processing system, *composite transmission*, that transmits all the detail and color information simultaneously over a single channel. This process of fitting the color values into the channel designed for monochrome service has involved many compromises that have impaired the quality of the image. Thus, a second major objective of HDTV development has been to remove these impairments. One way this may be done is by transmitting the color values in time sequence, rather than simultaneously. The time-sequential process is known as "time-multiplexed component transmission."

The methods employed to meet the needs of the eye more closely and to remove image impairments have required substantially more complex circuits, which have become available with the advent of digital video techniques (Chapter 4).

3.2 Visual Fields of Television

A central objective of the television service is to offer the viewer a sense of presence in the scene and of participation in the events portrayed. To meet this objective, the televised image should convey as much of the spatial and temporal content of the scene as is economically and technically feasible. One important limitation of the conventional services, now clearly recognized, is that their images occupy too small a portion of the visual field of view. Experience in the motion-picture industry (Section 1.3) has shown that a larger, wider picture, viewed closely, contributes greatly to the viewer's sense of presence and participation.

The current development of the HDTV service is directed toward the same ends. From the visual point of view, the term "high-definition television" is to some extent a misnomer in that the primary visual objective of the system is not to produce images having fine, sharp detail. The prime objective is to provide an image that occupies a larger part of the visual field. Higher definition is secondary; it need be no higher than is just adequate for the closer scrutiny of the image.

3.3 Foveal and Peripheral Vision

There are two areas of the retina of the eye to be satisfied by television: the fovea and the areas peripheral to it. The fine detail and edges of the image are perceived by a small central portion of the retina,[1] the *fovea*. Foveal vision extends over only about one degree of the visual field, whereas the total field to the periphery of vision extends about 160° horizontally and 80° vertically. Motions of the eye and head are necessary to assure that the fovea is positioned on that part of the retinal image where the detailed structure of the scene is to be discerned.

The portion of the visual field outside the foveal region provides the remaining visual information. Thus, a very large part of visual reality is conveyed by this extrafoveal region. The vital perceptions of extrafoveal vision, notably motion and flicker, have received only secondary attention in the development of television engineering. First attention has been paid to satisfying the needs of foveal vision. This is true because a system capable of resolving fine detail presents the major technical challenge: transmission channels that offer essentially no discrimination in the amplitude or time delay among the signals they carry over a very wide band of frequencies. The difficulty of extending the channel bandwidths used by the NTSC, PAL, and SECAM services has in fact been the principal impediment to the introduction of HDTV service.

Attention to the properties of peripheral vision has led to a number of constraints. Peripheral vision has great sensitivity to even modest changes in brightness and position. Thus, the bright portions of a wide image viewed closely are much more subject to flicker at its left and right edges than is the narrower image of the conventional systems. One result of this effect is that the 50-Hz image repetition rate of the PAL and SECAM services is inadequate for the HDTV versions of these systems. The rate adopted for their ATV and HDTV displays is 100 Hz.

3.4 Vertical Detail and Viewing Distance

Figure 3.1 illustrates the geometry of the field occupied by the televised image. The ratio of the picture width W to its height H is the *aspect ratio*. The viewing distance D determines the angle h subtended by the picture height. This angle (=2 tan-1 $H/2D$) is usually measured by the ratio of the viewing distance to the picture height D/H. The smaller this ratio, the more fully the image fills the field of view.

The useful limit to the viewing distance is that at which the eye can just perceive the finest details in the image. Closer viewing serves no purpose in resolving detail while more distant viewing prevents the eye from resolving all the detailed content of the image. The preferred value of the viewing distance, expressed in picture heights, is the *optimal viewing ratio*, commonly referred to as the "optimal viewing distance." It defines the distance at which the viewer with normal vision would prefer to see the image, when pictorial clarity is the criterion.

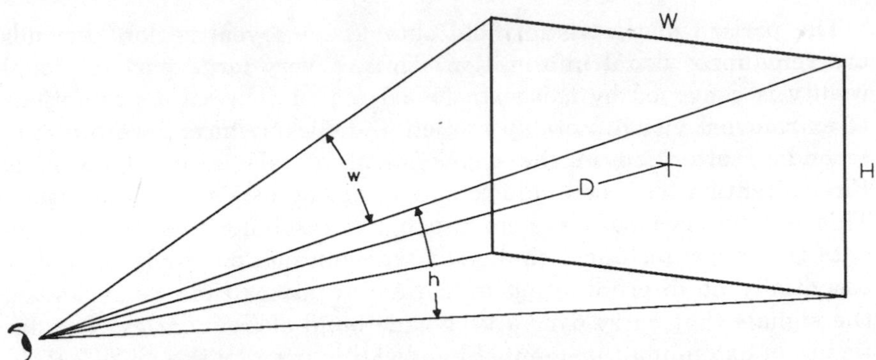

Figure 3.1 Geometry of field of view occupied by a television image.

The optimal viewing ratio is not a definite value, since it varies with the subject matter, the viewing conditions, and the acuity of the viewer's vision. Its nominal value does serve, however, as a convenient basis for comparing the performances of the conventional and the ATV and HDTV services.

The computation of the optimal viewing ratio depends on the degree of detail offered in the vertical dimension of the image, without reference to its pictorial content. The discernible detail is limited by the number of scanning lines presented to the eye and by the ability of these lines to present the details separately.

The smallest detail that can be reproduced in the image is known as a *picture element* (pixel). Ideally, each detail of the scene would be reproduced by one picture element, that is, each scanning line would be available for one picture element along any vertical line in the image. In practice, however, some of the details in the scene inevitably fall between scanning lines, so that two lines are required for such picture elements. Thus some vertical resolution is lost. Measurements of this effect show that only about 70 percent of the vertical detail is presented by the scanning lines. This ratio is known as the Kell[2] factor; it applies irrespective of the manner of scanning, whether the lines follow each other sequentially (progressive scan) or alternately (interlaced scan).

When interlaced scanning is used, as in all the conventional systems, the 70 percent figure applies only when the image is fully stationary and the line of sight from the viewer does not move vertically by even a small amount. In practice these conditions are seldom met, so an additional loss of vertical resolution, called the interlace factor, occurs under typical viewing conditions. This additional loss depends on many aspects of the subject matter and viewer attention, so there is a wide range of opinion on its extent. Under favorable conditions the additional loss reduces the effective value of vertical resolution to not more than 50 percent, that is, no more than half the scanning lines display the vertical detail of an interlaced image. Under unfavorable conditions, a larger loss can occur. The effective loss also increases with image brightness, as the scanning beam becomes larger.

Since interlacing imposes this additional detail loss (as well as associated degradations due to flicker along horizontal edges, known as "shimmer," and among objects aligned horizontally, known as "glitter"), it was decided by some system designers early in the development of HDTV to abandon interlaced scanning for image display. It has been necessary to design HDTV systems so that they can provide conventional receivers with interlaced scanning. In several proposed HDTV systems, the camera and the display are designed for progressive scanning. Progressively scanned displays can also be derived from interlaced transmissions by digital image-storage (see Section 2.3). Such conversion of scanning improves the vertical resolution by about 40 percent.

Table 3.1 compares the optimal viewing ratios and associated image characteristics of existing and proposed HDTV systems, as well as the NSTC, PAL, and SECAM systems with interlaced and progressive scanning. Two of the HDTV systems listed are conceptual only; they represent values consistent with design trends for terrestrial broadcast in 1989, prior to adoption of standards by the American and European authorities. Compatibility is assumed, with scanning line rates twice those of the conventional systems, no change in the field scanning rate, and an auxiliary channel (Section 2.12) one-half as wide as the conventional channel.

The basis for the figures presented for the optimal viewing ratio is an extensive study by Fujio and associates at the Japan Broadcasting Corporation (NHK).[3] They concluded that the preferred distance for viewing the 1125-line images of their system has a median value of 3.3 times the picture height, equivalent to a vertical viewing angle of 17°. This covers about 20 percent of the total vertical visual field. The other figures, given in Table 3.1, are based on this 3.3 figure, adjusted according to the value of vertical resolution stated for each system. The vertical resolutions in this table are 50 percent of the active scanning lines for interlaced scanning and 70 percent for progressive scanning.

Table 3.1 Spatial Characteristics of Television Systems (based on luminance or equivalent Y signal)

System	Total lines	Active lines	Vertical resolution	Optimal viewing distance	Aspect ratio	Horizontal resolution	Total picture elements	Field of view[a] *Vertical*	Field of view[a] *Horizontal*
HDTV-p USA	1050	960	675	2.5	16/9	600	720,000	23°	41°
HDTV-p Europe	1250	1000	700	2.4	16/9	700	870,000	23°	41°
HDTV-i NHK	1125	1080	540	3.3	16/9	600	575,000	17°	30°
NTSC-i	525	484	242	7.0	4/3	330	106,000	8°	11°
NTSC-p	525	484	340	5.0	4/3	330	149,000	12°	16°
PAL-i	625	575	290	6.0	4/3	425	165,000	10°	13°
PAL-p	625	575	400	4.3	4/3	425	233,000	13°	18°
SECAM-i	625	575	290	6.0	4/3	465	180,000	10°	13°
SECAM-p	625	575	400	4.3	4/3	465	248,000	13°	18°

p = progressive display; i = interlaced display; PH = picture height.
[a]At optimal viewing distance; video bandwidths (luminance): USA–8 MHz, Europe–9 MHz, NHK–20 MHz, NTSC–4.2 MHz, PAL–5.5 MHz, SECAM–6 MHz.

3.5 Horizontal Detail and Picture Width

Since the fovea is approximately circular in shape, its vertical and horizontal resolutions are nearly the same. This would indicate that the horizontal resolution of a display should be equal to its vertical resolution. Such equality is the usual basis of television system design, but it is not a firm requirement.

A study published in 1940 by Baldwin of Bell Telephone Laboratories[4] showed that the overall subjective sharpness of an image is greatest when the vertical and horizontal dimensions of the picture elements are the same (that is, with equal vertical and horizontal resolution). But, as shown in Figure 3.2, taken from his paper, considerable variation in the shape of the picture element produces only a minor degradation in the sharpness of the image, provided that its

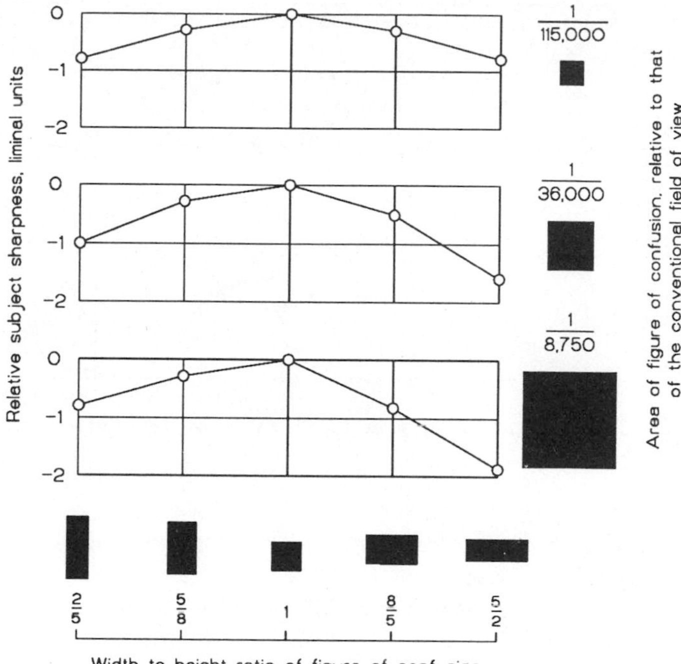

Figure 3.2 Visual sharpness as a function of the relative values of horizontal and vertical resolution. The liminal unit is the least perceptible difference.

area is unchanged. This seminal finding led to the conclusion that sharpness depends primarily on the *product* of the resolutions, that is, it depends on the total *number* of picture elements in the image.

Advantage of the freedom to extend the horizontal dimensions of the picture elements has been taken in wide-screen movies. For example, the Fox CinemaScope system[5] uses anamorphic optical projection to enlarge the image in the horizontal direction. Since the emulsion of the film has equal vertical and horizontal resolution, the enlargement lowers the horizontal resolution of the image. A limit was reached when the image became 2.35 times as wide as it is high (aspect ratio 2.35:1). Most current motion pictures have an aspect ratio of about 1.85:1.

When wide-screen films are televised with the conventional aspect ratio of 1.33:1 (= 4/3), the full width of the film image cannot be shown. This requires that the televised area be moved laterally when

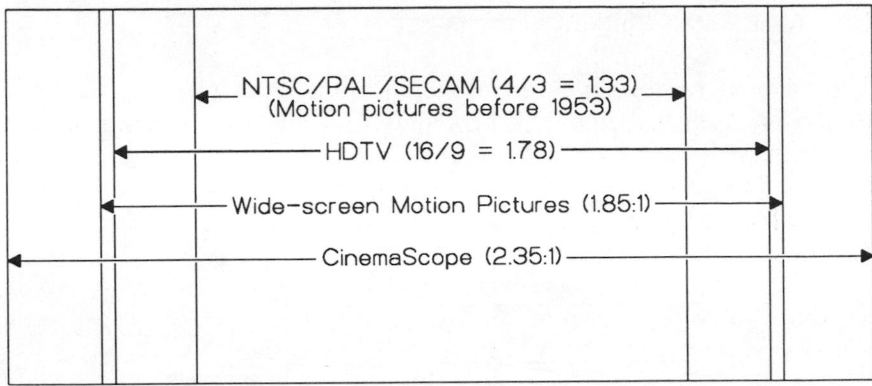

Figure 3.3 Comparison of the aspects ratio of television and motion pictures.

scanned at the studio, to keep the center of interest within the area displayed by the receiver.

The picture width chosen for HDTV service is 1.777 (= 16/9) times the picture height. This is a compromise with the wider ratio of the film industry, imposed by the constraints on the width of the HDTV channel. Other factors being equal, the video baseband increases in direct proportion to the picture width.

An aspect ratio greater than 1.33:1 was an early objective of HDTV system design. The NHK system was initially designed for a ratio of 1.666:1 (= 5/3), but in later work by other designers, the 16/9 ratio became first choice, and it now has been adopted by most other system proponents, including the NHK system. Figure 3.3 shows the relative widths of the aspect ratios of television and motion pictures. The 16/9 HDTV ratio covers nearly all the most popular wide-screen displays, except CinemaScope.

From the several aspect ratios we can determine the horizontal field of view from the horizontal angle w = 2 tan-1 $W/2D$ in Figure 3.1, at the optimal viewing ratio for each system. The NHK system, for example, with an optimal viewing distance of 3.3 times the picture height and 16/9 aspect ratio offers a horizontal viewing angle of 30°, about 20 percent of the total horizontal visual field. While this is a small percentage, it covers that portion of the field within which the majority of visual information is conveyed. The aspect ratios, resolutions, and fields of view of the HDTV and conventional systems are listed in Table 3.1.

3.6 Total Detail Content of the Image

The vertical resolution is equal to the number of picture elements separately present in the picture height, while the number of elements in the picture width is equal to the horizontal resolution times the aspect ratio. The product of the number of elements vertically and the number horizontally, the total number of picture elements in the image is an important parameter. Since all the elements must be scanned during the time between successive frame scans, the rate of scanning (and the concomitant video bandwidth) is directly proportional to the total number of elements (Table 3.1).

3.7 Perception of Depth

Perception of the third spatial dimension, depth, depends in natural vision primarily on the angular separation of the images received by the two eyes of the viewer. Attempts to produce a binocular system of television have been made, but their cost and inconvenience have outweighed their benefits. A considerable degree of depth perception is inferred in the flat image of television from the perspective appearance of the subject matter and from camera technique by the choice of the focal length of lenses and by changes in depth of focus. Continuous adjustment of focal length by the zoom lens provides the viewer with an experience in depth perception wholly beyond the scope of natural vision.

No special steps have been taken in the design of the HDTV service to offer depth clues not previously available. However, the wide field of view offered by HDTV displays provides significant improvement in depth perception, compared with that of conventional systems.

3.8 Contrast and Tonal Range

The range of brightness (contrast) that can be accommodated by television displays is severely limited compared to that of natural vision, and this limitation has not been avoided in HDTV equipment. Moreover, the display brightness of HDTV images is limited basically by the need to spread the available light over a large area.

Within the upper and lower limits of display brightness, the quality of the image depends on the relation between changes in brightness in the scene and the corresponding changes in the image. It is an accepted rule of image reproduction that the brightnesses should be directly proportional, that is, the curve relating input and output brightnesses should be a straight line. This continues to be the criterion for the HDTV service.

Since the output-versus-input curves of cameras and displays are, in many cases, not straight lines, intermediate adjustment ("gamma correction") is needed. Gamma correction in HDTV requires particular attention to avoid excessive noise and other defects that are particularly evident at close scrutiny of the display.

3.9 Luminance and Chrominance

The retina of the eye contains three sets of cones that are separately sensitive to red, green, and blue light. The cones are uniformly dispersed in the color-sensitive area of the retina, so that light falling on them produces a sensation of color that depends on the relative amounts of red, green, and blue light present. When the ratio of red:green:blue is approximately 30:60:10, the sensation produced is that of white light. Other ratios produce all the other color sensations. If only red cones are excited, the sensation is red. If both red and green intensities are present, the sensation is yellow; if red predominates, orange.

Given this process of color vision, it would appear that color television would be most directly provided by a camera that produces three signals, proportional respectively to the relative intensities of red, green, and blue light in each portion of the scene. The three signals would be conveyed separately to the input terminals of the picture tube, so that the tube would reproduce at each point the relative intensities of the red, green, and blue discerned by the camera. This separate utilization of three component-color signals occurs at the camera and display. In the transmission between them, a different group of three is used: *luminance* and two *chrominance* signals.

This form of transmission arose originally in the NTSC color system, when it was realized that compatible operation of the millions of existing monochrome receivers required a "black-and-white" signal, which became known as luminance and was given the symbol *Y*.[6] The additional information needed by color receivers was evidently the difference between a color value and its luminance component, at each point in the scene. The color value may have as many as three components: red, green, and blue (symbols R, G, and B), so the chrominance signal must embrace three values: red-minus-luminance, green-minus-luminance, and blue-minus-luminance (R-Y, G-Y, and B-Y, respectively).

At the receiver these "color-difference" signals are added individually to the luminance signal, the sums representing the individual red, green, and blue signals that are applied to the display device. The luminance component is derived from the camera output by combining, in proper proportions, its red, green, and blue signals. Thus, symbolically, $Y = R + G + B$.

In the PAL system the R-Y and B-Y color-difference signals are transmitted, at reduced bandwidth, with the luminance signal. The G-Y signal, equal to $-(R + B)$ is reconstructed at the receiver. In the NTSC system,[7] two different color-difference signals (I and Q) are used. They are combinations of the R-Y and B-Y signals that serve the same purpose.

The ability of the eye to discern fine details and sharpness of outlines ("*visual acuity*") is greatest for white light. Hence, the luminance signal must be designed to carry the finest detail and sharpest edges of the scene.

3.10 Chromatic Aspects of Vision

Two aspects of color science are of particular interest to television engineers in the development of HDTV. The first is the total range ("gamut") of colors perceived by the camera and offered by the display. The second is the ability of the eye to distinguish details presented in different colors. The color gamut defines the realism of the display while attention to color acuity is essential to avoid presenting more chromatic detail than the eye can resolve.

The gamut perceived by a viewer with normal vision (represented by the "C.I.E. Standard Observer of 1931") is shown in the C.I.E. diagram[8,9] in Figure 3.4. Color perceptions were measured by offering to viewers combinations of the three standard C.I.E. primary colors: spectrally pure red of 700 nm wavelength, green of 546.1 nm, and blue of 435.8 nm. These and all the other spectrally pure colors are located on the upper boundary of the diagram. The purple colors, which do not appear in the spectrum, appear on the line connecting the ends of the upper locus. The colors located within this outer boundary are perceived as being mixed with white light, that is, they become more pastel as the center of the diagram is approached. At the center of the diagram the color represented is white, and in the

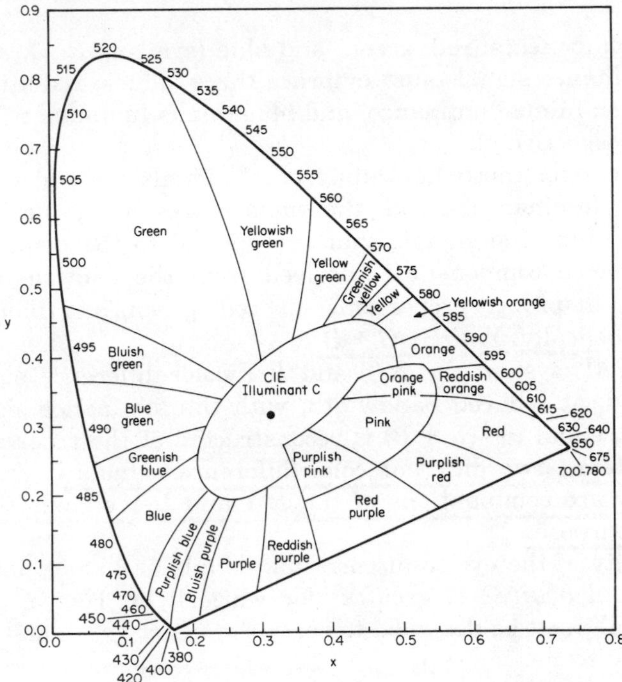

Figure 3.4 The C.I.E. Chromaticity Diagram, showing the color sensations and the triangular gamuts of the C.I.E. primaries compared with the FCC (NTSC) primaries. (From Reference 9.)

center of the white region is a particular white (C.I.E. Illuminant C) that matches sunlight from the northern sky.

Each point on the chromaticity diagram is identified by two numbers, its x and y coordinates, that define a particular color uniquely from all others. Three such points identify the primary colors used in a camera, and the triangle formed by them encloses the total gamut of colors which the camera can perceive. Another triangle limits the gamut of the display.

The chromaticity diagram[10] in Figure 3.5 compares the gamuts of interest in television. The largest triangle is that formed by the C.I.E. primaries. The smaller triangles are labeled to show the television system to which they apply. The television gamuts evidently cover a small portion of the range of colors visible to the eye.

This limited range is nevertheless wholly adequate for television. This is evident from the irregular area shown in the diagram that bounds the surface colors of the inks, dyes, and pigments[11] used in other media. These colors offer most viewers their concept of the

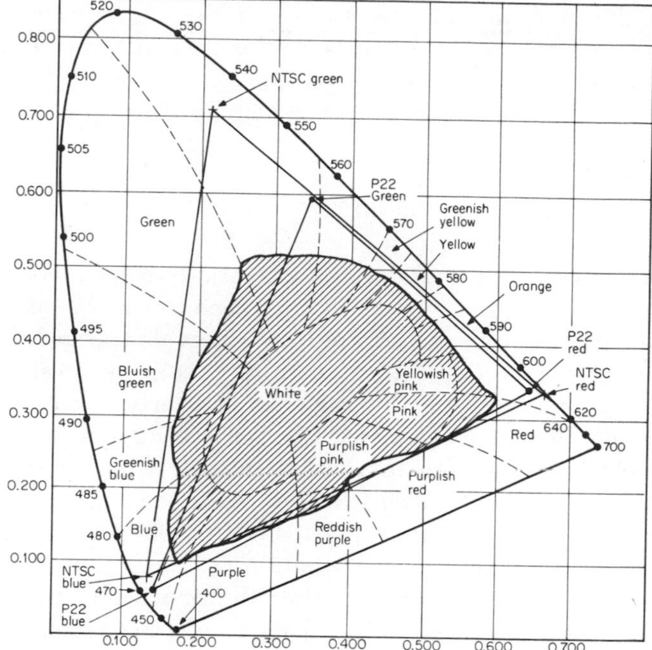

Figure 3.5 The range of surface colors exhibited by pigments, dyes, and ink compared with the NTSC and PAL/SECAM gamuts.

"real" colors of material objects, so their acceptance of the restricted gamut is readily explained. The slight differences in the primaries standardized for HDTV service are not intended to offer a wider gamut. They were chosen in view of the experience of the broadcasters and the receiver manufacturers.

3.11 Acuity in Color Vision

The second aspect of color vision is the acuity of the eyes for the primary colors. This was a basic factor in the development of the NTSC system, and it has so remained during the subsequent design of the PAL, SECAM, and HDTV systems. In the earliest work on color television[7] equal bandwidth was devoted to each primary color, although it had long been known by color scientists that the color acuity of the normal eye is greatest for green light, less for red, and much less for blue. The NTSC system made use of this fact by devoting much less bandwidth than luminance to the red-minus-luminance channel and still less to blue-minus-luminance. All color television services now employ this device in one form or another. Properly applied, it offers economy in the use of bandwidth without loss of resolution or color quality in the displayed images.

3.12 Temporal Factors in Vision

The response of the eye over time is essentially continuous, that is, the cones of the retina convey nerve impulses to the visual cortex of the brain in pulses recurring about 1000 times per second. If television were required to follow this design, a channel several kilohertz wide would be required in HDTV for each of nearly a million picture elements, equivalent to a video baseband exceeding a gigahertz.

Fortunately television can take advantage of a temporal property of the eye, *persistence of vision*, the retention of an image after the light is removed. The special properties of this phenomenon must be carefully observed to assure that motion in the scene is conveyed smoothly and accurately and that the illumination of the scene is free from the effects of flicker in all parts of the image. A conflict encountered in HDTV design is the need to reduce the rate at which some of the spatial aspects of the image are reproduced, thus conserving bandwidth, while maintaining temporal integrity, particularly the position and shape of objects in motion. To resolve such conflicts, it has become standard practice to view the television signal and its processing simultaneosuly in three dimensions: width, height, and time (see Chapter 5).

3.13 Temporal Aspects of Illumination

Basic to television and motion pictures is the presentation of a rapid succession of slightly different still pictures ("frames"). Between frames the light is cut off briefly. The visual system retains the image during the dark interval so that, under specific conditions, the image appears to be present continuously. Then, any difference in the position of an object from one frame to the next is interpreted as motion of that object.

For this process to represent visual reality two conditions must be met. First, the rate of repetition of the images must be high enough so that the motion is depicted smoothly, with no sudden jumps from frame to frame. Second, the rate must be high enough so that the persistence of vision extends over the interval between flashes. Here an idiosyncracy of natural vision[12] appears: the *brighter* the flash the *shorter* the persistence of vision. This is quite the opposite of what one would expect. The result is that bright pictures require rapid repetition. Otherwise, the visual system becomes aware that the light has been removed and the illumination is perceived as unsteady. Thus arises a persistent problem in television: flicker.

3.14 Continuity of Motion

Early in the development of motion pictures it was found that motion could be depicted smoothly at any frame rate faster than about 15 per second. This led to the establishment of 16 frames per second as a standard, a rate still used in home movie equipment. Experience in production for theaters showed that very rapid motion, so prevalent in the Wild West, could not be faithfully shown at 16 frames per second. So, despite an increase in the cost of film of 50 percent, the frame rate was changed to 24 frames per second. This remains the worldwide standard.

The repetition rate adopted for the European monochrome television service was 25 Hz, in accord with the 50-Hz power in that region, and this standard has persisted in the PAL and SECAM color systems. Since this rate does not match the 24-frame rate of motion pictures, it has become standard practice to telecast film in the PAL and SECAM countries at 25 frames per second, thus raising the pitch of the accompanying sound by 4 percent and shortening the performance by the same amount. These discrepancies have become fully accepted, and only those possessing the rare sense of perfect pitch are known to object—and then only during musical productions.

The frame rate for the NTSC system was initially chosen at 30 per second, not then to reduce flicker, but because the 60-Hz power used would otherwise cause disturbances in scanning and signal processing. The electric and magnetic fields then causing these problems are now under such strict control that the 30–60 Hz ratio is no longer required. To illustrate this fact, when the NTSC color service was introduced in 1953, the 30-Hz frame rate of the NTSC monochrome service was changed by 0.1 percent to 29.97 Hz, to maintain the visual-aural carrier separation at precisely 4.5 MHz.[13] NTSC telecine scanning equipment presents film images at the 24-Hz standard, but transposes to the 29.97-Hz scanning rate.

The 24-frame rate of motion pictures does not solve all the problems of reproducing fast action. Restrictive production techniques, such as limited camera angles and restricted rates of panning, must be observed when rapid motion is encountered. A recent proposal to change the film standard to 30 frames per second, to alleviate the rapid-motion problem and to ease film scanning conversion in the NTSC and compatible HDTV systems, has met strong resistance because of the heavy investment in 24-frame projectors in theaters.

3.15 Smear and Related Effects

The acuity of the eye viewing objects in motion is impaired by the fact that the temporal response of the fovea is slower than that of the surrounding regions of the retina. Thus, a loss of sharpness in the edges of moving objects is an inevitable aspect of natural vision, and use has been made of this fact in the design of television systems.

Much greater losses of sharpness and detail, under the general term *smear,* occur whenever the image moves across the sensitive surface of the camera. Each element in the surface then receives light not from one detail but from a succession of them. The signal generated by the camera at that point is thus the sum of the passing light, not that of a single detail, and smear results. As in photography, this effect can be reduced by using a short exposure, provided that there is sufficient light relative to the camera's sensitivity. Electronic shutters are used to limit the exposure to 1/1000 second or less when sufficient light is available.

Another source of smear occurs if the camera response carries over from one frame scan to the next. The retained signal elements from the previous scan are then added to those of the current scan, and any change in their relative position causes a misalignment and consequent loss of detail. Such "carry-over" smear occurs when the exposure occupies the full scan time, i.e., under conditions of poor illumination. A similar carry-over smear can occur in the display, when the light given off by one line persists long enough to be present appreciably during the successive scan of that line. Such carry-over helps reduce flicker in the display, and there is room for compromise between flicker reduction and loss of detail in moving objects.

3.16 Flicker

The peculiarity of the eye that demands rapid repetition of pictures for continuity of motion demands even faster repetition to avoid flicker. As theater projectors became more powerful and the images correspondingly brighter, flicker at 24 frames per second became a serious concern. To increase the frame rate without otherwise subjecting the system to change, a two-bladed rotating shutter was added to the projector, between the film and the screen. When this shutter was synchronized with the film advance, it shut off the light briefly while the film was stationary. The flash rate was thus increased to 48 per second, allowing a substantial increase in screen brightness. No adverse effect was produced on the appearance of moving objects since the frame rate remained at 24 frames per second. In due course, wider and brighter pictures became available and a 3-bladed shutter is now available, increasing the flash rate to 72 Hz.

3.17 Defects of Interlaced Scanning

No such simple interruption of the image suffices for television. To obtain two flashes for each frame, it was arranged from the beginning of public television service in 1936 to employ the technique of interlaced scanning, which divides the scanning pattern into two sets ("odd" and "even") of spaced lines that are displayed sequentially, one set fitting precisely into the spaces of the other. Each set of lines is called a *field* and the interlaced set of the two lines is a *frame*. The field rate for PAL and SECAM is 50 Hz, for NTSC, 59.94 Hz.

This procedure is the source of several degradations of image quality. While the total area of the image flashes at the rate of the field scan, twice that of the frame scan, the individual lines repeat at the slower frame rate, and this gives rise to several degradations associated with the effect known as *interline flicker*. This causes small areas of the image, particularly when they are aligned horizontally, to display a shimmering or blinking visible at the usual viewing distance. A related effect is unsteadiness in extended horizontal edges of objects, as the edge is portayed by a particular line in one field and by another line in the next. These effects become more pronounced the higher the vertical resolution provided by the camera and its image enhancement.

Interlacing also introduces aberrations in the vertical and diagonal outlines of moving objects. This occurs because vertically adjacent picture elements appear at different times in successive fields. An element on one line appears 1/50 second (or 1/59.94 second) later than does the vertically adjacent element on the preceding field. If the objects in the scene are stationary, no adverse effects arise from this time difference. But if an object is in rapid motion, the time delay causes the elements of the second field to be displaced to the right of, instead of vertically or diagonally adjacent to, those of the first field. Close inspection of such moving images shows that their vertical and diagonal edges are not sharp, but display a series of stepwise serrations, usually coarser than the basic resolution of the image. Since the eye loses some acuity as it follows objects in motion, these serrations are often overlooked. But they represent an important impairment compared with motion picture images, for which HDTV is now available as a production tool. All the picture elements in a film frame are exposed and displayed simultaneously so the impairments due to interlacing do not occur.

As previously noted, the defects of interlacing have been an important target in HDTV development. To avoid them, scanning in the camera must be progressive, using only one set of adjacent lines per frame. At the receiver, the display scan must match that at the camera.

Since conventional receivers employ interlaced scanning, they cannot respond directly to a progressively scanned image. Rather the HDTV transmission must offer a signal which, with the intervention of a converter, can be transposed from progressive to interlaced scanning. This can be arranged by retaining the progressively scanned signal in a digital-store, from which signals representing the average of two successive progressively scanned lines are formed. When this is done, much of the degradation associated with interlacing can be avoided, particularly the serrated edges of objects in motion.

The fact that the field and frame rates of PAL and SECAM (25 and 50 Hz) and those of NTSC (29.97 and 59.94 Hz) are different has affected the design of the HDTV service in two important ways. One is their relative susceptibility to flicker. The higher rates of NTSC allows its images to be about six times as bright for a given flicker limit. This difference has carried over into the HDTV systems that are designed to be compatible with the 50-Hz systems. Also, the wider visual field of HDTV images make them more prone to flicker, so in planning for HDTV displays in the PAL and SECAM services consideration is being given to the use of a field rate of 100 Hz.

The second result of the field and frame rate differences is their impact on worldwide standardization of the HDTV service. Whereas standardization within the broadcast plant (Section 1.11) may be achieved, the prospect is less certain in the transmissions to the viewing audience. If compatibility (Chapter 6) becomes the predominant economic and political force affecting the choice of HDTV standards, then the 50-Hz versus 59.4-Hz difference in the conventional services will be difficult to avoid in the rates adopted for HDTV. The plans currently announced by leading developers of HDTV technology strongly suggest that this will be the case. In that event, the hoped-for world HDTV standard would appear to be unattainable until a wholly new approach is taken decades in the future.

3.18 Visual Basis of Video Bandwidth

The time consumed in scanning fields and frames is measured in hundredths of seconds. Much less time, less than a tenth of a millionth of a second, is available to scan a picture element in HDTV. The visual basis lies in the very large number of picture elements (600,000 to 900,000, Table 3.1) that are required to satisfy the eye, when the image is under the close scrutiny offered by the wide-screen display. The eye further requires that signals representative of each of these elements be transmitted during the short time that persistence of vision allows for flicker-free images, not more than 1/25 second. It follows that the rate of scanning the elements ranges up to 22.5 million per second. This challenging requirement is directly derived from the properties of the eye. When translated into engineering terms, the rates of scanning picture elements are stated in video frequencies. At best, one cycle of video frequency can represent just two horizontally adjacent picture elements, so the scanning rate of 22.5 million picture elements per second requires video frequencies up to about 11 MHz. Finding the means to accommodate such wide bands of frequencies is the outstanding achievement in the development of the HDTV service. Specifics have been treated in Chapter 2.

Table 3.2 lists the temporal aspects of the several systems whose spatial features are shown in Table 3.1.

3.19 Sharpness of Images in Television and Motion Pictures

Comparisons of the ability of television and motion pictures to present sharply defined pictures have occupied much attention during the development of HDTV. A well-regarded summary[14] of the opinions of qualified specialists appears in Figure 3.6. This diagram relates two measures of sharpness (definition as seen subjectively): numerical and verbal. On the curve relating them are located the regions of sharpness offered by various types of color film and by 525-

Table 3.2 Temporal Characteristics of Television Systems

System	Total channel width	Video basebands			Scanning rates		
		Y	*R-Y*	*B-Y*	Camera	HDTV display	Conventional display
HDTV USA	9.0	10.0	5.0	5.0	59.94-p	59.94-p	59.94-i
HDTV Europe	12.0	14.0	7.0	7.0	50-p	100-p	50-i
HDTV Japan	30.0	20.0	7.0	3.0	60-i	60-i	NA
NTSC	6.0	4.2	1.0	0.6	59.94-i	NA	59.94-i
PAL	8.0	5.5	1.8	1.8	50-i	NA	50-i
SECAM	8.0	6.0	2.0	2.0	50-i	NA	50-i

and 1000-line color television images, all at a viewing distance of five times the picture height. These expert estimates stand behind the widely stated claim that HDTV systems offer performance equal to that of 35-mm film as projected in theaters.

3.20 Aural Component of Visual Realism

The realism of a television image depends to a great degree on the realism of the accompanying sounds.[15] Particularly in the close viewing of HDTV images, if the audio system is monophonic, the sounds appear to be confined to the center of the screen. The visual and aural senses then convey conflicting information. So, from the beginning of HDTV system design, it has been clear that stereophonic sound must be used. The NTSC sound standard was converted to stereo in 1985, and by 1988 approximately half of the terrestrial broadcast stations then operating in the United States and many cable systems could offer stereophonic reproduction when appropriate program material was available. When the FCC set the standard, provision was made for a second sound channel intended primarily for offering a second language (in particular, Spanish). The multilanguage problem is especially severe in Europe. The direct satellite

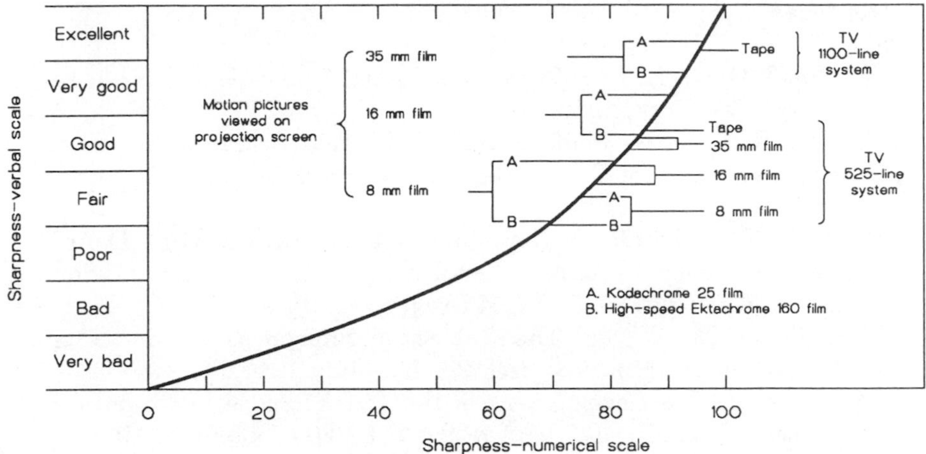

Figure 3.6 Comparative sharpness (definition as seen subjectively) of motion-picture film projection and television systems when the images are viewed at five times the picture height. (From Reference 14.)

broadcast service there provides channels for two languages, with provision for expansion.

The quality of the sound accompanying HDTV images should be as realistic as the state of the art allows. The quality standard for high-fidelity sound has been set by the digital compact disk. This medium covers the audio frequencies from below 30 to above 20,000 Hz, beyond the range of most ears, and the dynamic range from silence to fortissimo exceeds 90 dB (a sound pressure range of better than 30,000 to 1).

Cost factors and cabinet space have limited the sound performance of conventional receivers to levels far below these limits. The experience of the "simulcast," in which the accompanying sound is provided by an fm stereo broadcast reproduced on high-fidelity equipment, gives ample evidence of the effect of sound quality on the impact of the visual display.

While the factors that limit sound quality in conventional receivers apply to some extent to HDTV receivers, the intent of HDTV system designers is to approach the performance of the compact disk as closely as the limits of the distribution system permit. This involves transmitting HDTV sound digitally, while maintaining compatible sound in the analog form specified for the the NTSC, PAL, and SECAM services.

References

1. Hubel, David H. *Eye, Brain and Vision.* Scientific American Library, New York, 1988.
2. Kell, R. D., A. V. Bedford and M. Trainer. Scanning Sequence and Repetition of Television Images. *Proc. IRE* 24: 559 (April 1936).
3. Fujio, T., J. Ishida, T. Komoto and T. Nishizawa. High Definition Television Systems—Signal Standards and Transmission. *SMPTE Journal* 89: 579–584 (August 1980).
4. Baldwin, M. W., Jr. The Subjective Sharpness of Simulated Television Images. *Proc. IRE* 28: 458 (July 1940).
5. Belton, J. The Development of the CinemaScope by Twentieth Century Fox. *SMPTE Journal* 97: 711–720 (September 1988).
6. Benson, K. Blair (Ed.). *Television Engineeering Handbook.* McGraw-Hill, New York, 1986, pp. 21.57–21.72.
7. Fink, D. G. Perspectives on Television: The Role Played by the Two NTSC's in Preparing Television Service for the American Public. *Proc. IEEE* 64: 1322–1331 (September 1976).
8. Judd, D. B. The 1931 C.I.E. Standard Observer and Coordinate System for Colorimetry *J. Opt. Soc. Am.* 23: 359–374 (1933).
9. Kelly, K. L. Color Designation of Lights. *J. Opt. Soc. Am* 33: 627–632 (1943).
10. Benson, K. Blair (Ed.). *Television Engineering Handbook.* McGraw-Hill, New York, 1986, p. 21.53.
11. Pointer, R. M. The Gamut of Real Surface Colors. *Color Res. App.* 5: 145–155 (1945).
12. Porter, J. C. *Proc. Roy. Soc. (London)* 86: 945 (1912).
13. Fink, D. G. *Color Television Standards.* McGraw-Hill, New York, 1955, pp. 108–111.
14. Fink, D. G., et. al. The Future of High-Definition Television. *SMPTE Journal* 89: 89–94, 153–161 (February and March 1980). This diagram was prepared by R. M. Wilmotte (consultant), A. Rose of RCA, and C. N. Nelson of Eastman Kodak.
15. Suitable Sound Systems to Accompany High-Definition and Enhanced Television Systems—Report 1072. Recommendations and Reports of the CCIR, 1986. Broadcast Service—Sound. International Telecommunications Union, Geneva, 1986.

Chapter

4

Digital Operations in Video Systems

4.1 Digital Functions in Television

As pointed out in previous chapters, digital techniques have not only established themselves in the transmitters and receivers of the conventional services, but they have become a central feature in the design and implementation of advanced and HDTV systems. This chapter is devoted to the fundamental processes involved in digital video.

Broadly speaking, digital methods in television provide the three "p's" of signal technology: processing, protection, and preservation. Digital signal *processing* is used, as previously noted, in functions following the second detector of receivers; in standards conversion among the NTSC, PAL, and SECAM services; in scanning conversion between the interlaced and progressive modes; and in bandwidth compression for the MAC and MUSE direct broadcast satellite systems.

Digitization of signals provides *protection* against noise, reflections, and accumulated errors in networks and in successive generations of recordings. Finally, storage of digitized signals in memory banks permits their *preservation* for such transient purposes as still-frame and picture-within-picture displays, or in permanent storage on digital tape or disks.

Common to all these uses is the conflict between the advantages provided by digital signals and the wide bandwidths they occupy. An example is the format prescribed in the CCIR international standard for digital operations within studios. This standard calls for 216 million binary digits per second, occupying a bandwidth up to 108 MHz, more than ten times the spectrum required by the older (and still prevalent) analog circuits in studios. Such bandwidths are not now available in any of the media that serve the public, although they are offered in the fiber-optic networks of the telephone companies and in microwave radio relay circuits. In studio practice, digital processing is confined to individual items of equipment or assemblies with short interconnections.

4.2 Digital Entities: Bits, Bytes, and Words

The digital video signal is made up of binary digits or *bits,* signals that are either "present" or "absent." The bit (b) is often shown as a rectangular waveform, but it may have any form that suffices to distinguish presence from absence. In electrical terms the bit has three forms: "on" versus "off," "plus" versus "minus," and "1" versus "0." Figure 4.1 shows that these are different in that a downward pulse can represent "off" in the "plus-minus" version. A shift in the d-c level of the "on-off" signal also provides a "plus-minus" variation.

In computer science a sequence of 8 bits is a "byte." This number of bits is sufficient to represent all of the characters (lowercase and capital letters, numbers, punctuation marks, etc.) defined in the American Standard for Computer Information Interchange (ASCII) standard. The memory capacity of integrated circuits is stated in bytes. Thus, a "256-kilobyte" random-access memory (RAM) chip can store 262,144 bytes or 2,097,152 bits.

The general term for a collection of bits is the *word,* usually of stated length. In computers the word length is the bit capacity of each memory position; it varies from 16 bits in a personal computer to 60 bits in a supercomputer. In television, the common word lengths are 7, 8, or 9 bits, the 8-bit word being most commonly used. Figure 4.1 shows three versions of an 8-bit word, on-off and plus-minus, with the corresponding designations of 1's and 0's.

Since each bit in an 8-bit word can be either a 1 or a 0, there are 256 (= 2^8) different ways that the 1's and 0's can be arranged. Thus, an 8-bit word can represent each of 256 different signal levels, with each level representing 1/255 = 0.39 percent of the total range. The error in amplitude is approximately 0.4 percent after quantization

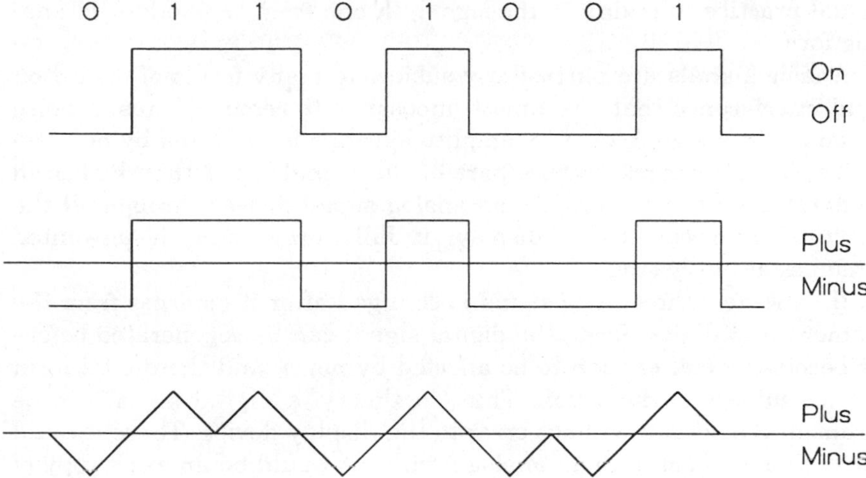

Figure 4.1 An 8-bit digital word with on-off versus plus-minus indications of a 1-bit's presence or absence.

(Section 4.5). This fine distinction among amplitudes is sufficient for high-quality representation of the video signal, and the 8-bit word was accordingly standardized by the CCIR. The 7-bit word, primarily used in receiver processing, has 128 levels, i.e., an uncertainty in level of about 0.8 percent. The 0.4 percent uncertainty is equivalent to 48 dB below the maximum signal level, the 0.8 percent to 42 dB. This 42- to 48-dB range in the signal circuits compares with the 48-dB signal-to-noise ratio commonly accepted as good performance at the output of a color camera. The 9-bit word is used for additional precision in processing within digital video equipment.

4.3 Analog and Digital Signals

The signal produced by the camera is an analog of the light falling on its sensitive surfaces, distributed in width and height, and discerned by the scanning beams, in time, as a sequence of successive picture elements arranged in fields and/or frames. The camera output is three continuously varying voltages which, from time to time, may have any value within the operating range. The same type of continuously varying voltages is required at the three input terminals of the picture tube. Between camera and display it has been the usual practice to maintain the signal, in one form or another, in analog form.

Analog signals are particularly subject to many forms of distortion and interference that are almost impossible to remove. Thus, if noise is added to a signal, or if its amplitude range is distorted by nonlinearity, these changes become part of the signal, as if they had been generated by the camera. As an analog signal passes through all the circuits from camera to display, it falls prey to such unwanted changes at every stage.

If, however, the analog signal is changed after it emerges from the camera into digital form, the digital signal can be regenerated before it becomes weak enough to be affected by noise, and the digital form is not subject to distortion. Thus, in theory, a digital signal can be transmitted totally without error to the display device. There it could be converted back into an analog form that would be an exact copy of the camera output.

In practice, there are types of interference and other forms of damage to which digitized signals are subject, but protection against them can be obtained by adding extra bits, not necessary for signal content, but to provide redundancy and error correction. To carry out this process, the analog signal must first be converted into digital form (analog-to-digital, or A-to-D conversion) and after transmission and processing back into analog form (digital-to-analog, or D-to-A) conversion).

4.4 Analog-to-Digital Conversion

To convert from the analog to the digital form, it is necessary to create a succession of digital words that comprise only two discrete values, 0 and 1. Figure 4.2 shows the essential elements of the analog-to-digital converter. The input analog signal must be confined to a limited spectrum (to avoid spurious components in the reconverted analog output, Section 4.6), so a low-pass filter precedes the converter. The converter proper first *samples* the analog input, measuring its amplitude at regular, discrete intervals of time. These individual amplitudes are then matched, in the *quantizer,* against a large number of discrete levels of amplitude (256 levels to convert to 8-bit words, Section 4.2). Each one of these discrete levels can be represented by an 8-bit word, its digital equivalent. The process of matching each discrete amplitude with its unique word is carried out in the *encoder,* which in effect scans the list of words and picks out the one that matches the amplitude then present. The encoder passes

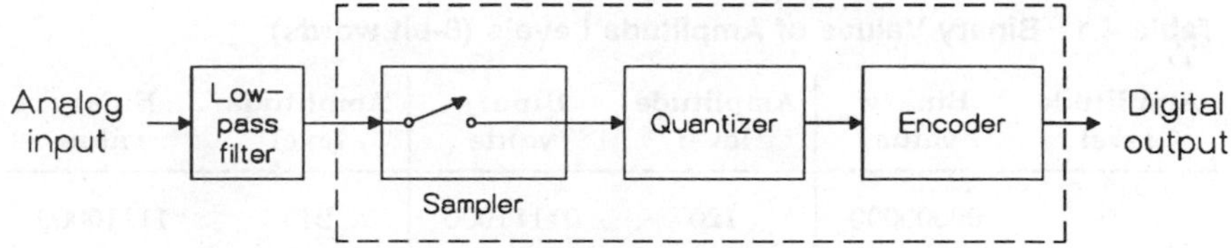

Figure 4.2 Elements of an analog-to-digital converter.

out the series of code words in the sequence corresponding to the sequence in which the analog signal was sampled. This "bit stream" is the digital version of the analog input.

The list of digital words corresponding to the sampled amplitudes is known as a *code*. Table 4.1 represents a simple code showing amplitude levels and their 8-bit words in three ranges: 0–15, 120–135, and 240–255. Signals encoded in this way are said to be *pulse-code-modulated* (PCM). Table 4.1 is compiled by adding the weights ("place values" in decimal form) of each 1 bit in a word. Reading from right to left, these values are 1, 2, 4, 8, 16, 32, 64, and 128. The amplitude level is equal to the sum of those place values where 1's appear. The amplitude level 120 in the table has a binary value 01111000, with 1's in the 8, 16, 32, and 64 places. Its amplitude value is thus 8 + 16 + 32 + 64 = 120.

While the PCM code is sometimes used, more elaborate codes, having many additional bits per word, are generally used in transmitting through equipment or over circuits where errors may be introduced into the bit stream (Section 4.8). Figure 4.3 shows a typical video waveform and several quantized amplitude levels based on the PCM code scheme of Table 4.1.

The great speed at which the converter must match the analog samples to the digital code words (hundreds of millions of bits per second) cannot be handled by circuits that operate sequentially. Instead, parallel channels are provided, one for each of the 256 code words in 8-bit sampling. This is a prime example of the great dependence of digital video techniques on integrated circuits. An evident complication is that each of these parallel channels must operate in precise time synchronism. This is arranged by a common time source, the *clock*. All converters require a clock to set the sampling rate and to control the parallel channels.

A simplified diagram of such a parallel A-to-D converter[1] (a *flash-converter*) is shown in Figure 4.4. A fixed reference voltage is applied across 255 identical resistors in series. Taps are taken off from each resistor. The bottom tap is zero voltage (the 256th level), and each

Table 4.1 Binary Values of Amplitude Levels (8-bit words)

Amplitude level	Binary value	Amplitude level	Binary value	Amplitude level	Binary value
0	00000000	120	01111000	240	11110000
1	00000001	121	01111001	241	11110001
2	00000010	122	01111010	242	11110010
3	00000011	123	01111011	243	11110011
4	00000100	124	01111100	244	11110100
5	00000101	125	01111101	245	11110101
6	00000110	126	01111110	246	11110110
7	00000111	127	01111111	247	11110111
8	00001000	128	10000000	248	11111000
9	00001001	129	10000001	249	11111001
10	00001010	130	10000010	250	11111010
11	00001011	131	10000011	251	11111011
12	00001100	132	10000100	252	11111100
13	00001101	133	10000101	253	11111101
14	00001110	134	10000110	254	11111110
15	00001111	135	10000111	255	11111111

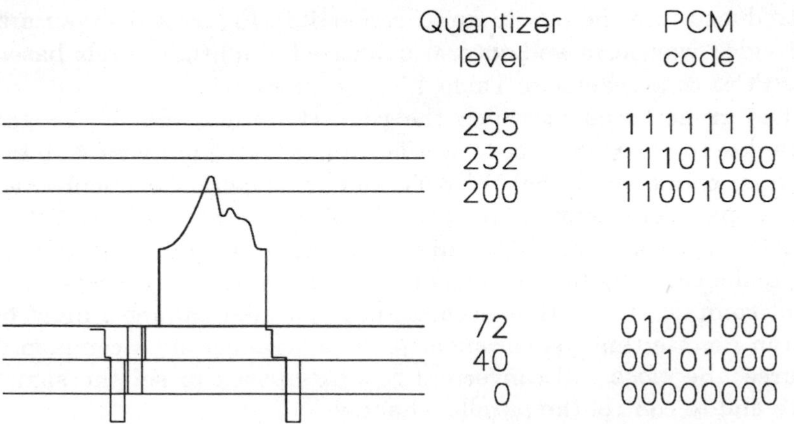

Figure 4.3 Video waveform quantized into 8-bit words (compare amplitudes and words with Table 4.1).

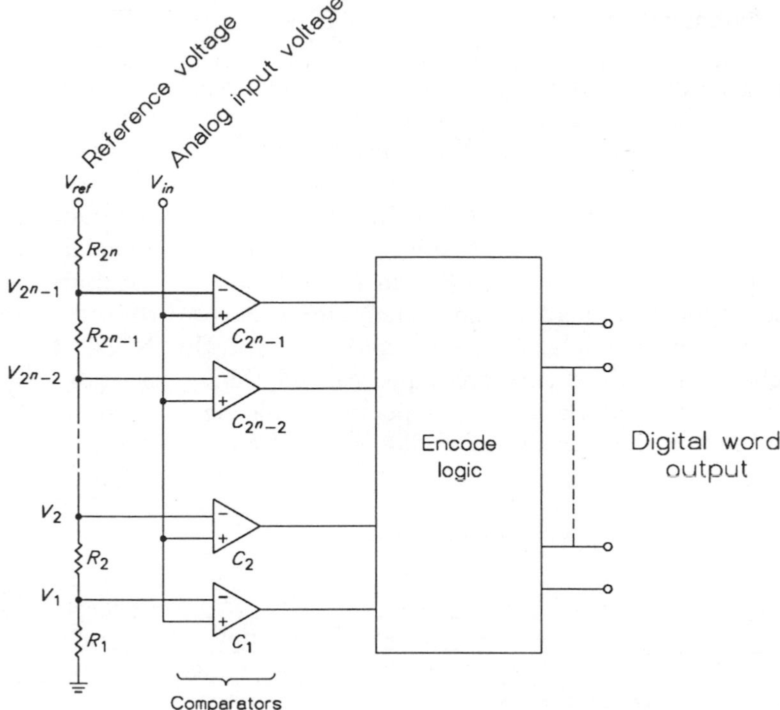

Figure 4.4 Parallel operation of "flash" type of an analog-to-digital converter using resistors in series to establish quantized levels.

successively higher tap presents a higher voltage than the one below it, by 1/255 of the reference voltage. Each tap leads to one input of an individual comparator. Its other input is connected to the analog input signal. As the analog signal varies, each comparator determines, at each sample time, if the analog value is above or below the reference voltage applied to it. If above, the comparator generates a "0," if below, a "1." The comparator's outputs are fed to a converter, which is timed to accept the comparator digits in parallel and convert them into serial form, producing an 8-bit word for each sample.

Even so elaborate a device as the flash converter cannot operate at the hundreds of millions of bits per second required by the CCIR studio standard, so two flash converters may be used in cascade, one in effect handling the upper range of reference voltages, the other the lower range, the two outputs being compared to determine where the first and last 0's and 1's appear.

4.5 Quantization Error

Analog-to-digital conversion is not an exact process, since the comparison between the analog sample and a reference voltage is uncertain by the amount of the difference between one reference voltage and the next. The uncertainty amounts to plus or minus one half that difference. When words of 8 bits are used, this uncertainty occurs in essentially random fashion, so its effect is equivalent to the introduction of random noise ("quantization noise"). Fortunately such noise is not prominent in the analog signal derived from the digital version. For example, in 8-bit digitization of the NTSC 4.2-MHz baseband at 13.5 megasamples per second (Ms/s), the quantization noise is about 60 dB below the peak-to-peak signal level, far lower than the noise present in the analog signal from the camera.

4.6 The Nyquist Limit and Aliasing

A definite rule must be observed in sampling an analog signal if it is to be reproduced without spurious effects known as *aliasing*. That rule is that the time between samples must be short compared to the rates of change of the analog waveform. Nyquist[2] first stated this rule in 1924. In video terms, the sampling rate in megasamples per second must be at least twice the maximum frequency in MHz of the analog signal. Thus, the 4.2 MHz maximum in the luminance spectrum of the NTSC baseband requires that the NTSC signal be sampled at 8.4 megasamples per second, not less. Conversely, the 13.5 megasamples per second rate specified in the CCIR studio digital standard can be applied to a signal having no higher frequency component than 6.75 MHz. If studio equipment does exceed this limit, and many cameras and associated amplifiers can do so, a low-pass filter must be inserted in the signal path before the conversion from analog to digital form takes place. A similar band limit must be met at 3.375 MHz in the chrominance channels before they are digitized.

If the sampling occurs at a rate lower than the Nyquist limit, the spectrum of the output analog signal contains spurious components, which are actually higher-frequency copies of the input spectrum that have been moved down so that they overlap the desired output spectrum. When this output analog signal is displayed, the spurious information shows up in a variety of forms, depending on the subject matter and its motions. Moiré patterns are typical, as are distorted and randomly moving diagonal edges of objects. These aliasing effects often cover large areas and are visible at normal viewing distances, so care must be taken to meet or exceed the Nyquist limit in all analog-to-digital conversions.

Aliasing may occur not only in digital sampling, but whenever any form of sampling of the image occurs. An example long familiar in moving pictures is the appearance of the wheels of vehicles that seem to move backward. This occurs because the image is sampled by the camera at 24 frames per second. If the rotation of the spokes of the wheel is not precisely synchronous with the film advance, another spoke takes the place of the adjacent one on the next frame, at an earlier time in its rotation. The two spokes are not separately identified by the viewer, so the spoke motion appears reversed.

There are many other examples of image sampling that occur in television. The camera samples the scene vertically during every field, or every frame in progressive scan, and there are minor differences in successive scans that extend over as many as four fields (roughly 1/15 second in the NTSC system) before the sequence repeats. The display similarly offers a series of samples in the vertical dimension, with results that depend not only on the time-versus-light characteristics of the display device but also, more importantly, on the time-versus-sensation properties of the human eye. These aspects of sampling have an important bearing on the design of advanced and HDTV systems. They are treated in Chapter 5, which deals with the spatial and temporal analysis and synthesis of the imaging process.

4.7 Digital-to-Analog Conversion

The digital-to-analog converter for 8-bit words has eight inputs, one for each bit in the word. Its function is to translate each word into its corresponding quantized amplitude level, i.e., proceeding from the 8-bit word in the "binary value" column of Table 4.1 to the "amplitude level" immediately to its left. In this way, the succession of words in the bit stream produces a succession of amplitudes which trace out the analog waveform. The amplitudes thus generated are not identical in size to the input amplitude samples, because of the quantization error (Section 4.5), so the analog output wave displays a series of minute steps. The spectral content of these steps is outside the spectrum of the input analog signal, so they are readily removed by passing the D-to-A converter output through a low-pass filter like that at the input of the A-to-D converter (Figure 4.4).

The D-to-A converter is faced with the same requirement for high-speed operation as the A-to-D converter, since its input words may arrive at rates in the millions of bits per second. It operates by determining the place value of each 1 bit in the word (Section 4.4). The leftmost bit contributes 128 to the amplitude, the rightmost 1, and the intermediate bits accordingly in the manner represented in Table 4.1.

Figure 4.5a shows a typical D-to-A converter that operates by creating currents that match the place value of each 1 bit in the word and adding the currents to create the corresponding amplitude level, which serves as the output amplitude for that word. One form of current generator is shown in Figure 4.5b. It consists of identical resistances in parallel groups of 1, 2, 4, 8, 16, 32, 64, and 128 each. When a reference voltage is applied across all the resistors, the currents through the parallel groups range from 1 to 128, respectively. Each current value is selected by one of eight switches, which responds to the 0 or 1 bit it receives, switching on when a 1 bit is applied, remaining off for a 0 bit. The sum of the switched-on currents then represents the output amplitude corresponding to the input word. The current is converted to voltage by passage through a stabilized operational amplifier.

4.8 Protection of Digital Signals in Storage and Transit

By regeneration of the digital bit streams at regular intervals, most of the disturbances to which analog signals are exposed may be avoided. But errors can be introduced at any stage in the transmission, storage, or retrieval of digital signals by errors in timing, by the introduction of bits from extraneous sources, or by their omission as a result of failure in counting and recognition. To minimize the effect of such inaccurate transmission and processing functions, much effort has been spent on methods to conceal or correct errors in digital functions. In general, the protection and correction of digital signals has involved the introduction of at least 1 and often as many as 4 bits to an 8-bit word. This proportionately increases the bandwidth occupied at a given sampling rate, but the signal protection is deemed well worth the spectrum cost, particularly when long-distance transmission is not involved.

The word with its extra bits is known as a *block*, identified by the total bits and the bits carrying signal information. Thus, a (10,8) block embodies 10 bits: 8 bits for information and 2 for error protection. A simple code employs (9,8) blocks, the extra bit being known as a "parity bit." The encoder counts the number of 1's in the 8-bit

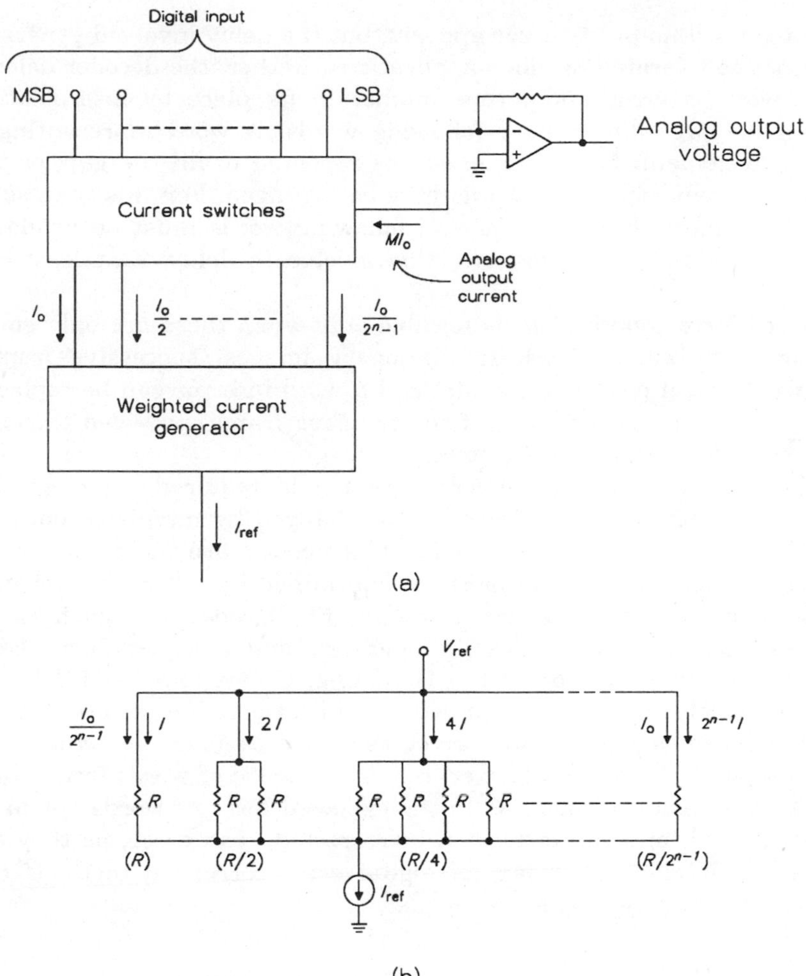

Figure 4.5 Parallel operation of a digital-to-analog converter, which adds currents corresponding to the places of the 1 bits in the word. (a) Elements of the converter; (b) detail of current generator consisting of sets of identical resistors in parallel.

word. If this count is an odd number, it appends a 1 to make the 9-bit word; if an even number, a 0 is added. Thus, all the 9-bit words contain an even number of 1's. The decoder similarly counts the number of 1's in each 9-bit word and if that number is odd, that word is known to be incorrect in at least one of its bit positions. The decoder might then wait for that particular 9-bit word to be retransmitted until no parity error appears, but the delay involved generally cannot be tolerated in video applications. Rather, the decoder deletes the word in error and forms another in its place by interpolation from the preceding and/or following words. A word representing a picture element, for example, can be repeated to fill the gap, or the preceding and following words may be averaged. For this process to be performed, the preceding and following words must be available in storage and the error correction involves a delay of at least one line scan.

Such *error concealment* is feasible only when there are only small changes between words. In stationary images, successive frames have identical picture elements, and a word in error can be replaced by the corresponding word of the previous frame, provided that the previous frame's words are stored.

Much more elaborate error correction is required when several successive bits in a word are lost or changed by impulsive noise or other interference. The correction bits needed are more numerous, and their position in the block is determined by a complex scheme, known as a *forward error correction* (FEC) code. One such set of codes was devised by Bose, Chaudhuri, and Hocquenghem (BCH codes), a generalization of the Hamming, Golay, and Reed-Solomon codes widely used in digital video applications. These are complicated schemes which can correct as many as four bit errors in a 12-bit block. To do so, however, requires storage of words from which predictions can be made as to the expected form of words yet to be transmitted, and the latter can be corrected, if in error, as they appear. A detailed treatment of digital error correction methods has been prepared by Tarnai.[3]

4.9 Post-Detector Digital Processing

Essentially all of the signal processing following the second detector in television receivers is now performed in integrated circuits. The majority of these operations are performed with analog signals, but digital methods have recently been introduced. Figure 4.6 shows the digital elements in a typical receiver. The analog signal from the second detector is sampled at 13–14 Ms/s (13.5 Ms/s in a typical case) into words of 7 bits each. The digitized signal feeds two processors, for scan synchronization and for luminance and chrominance processing.

The sync block contains a clock operating at a multiple of the subcarrier frequency. This clock controls the sampling rate in the A-to-D

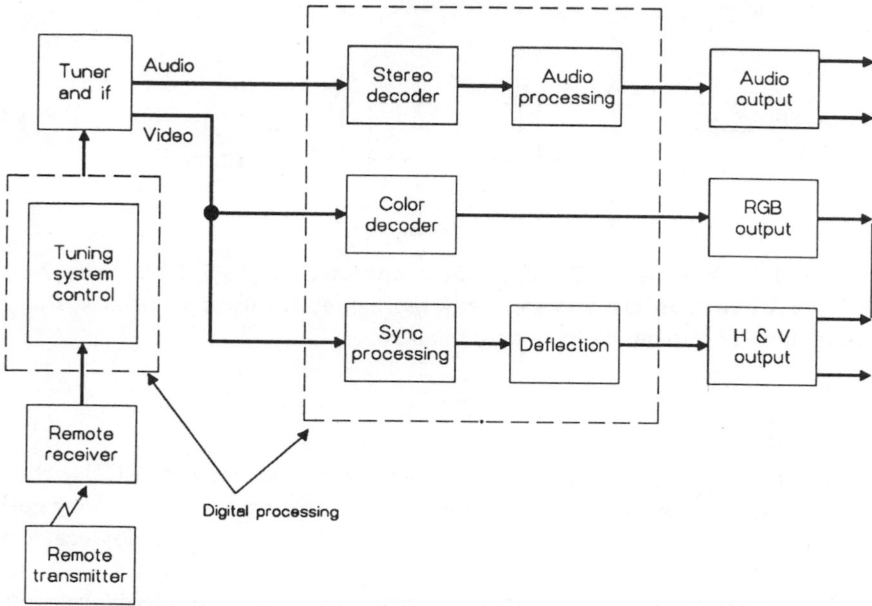

Figure 4.6 Digital functions following the second detector of a conventional receiver.

converter and all the other digital functions of the receiver. The code words corresponding to the analog vertical and horizontal synchronization signals are regenerated to remove the effects of noise or other impairments and are used to control the sweep circuits. The words corresponding to the color synchronization, luminance, and chrominance are passed through a comb filter to separate the respective sidebands.

To produce a one-line delay, this filter must temporarily store 6006 bits in the NTSC system, sampled at 13.5 Ms/s, 7 bits per word. Such large numbers are readily attainable in integrated circuits. These digital comb filters are connected as shown in Figure 4.7.

After filtering, the luminance signal *Y* and the color-difference signals *I* and *Q* are processed separately to form the digital words for *Y*, *R-Y*, and *B-Y*. These are then matrixed in digital form to create the words for *R*, *G*, and *B*. At this stage three D-to-A converters produce the analog versions for the display. Analog signals from other sources, such as tape recorders, are fed in at this point. The A-to-D and D-to-A converters employ the principles described in Sections 4.4 and 4.7.

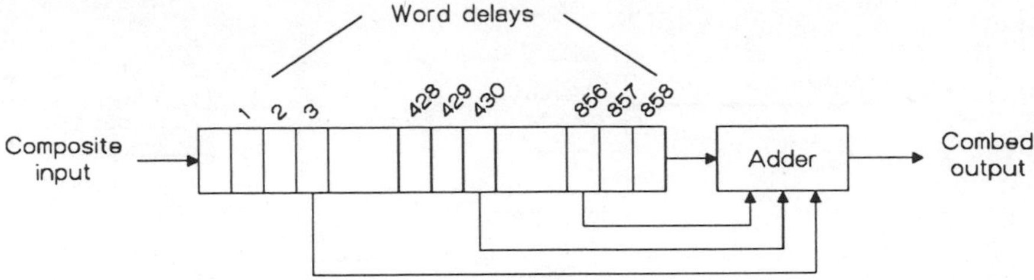

Figure 4.7 One-line CCD comb filter operated at 13.5 Ms/s with 7-bit words. Three taps are shown; many more may be used to secure sharp separation of luminance from chrominance.

The digital comb filter employs *charge-coupled devices* (CCDs) that pass charges (representing the successive bits) from one memory cell to the next as each bit enters the first cell. These CCD devices are described in Section 4.10.

More elaborate digital functions are performed in receivers that not only store the bit content of one line at a time, but also have storage capacity for all the 7-bit words in a field scan (field-store) or a frame scan (frame-store). Use of a field-store to convert from interlaced to progressive scanning was described in Section 2.3.

Another important use of field and frame storage is noise reduction by comparing the digitized signals of two successive scans of the same line. If the digital content of a frame-store is compared, word by word, with that of the next frame scan, any difference between corresponding words can be corrected by error concealment. This process can offer as much as 10 dB improvement in the signal-to-noise ratio of stationary images, but since the corresponding words differ in the parts of an image in motion, the comparison process must be disabled in moving parts of the scene.

Storage of fields or frames can also be used to display still pictures ("freeze frames"). After one frame is stored, the signal source is disconnected, and the frame-store is read out repetitively, the bit stream being presented to the processors shown in Figure 4.6. By timing the readout to occur during a portion of the conventional picture scan, a smaller picture derived by storage from another channel may be inserted in the display. By restricting the number of bits in the inserted images, memory may be available for several different such *picture-in-picture* displays.

4.10 Digital Memory Cells

The storage, delay, or transfer of digital words requires mechanisms for placing the corresponding bit signals into known locations and later retrieving them. The individual location where a bit is stored is known as a *memory cell.* The quantity stored is electrical charge, trapped in a capacitance in, or associated with, a semiconductor device. Enough charge must remain to be recognized when the information is retrieved, so the memory cell must be designed to avoid excessive leakage during the storage time. Moreover, at the time the bit storage is "read" to determine whether a 0 or a 1 is present in the memory, the stored charge must not be removed. If the charge is removed ("destructive readout") it must be replaced so that the memory pattern is preserved.

4.10.1 Charge-Coupled Cells

The charge-coupled device (CCD) is designed to pass the stored charge representing a bit from one memory cell to the next under the timing control of the system clock. A bit signal entering the CCD thus emerges from it after a delay equal to the number of cells times the clock interval. One example (Section 4.9) is the digital comb filter that delays the bit stream for the duration of a line scan. Such a CCD, operating in the NTSC system for the full line scan time at the usual clock speed of 13.5 Mb/s, must have room to accommodate 858 words of 7 bits each, which requires no less than 6000 individual memory cells.

Figure 4.8 shows four of the memory cells of a "surface-channel" CCD. Each cell is field-effect transistor (FET). The charge is stored at the surface of the *p*-type semiconductor, below the storage plate. Clock pulses are applied to the storage plates and to the transfer plates between them. When the clock voltage on the transfer plates drops below that on the storage plates, the stored charge moves to the right, occupying the surface under the next storage plate. This shift recurs with each clock cycle. In this way the succession of bits entering the input (leftmost cell of the whole CCD) appears at the output (rightmost). Leakage from the cells can occur, so the switching from cell to cell must occur rapidly. The clock rates employed in digital video are high enough to avoid this problem.

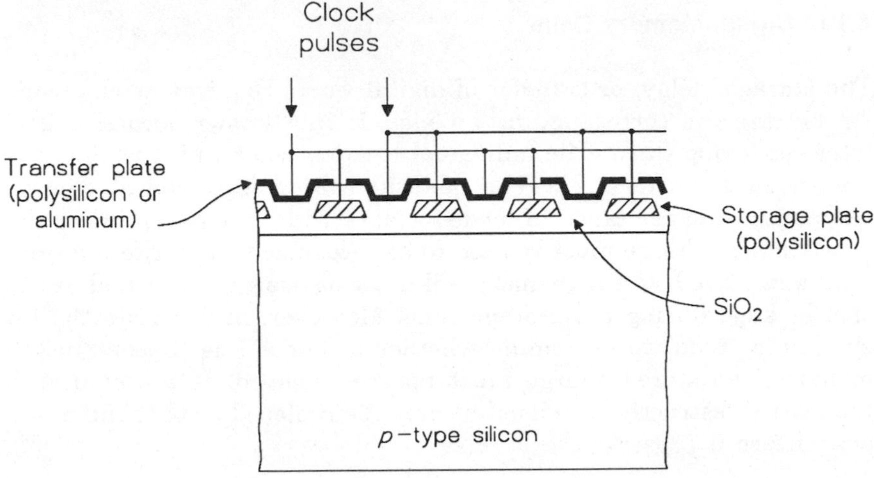

Figure 4.8 Four memory cells of a charge-coupled device showing trapped charge, storage, and transfer plates, and the clock voltage that moves the stored charge one cell to the right for each clock pulse.

The digital CCD is often referred to as a *digital shift register*. The principle of charge-shifting also can be employed with samples of analog signals in a *bucket brigade device* (BBD). One such application is in CCD color cameras.

4.10.2 Dynamic Access Cells

When the content of a memory cell is to be introduced and later retrieved, one cell at a time, it is necessary that each cell have an *address* at which it may be found, and a means of inserting the charge to be stored and of determining at a later time whether charge is stored. Thus, at a minimum, each memory cell must have two connections: one (the *word-line*) to address the position of the cell in the appropriate row of the memory array and the other (the *bit-line*) to introduce the charge and to determine its presence or absence.

Figure 4.9 shows the elementary form of the *dynamic memory cell* with its word-lines and bit-lines. It consists of a field-effect transistor T and a capacitor C. The word-line opens the transistor, permitting the bit-line to add charge to, or remove it from, the capacitor. In the integrated circuit, the area of the capacitor is very small and the

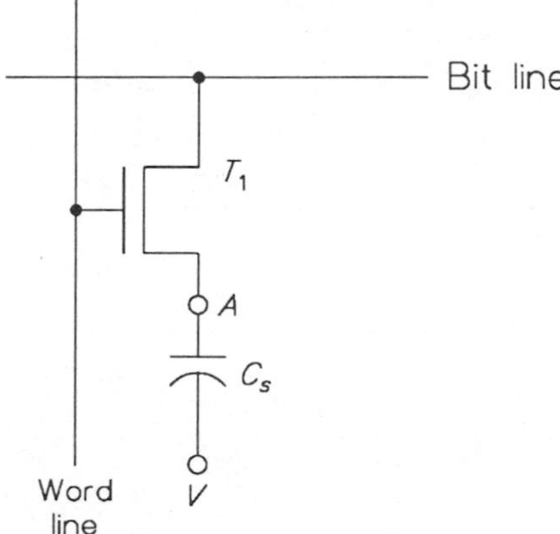

Figure 4.9 Elementary form of the dynamic memory cell, consisting of a transistor controlled by the word-line and a capacitor charged or discharged through the bit-line.

charge so minute that amplification is needed in the readout process. This is provided by an auxiliary transistor at each cell.

This type of memory cell is called "dynamic" because when the charge is read out, it is removed from the capacitor and the stored bit is lost. This lost charge must be replaced (*refreshed*) at once by an auxiliary circuit. This requirement increases the complexity of the memory circuits, but the simplicity of the basic cell outweighs this detriment, so large memories usually employ cells of the dynamic type.

4.10.3 Static Access Cells

A more complicated circuit has the property of retaining its stored bit during the readout process. Figure 4.10 shows such a *static memory cell.* It consists of field effect transistors connected in criss-cross fashion. When one transistor is not conducting, its opposite neighbor conducts, and this condition is maintained indefinitely until the first transistor is forced into conduction. Then the two transistors exchange roles, and the exchanged condition again remains fixed until another exchange is imposed. The conductive condition of the transistors is taken to represent the presence of a 1 bit. When the word-line opens the bit-line connections, each bit-line can "write in" a bit (cause the conduction condition to occur in its associated transistor)

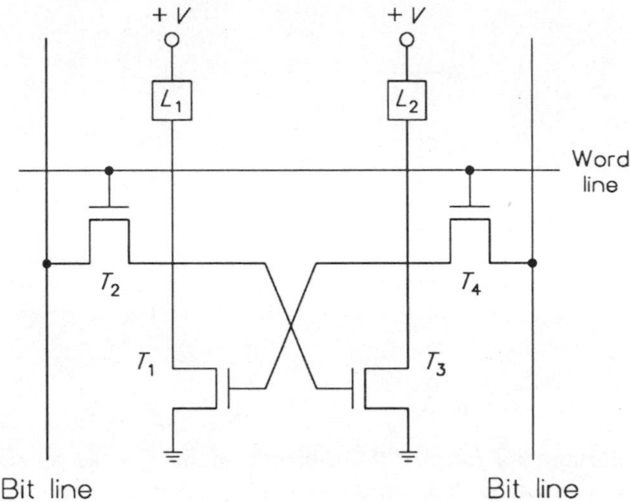

Figure 4.10 Static memory cell, with transistors in crossed connection ("flip-flop" circuit) and associated word-line and bit-lines.

or "read out" a bit (reveal what that condition is). The storage of a read-in bit is maintained after the word-line closes the bit-line connections. The static access cell, while straightforward from an operational point of view, involves larger areas in the integrated circuit than does the dynamic cell and its refresh circuitry.

4.11 Memory Arrays

To handle the vast amount of information typical of digital video applications, very large arrays of memory cells must be available and each cell in the array must be reached, without error, for the storage or retrieval of a bit. Two forms of such arrays are used, for handling digital data and for storing instructions, respectively. The first must be highly flexible, i.e., every bit in the array must be capable of being examined and of being changed individually. The second is wholly inflexible; the bit content of each of its cells remains unchanged at all times. They are known, respectively, as RAM and ROM arrays.

4.11.1 Random-Access Memories

In many digital applications, it is necessary to write in and read out the content of memory cells at particular locations in the memory array, without affecting the contents of other cells. This process is known as *random access*, and the array that provides it is a *random-access memory* (RAM). The DRAM uses dynamic cells (requiring refreshing); the SRAM uses static cells. DRAMs are used almost exclusively in video work, because they are smaller, cheaper, and use less power.

Since every memory cell (Section 4.10, above) has a word-line and a bit-line, these lines must be carefully organized to permit fast and accurate storage and retrieval to and from many thousands of cells. It is a remarkable achievement that the 2,097,152 cells in a 256-kilobyte RAM chip can be addressed and read individually via 16 external connections on the component package.

The arrangement of word-lines and bit-lines is illustrated in the right-angle array of Figure 4.11. When a particular word-line is activated, all the cells on that line are "open" to receive a bit or to reveal a bit's presence during the write and read functions, respectively. Only one word-line is activated at a time, and the order in which they are opened is determined by the device at the left of the array, the *address decoder*. This accepts at its input the *address word* that specifies the locations at which the bits of the *data word* are to be found or stored.

The reading and writing of the stored bits is performed by the device at the bottom of the array, the *memory buffer*, which provides a two-way connection to each of the bit-lines. During the memory input (write) phase, it delivers the appropiate bit to each bit-line to form the word to be stored in memory. During the output (read) stage, it senses the bit stored at each of the open cells, thus recovering the stored word. When the next word-line is opened by the address decoder, the processing of writing or reading bits is repeated for the next data word by the memory buffer. The address decoder operates with words large enough to select among many thousands of word-lines; and the memory buffer may store sequentially, in a charge-coupled register, many data words, feeding them into or out of the memory in the proper sequence.

4.11.2 Read-Only Memories

Some parts of a digital memory system are not required to change. This is true, for example, when a fixed program of digital instructions is required to control the functions of other parts of the system. Such instructions are stored digitally in a read-only memory (ROM). Structurally, the ROM resembles the RAM, but the storage of charge in the ROM memory cells is fixed by connections built into

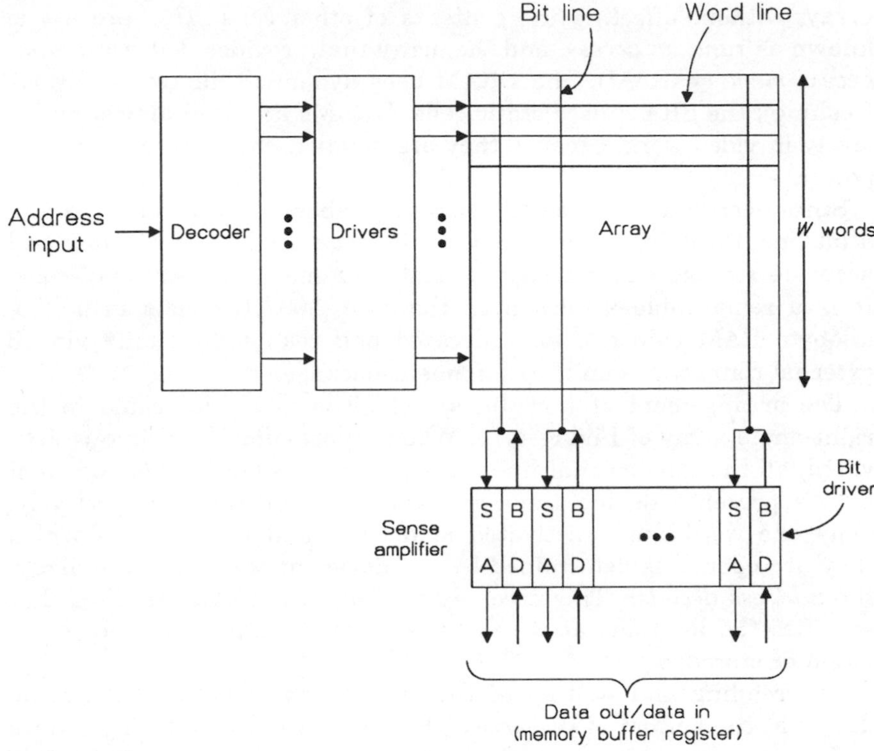

Figure 4.11 Array of dynamic memory cells in a random-access memory, showing row selection by word-lines and bit writing and reading via the bit-lines.

the integrated circuit during manufacture. Thus, when a particular combination of words is fed to the word-lines of the ROM, output words are produced at the bit-lines that correspond uniquely to the input words. This correspondence between input and output words remains fixed throughout the life of the ROM, unless changes in its strucure are made by external means. Such a changeable ROM is known as *programmable read-only memory* (PROM). The external influence may be exercised by heat (which opens fusible connections). In another ROM device ("electrically programmable," or EPROM), the stored charge in the memory cells is trapped behind a layer of the oxide in the semiconductor material. By applying a pattern of ultraviolet light, the trapped charge may be released at particular points in the array, altering the memory structure. In digital video applications, such after-the-fact reprogramming is needed during the

Table 4-2 Digital Video Storage Data [8-bit words at 13.5 ms/s for luminance (L) and 6.75 Ms/s for chrominance (C)]

		Line scan				Field scan			
		Total		Active		Total		Active	
System		Time (μs)	Words	Time (μs)	Words	Time (μs)	Words	Time (μs)	Words
NTSC	(L)	63.555	858	52.456	708	16.68	225,220	15.34	207,050
	(C)	63.555	429	52.456	354	16.68	112,610	15.34	103,525
PAL/ SECAM	(L)	64	864	51.7	698	20	270,000	18.39	248,224
	(C)	64	432	51.7	349	20	135,000	18.39	124,112

development of digital systems to determine the pattern of the ROM needed in mass production.

4.12 Line, Field, and Frame Stores

The memory arrays used to store video signals are based on the scan parameters: the line, the field, and the frame. Two functions must be distinguished: *delay* and *recall.*

4.12.1 Delay Storage

In the delay function the array simply delivers in sequence the stored words unchanged at its output. An example is the charge-coupled line-store used as a comb filter (Section 4.10). In many applications a delay of a whole field or frame may be required. This is arranged by connecting as many line-stores in series as there are lines in the field or frame. Table 4.2 shows the number of words involved in a field scan in each of the conventional systems (the numbers are doubled for a frame scan). In the NTSC system, with 8-bit words sampled at 13.5 Ms/s, 225,220 words are required to represent the luminance signal during the full field interval. The field-store array thus contains 8 × 225,220 = 1,801,760 memory cells, each of which must pass charge to its neighbor without error during the operating life of the equipment. This is routine performance in integrated circuits.

4.12.2 Recall Storage

The more general use of stores in scanning is to permit words to be recalled from storage for processing, correction, or display. The processes of recall are usually under the control of a program stored in a ROM array, but the data words representing the video content of the line, field, or frame are stored in RAM arrays, from which they are recovered by the processes described in Section 4.11. The number of cells is the same as in a storage array (nearly two million bits in the storage field array mentioned above), but for random access, each cell must have its own word-line and bit-line, and the word decoders and bit registers must cope with hundreds of thousands of address words and stored words, respectively. This more complicated function is also routine in the operation of ROM and RAM memories.

RAM arrays for digital video at the field- and frame-store level are, in 1990 practice, assemblies of several RAM chips developed for the computer industry. Ten 64-kilobyte RAM chips will suffice for an NTSC frame store of approximately half a million 8-bit words with its reading and writing functions. The cost of memory chips has been reduced by a factor of 10 in 10 years, and their bit capacity increased a factor of more than a thousand. Nevertheless, their cost is still high enough to add appreciably to the price of receivers employing field or frame storage.

4.13 Digital Filters

The charged-coupled delay line can serve as a general-purpose filter for analog or digital signals. Figure 4.12a shows one version, the *nonrecursive* filter. Taps are taken at the input and between each of the delay cells. The signals from these taps, after amplification, are fed to an adder circuit, which adds them to produce the output of the filter. The more tapped signals are added, the more sharply separated are the spectral components of the filter output. This type of filter can be recognized as the general form of the line-comb filters previously described, which use only one or two delay cells.

The second type of delay circuit, the *recursive filter,* is shown in Figure 4.12b. In this arrangement the adder circuit has two sets of delayed inputs. To the left of the adder, the arrangement is the same as that of the nonrecursive filter. To the right, however, the *output* of the filter is fed to another set of charge-coupled delay cells from which taps are taken. Signals from these taps are fed to the adder, but in reverse polarity. In this way, delayed versions of the output

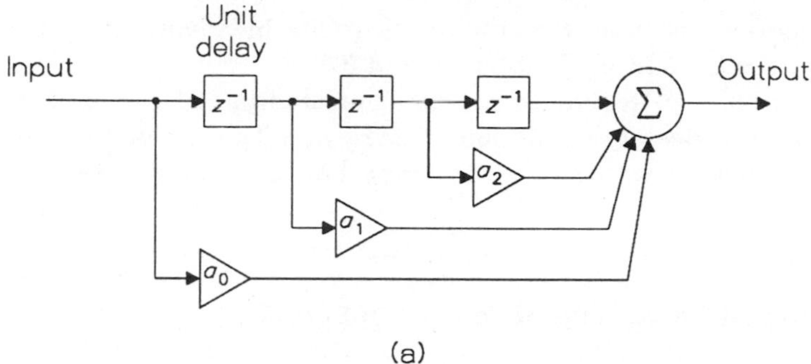

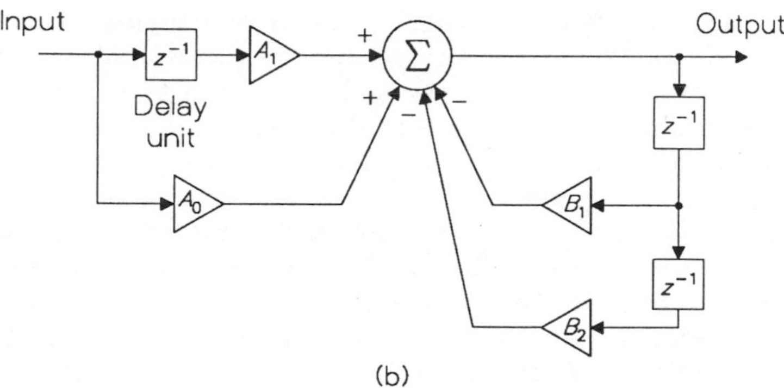

Figure 4.12 Digital filters: (a) nonrecursive, with delayed-signal taps from input only; (b) recursive, with taps from input and output.

signal are subtracted from delayed versions of the input. In most cases, the desired filter characteristic can be obtained with fewer delay cells in the recursive form. But large numbers of cells are readily obtained in integrated circuits in any event. In demanding applications, very many cells may be used. Examples are the filters used in the ACTV-1 compatible single-channel system (Sections 2.11 and 5.5). In the computer simulation of this system, one filter has 99 taps and two others have 15 taps each.

4.14 Digital Equipment and Systems

The foregoing treatment of digital video principles has been limited to the principles underlying the equipment and systems used in television engineering. Specific applications in advanced and HDTV systems are treated in Chapters 8 and 9.

An extensive collection of papers[4,5] on digital video has been published by the Society of Motion Picture and Television Engineers in four volumes, to which reference may be made for additional information.

4.15 Digital Processing of Television Sound

The outstanding performance of the digitally recorded compact disk audio system has encouraged the designers of HDTV systems to employ similar methods in transmitting the accompanying sound (Section 3.20). In the compact-disk system, the audio analog signal is sampled at 44.1 kHz, limiting the audio spectrum to 22,050 Hz and permiting an upper limit of the reproduced sound in excess of 20,000 Hz. The samples are quantized in 16-bit words in each of the stereo channels. This is equivalent to $2^{16} = 65{,}536$ levels, each occupying less than 0.002 of the dynamic range of the sound, equivalent to 96.3 dB from silence to peak fortissimo. To carry such a detailed digital signal, the overall channel bit rate exceeds 4.3 million bits per second. The bits impressed on the disk are illuminated by a solid-state laser and read off by a photodiode. The transmission paths are short, contained in a few cubic inches.

These specifications are a severe challenge to the television engineer, who must arrange for the digital bit stream to be carried over long distances in media subject to many types of error, both of omission and commission, in the signal path. So the digital audio specifications proposed for advanced television and HDTV systems are comparatively modest. Typical are the values proposed for the D2-MAC satellite service for Europe (Section 2.7). In accordance with the CCIR recommendation for digital audio interchange, the MAC signals are sampled at 32 kHz, limiting the frequency range at the loudspeakers to about 15 kHz. The samples consist of 14-bit words, that is, $2^{14} = 16{,}384$ quantization levels are used to provide a dynamic range of 84.3 dB, 12 dB less than offered by the compact-disk system. Since the minimum sound levels in the home are well above silence, and loudspeakers seldom operate at a peak power level above 20 watts, the 84 dB dynamic range appears to be wholly adequate for domestic use.

References

1. Peterson, J. G. A Monolithic Fully Parallel, 8-Bit A-to-D Converter. *ISSCC Digest* 1979, pp. 128–129.
2. Nyquist, H. Certain Factors Affecting Telegraph Speed. *Bell System Tech. J.* 3: 324–346 (March 1924).
3. Tarnai, E. J. Digital Television, In *Television Engineering Handbook*. Benson, K. Blair (Ed.), McGraw-Hill, New York, 1986, pp. 18.7–18.22.
4. *Digital Video*, Vol. 1 (1977), Vol. 2 (1979), Vol. 3 (1980). Society of Motion Picture and Television Engineers, White Plains, NY.
5. *Digital Television and Tape Recording*. Society of Motion Picture and Television Engineers, White Plains, NY, 1986.

Chapter

5

Space and Time Components of Video Signals

5.1 The Dimensions of Video Signals

The images picked up by a television camera are sampled in space and in time. The line structure represents sampling in the vertical axis. In digital processing, each line is sampled by the analog-to-digital converter, thereby producing samples in the horizontal axis of the image. These perpendicular sampling processes provide the spatial analysis. The field and frame repetitions provide the temporal sampling of the image. The signal content of all three dimensions must be exploited by the system designer to take full advantage of the channel's capacity to carry information. This approach, *spatio-temporal analysis*, has become an essential tool in advanced television system development.

Two of the dimensions are divided into specific intervals: the vertical distance between the centers of the scanning lines and the time occupied by each field scan. The third dimension, in the analog operation of the system, is not so divided since scanning the picture elements is then a continuous process.

The intervals in the vertical and field-time dimensions result from *sampling* of distance and time, respectively. The Nyquist limit applicable to sampled quantities (Section 4.6) imposes limits on their

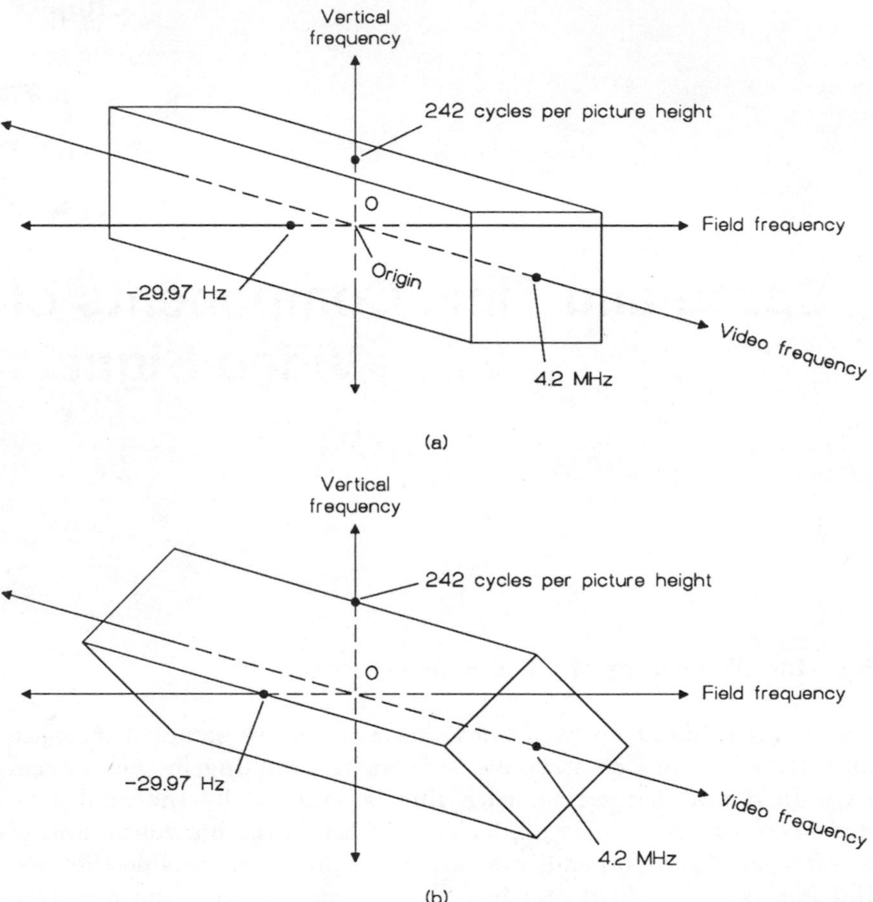

Figure 5.1 Nyquist volumes of the luminance signal for progressive (a) and interlaced (b) scanning.

rates of change and these limits must be observed in processing the signals resulting from the repetition of the scanned lines and fields.

The three-dimensional nature of the video signal requires that it be represented as a solid figure, known as a *Nyquist volume*. Examples of this figure for progressive (Figure 5.1a) and interlaced scanning (Figure 5.1b) of the luminance signal appear in Figure 5.1. The cross section of this volume is bounded by the maximum frequencies at which the vertical and time information may be transmitted, while its long dimension is bounded by the maximum frequency at which the picture elements may be scanned. Negative as well as positive frequencies (arising from the mathematical analysis of the signal space) are shown. They may be thought of as arising from the choice of the origin of the axis in each dimension.

Within the Nyquist volume reside all the permitted frequencies in the luminance channel. Study of the volume's contents thus reveals which frequencies are occupied by a given signal format, where additional signal information might be added, the extent of the interference ("crosstalk") thereby created, and where empty space may be created, by filtering, to reduce or eliminate the crosstalk. The spatiotemporal analysis[1,2] of signal content has been particularly valuable in the design of the single-channel advanced television systems, such as the ACTV-I system (Section 2.11).

5.2 Dimensions of the Nyquist Volume

To determine the specific dimensions and contents of the Nyquist volume, it is appropriate first to examine its cross section. This can be approached in two steps, shown in Figures 5.2 and 5.3. The first figure represents the intervals between samples, the second the corresponding frequency limits.

The intervals in Figure 5.2 are the center-to-center separations between lines in the vertical dimension of the image, shown on the vertical axis, and the field-to-field time intervals, on the horizontal axis. Figure 5.2a relates to progressive scan, in which the lines are adjacent to one another. Taking the number of active lines as 484 (NTSC system, Table 3.1*), the line spacing is found to be 1/484 of the picture height. The NTSC field-to-field interval is 1/59.94 second.

Positive and negative intervals† are shown on both axes. The axes thus represent the indefinite progression of the line and field scans, with one position and one time selected for the origin. The plot is thus covered by an indefinitely large number of separate points where a distance and a time coincide. These are marked on the figure as dots. It will be noted that successive intervals of 1/484 of the picture height and 1/59.94 second enclose the rectangular figure shown, and that the area of the plot is covered without gaps by such rectangles.

---

*The applicable values of intervals and rates for other systems are listed in Table 5.1.
†The mathematical analysis of the plots in Figures 5.2 and 5.3 involves Fourier series and matrix algebra. The simplified presentation given here arrives at the conclusions of the analysis in a form appropriate for this book.

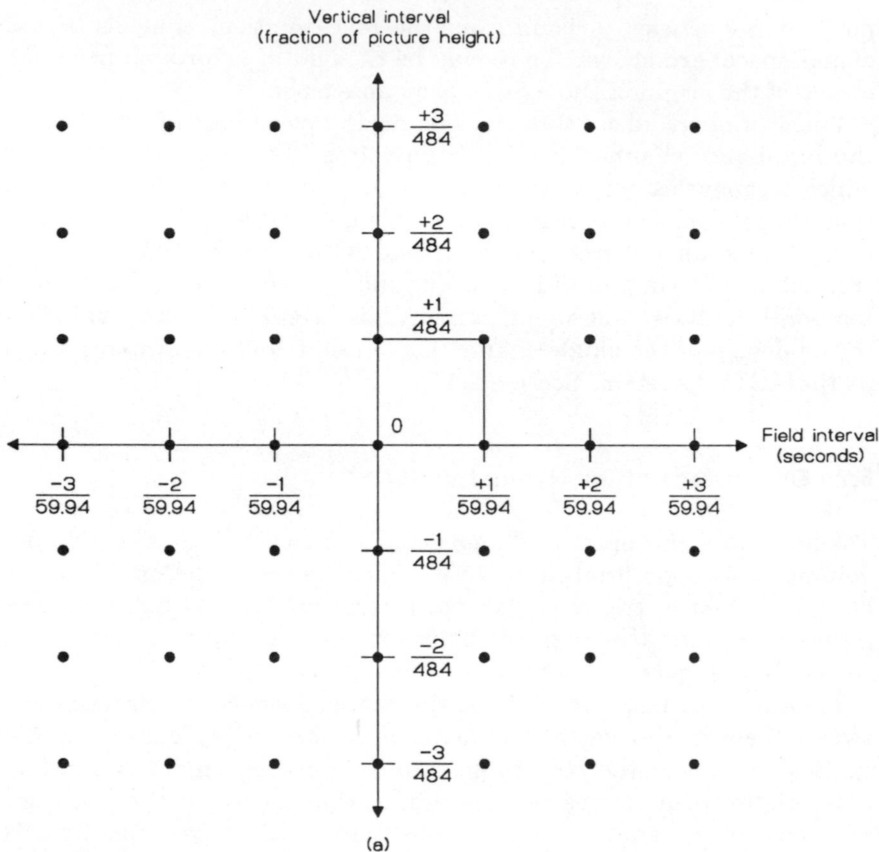

Figure 5.2a Sampled intervals in the vertical-picture and field-time axes for progressive scannings.

When the scanning is interlaced (Figure 5.2b), a different arrangement of the line intervals occurs. Vertically above each field interval, only half the line intervals are occupied, since in interlaced scanning every other line is omitted. Above the next field time to the right the same omissions exist, except that the occupied intervals lie opposite the omitted intervals to the left, i.e., the line intervals are interlaced. In this plot, we find that the figure connecting adjacent intervals is diamond-shaped. This figure is replicated throughout the area of the plot.

To convert the distance-versus-time intervals to the corresponding sampling rates (Figure 5.3), the axis numbers are inverted. Thus 1/484 of the picture height becomes 484 cycles per picture height

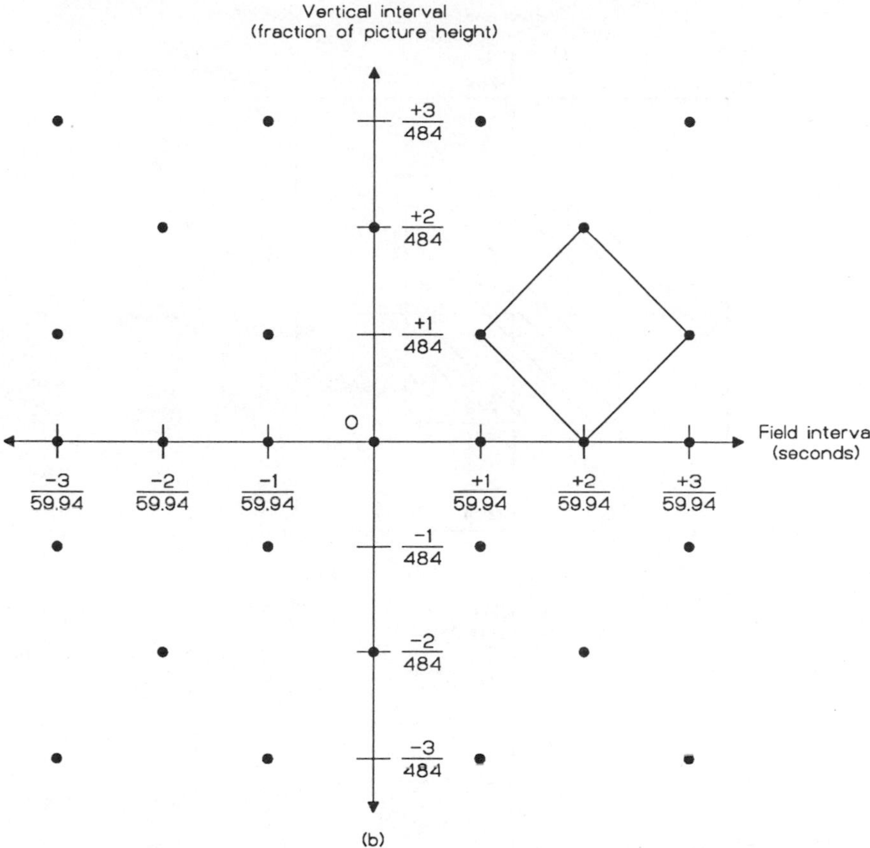

Figure 5.2b Sampled intervals in the vertical-picture and field-time axes for interlaced scannings.

(c/ph) and 1/59.94 second becomes 59.94 Hz. Figure 5.3a shows the inverted values over the rectangular region bounded by ±484 c/ph and ±59.94 Hz. These are the sampling rates for progressive scanning. The maximum frequencies for lines and fields, under the Nyquist limit, are one-half these values, 242 c/ph and 29.97 Hz, respectively. These limits bound an inner rectangle, shown shaded, in which all the frequencies of the sampled quantities must reside if aliasing is to be avoided. This rectangle is the cross section of the Nyquist volume in Figure 5.1a.

The corresponding plot for interlaced scanning is shown in Figure 5.3b. The area bounded by ±484 c/ph and ±59.94 Hz in this case is a diamond-shaped figure. When the Nyquist limit is applied to these

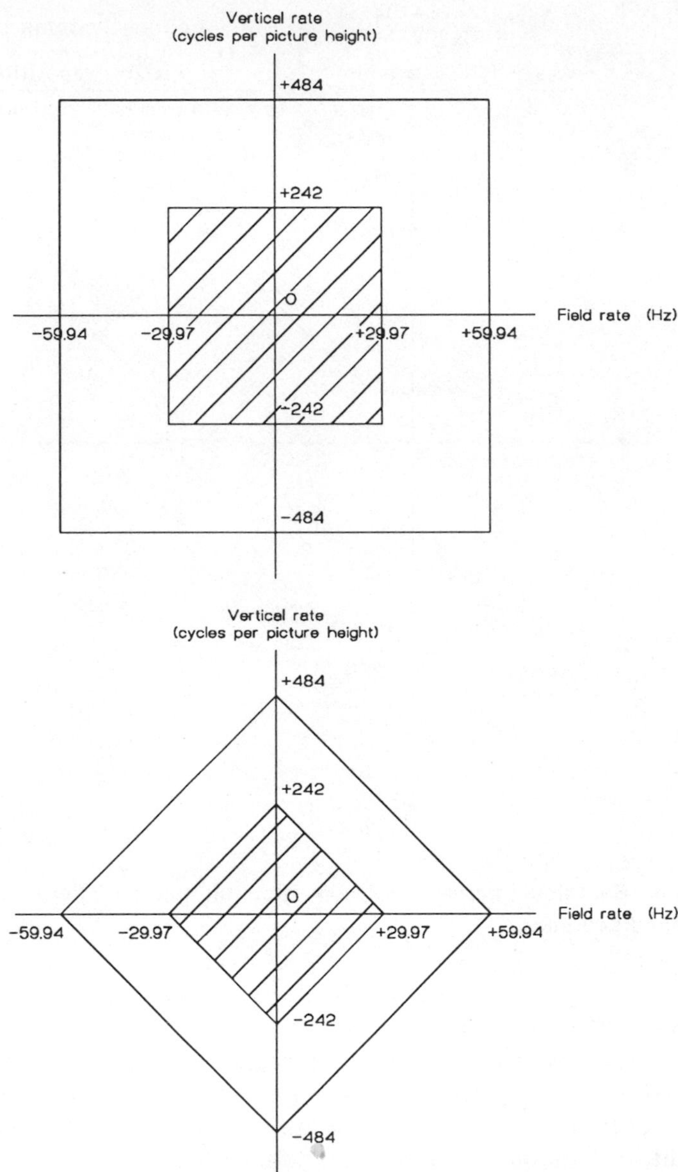

Figure 5.3 Sampling rates corresponding to the intervals in Figure 5.2.

values, the central diamond figure (shaded) is the area contained within ±242 c/ph and ±29.97 Hz. This area bounds the frequency range allowable in interlaced scanning, and is the cross section of the Nyquist volume shown in Figure 5.1b.

Table 5.1 Sampling Intervals and Rates of Conventional Systems

System		Vertical dimension		Picture repetition	
		Line interval (of ph)	Line rate (c/ph)	Field interval (ms)	Field rate (Hz)
NTSC	(L)	1/484	484	16.683	59.94
	(C)	1/121	121	66.733	14.985
PAL	(L)	1/575	575	20	50.
SECAM	(C)	1/143.75	143.75	80	12.5
NHK	(L)	1/1080	1080	16.667	60
HI-VISION	(C)	1/270	270	66.667	15

L = luminance; C = chrominance; ph = picture height.

5.3 Significant Frequencies in the Nyquist Volume

On and within the Nyquist volume, a number of subcarrier frequencies and their sideband regions can be identified. Figure 5.4 shows examples in the NTSC system (see Table 5.1 for other systems). The horizontal dimension extends to the limit of the video baseband, ±4.2 MHz in Figure 5.4a. The color subcarrier occupies a slice in Figure 5.4b through the volume, intersecting the horizontal axis in Figure 5.4a at ±3.58 MHz. Since the subcarrier repeats its full cycle only after four fields, its picture repetition rate is 14.985 Hz, and its line repetition rate is reduced to 121 c/ph. When these values are laid out within the Nyquist volume (Figure 5.4b), the subcarrier frequency appears on its surface, at the positions marked by the dots. About these dots as origins, Nyquist volumes are erected to represent the chrominance frequency space.

The spaces occupied in common by the luminance and chrominance frequency volumes are shown in Figure 5.5. Within this common space, crosstalk between the luminance and chrominance signals occurs. This crosstalk can be reduced or eliminated, if the common space is removed from the luminance volume by filtering[3] the luminance signal at the source and again prior to the display (see Section 2.6).

The 3.58-MHz subcarrier frequency is one of a set of freqencies that lie on the surface of the Nyquist volume. They are the odd multiples of half the line-scan frequency, subcarriers of alternating phase on successive scans of each line. One such multiple, 455, sets the color subcarrier in the NTSC system.

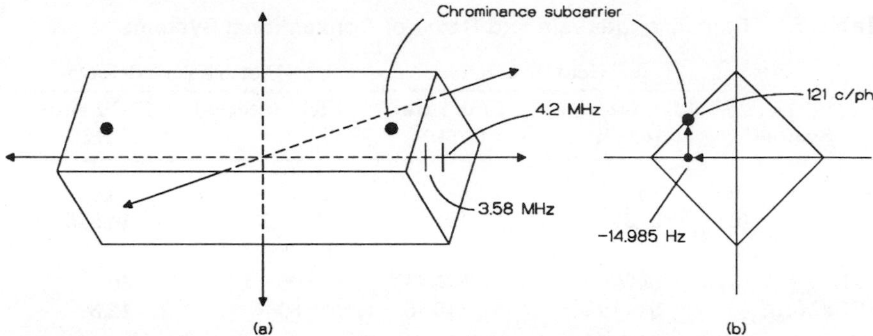

Figure 5.4 The Nyquist volume of the NTSC signal space: (a) location of the chrominance subcarrier; (b) cross section showing field- and line-frequency components.

The locus of these subcarrier frequencies is a horizontal line along the center of one of the Nyquist volume surfaces, marked SC-1 in Figure 5.6. Another such locus of subcarrier frequencies, SC-2, lies along the center of the adjacent face of the volume. Such a subcarrier can be formed by inverting the phase of an SC-1 type subcarrier on alternate lines. The locations of the SC-1 and SC-2 subcarriers, shown in the cross section (Figure 5.6b), show that they can coexist in the signal space. So an additional channel, available for auxiliary information, is provided by SC-2 subcarriers. These are known as

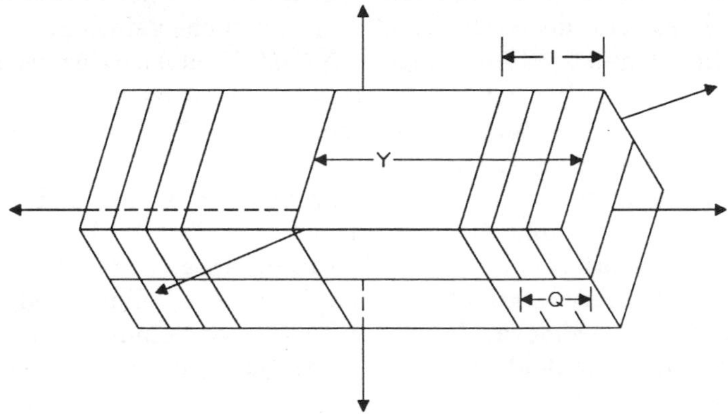

Figure 5.5 Nyquist volumes of the chrominance space positioned within the luminance volume.

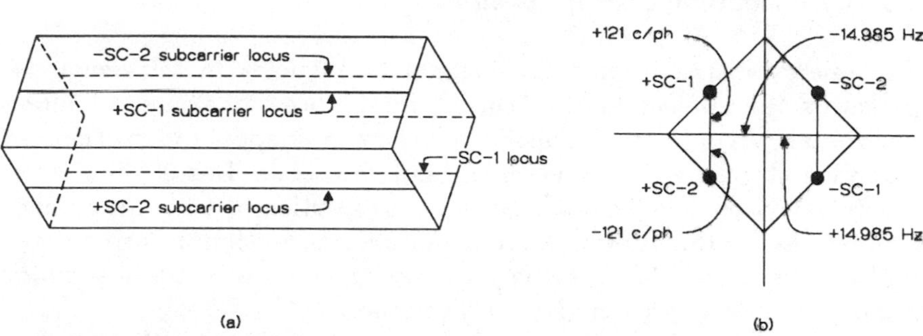

Figure 5.6 Subcarriers located on the faces of the Nyquist volume.

phase-alternate-frequency (PAF) carriers. Such a subcarrier carries auxiliary information in the ACTV-1 system, at the multiple 395 (Section 2.11).

The ability of SC-1 and SC-2 subcarriers to coexist can be shown by comparing the patterns produced by them on a display in the absence of the chrominance sidebands. In the SC-1 type of display, the pattern is a checkerboard whose individual elements appear to move upward. As the eye follows this motion, a given element moves from the bottom to the top of the image in about 8 seconds. When an SC-2 type of subcarrier is substituted, an otherwise identical pattern appears, but the motion is downward, completing itself in the same time. When both subcarriers are fed to the monitor, no vertical motion is seen since the two opposed motions are averaged by the eye. When appropriate modulation is used, the sidebands of the two subcarriers can carry different sets of signal information which are independent.

5.4 Signal Distinctions in the Nyquist Volume

The Nyquist volume is indicative only of the frequency content of a signal. It does not reveal how signals may otherwise be distinguished. To separate or protect signals sharing the frequency space, several characteristics or processes may be used. These rely (1) on the relative amplitude (energy content) of conflicting signals; (2) on changes with the time of signal occupancy, determined by the modulation used; and (3) on the distribution of the frequency space by imposing barriers to the signal spectra by filtering.

5.5 Distinction by Energy Level

The energy distribution among signal frequencies in television depends on the subject matter: colors; luminances; sharpnesses along horizontal, vertical, and diagonal edges; the shapes and motions of extended objects; and so on, throughout the scene. Despite this variety in televised images, some general properties of the energy concentrations in the luminance and chrominance signals are known and can be used to define the relative importance of the frequency regions they occupy in the Nyquist volume.

The energy concentration in luminance is greatest at the low frequencies, by which extended shapes are defined. The high frequencies serve to define edges; above 2 MHz, most of the signal energy is of low amplitude.

The energy in the chrominance signal is also greatest at the low frequencies, and it is largely removed in the Q component, to meet the NTSC standard, above 0.5 MHz. The I component extends to about 1.5 MHz.

When the luminance and chrominance signals are combined in the NTSC signal spectrum, the high-energy chrominance surrounds the 3.58-Mhz subcarrier, which is superposed on the low-energy region of the luminance spectrum. The crosstalk is thus minimized, but is still severe enough to affect adversely the performance of the great majority of receivers (Section 2.2). Reduction of this crosstalk between luminance and chrominance, as well as with the auxiliary signals of advanced systems, has been guided by their identification in the frequency space of the Nyquist volume. Examples appear in the descriptions of advanced and HTDV systems in Chapters 8 and 9.

5.6 Distinction by Modulation Method

A method widely used to permit two distinct signals to occupy the same frequency space is *quadrature modulation*. This is used in the NTSC and PAL systems to impose the two color-difference signals on the chrominance subcarrier. The Nyquist volumes of the I and Q sidebands of the NTSC system are shown in Figure 5.5 as they are located within the luminance volume.

While crosstalk occurs between luminance and chrominance, none is present between the I and Q components. This is accomplished by the use of quadrature modulation. The signals are separated by combining two color subcarriers, whose oscillations are out of phase by 90°, separately modulated by the I and Q signals. In the Nyquist volume, each point in the frequency space occupied by the subcarrier sidebands is filled first by the I sideband and next by the Q component, but in stationary images neither occupies the space at the same time.

In the NTSC and PAL systems the subcarrier is suppressed and only the sidebands are transmitted. To recover the I and Q sideband components at the receiver, the color burst synchronization signals are used to recreate the subcarrier, which is combined with the I and Q sidebands.

Quadrature modulation and demodulation find many uses in advanced and HDTV systems. In the ACTV-I single-channel system (Section 2.11), the PAF subcarrier at 3.107 MHz is quadrature modulated, and another auxiliary signal (the "helper" signal that restores the vertical detail of moving objects) is quadrature modulated with the picture carrier. Other examples of sequential occupancy of points in the frequency space are provided by the phase-alternate subcarriers of the SC-1 and SC-2 types (Section 5.3).

Conventional amplitude modulation of carriers and subcarriers does not provide such sequences, i.e., the am sidebands occupy the frequency space continuously. With such modulation, filtering is the means of separating signals occupying the same space.

5.7 Distinction by Filtering

To make room for chrominance and other auxiliary signals, some of the frequency space in the Nyquist volume must be emptied by filtering the luminance signal. The unoccupied regions are known as *channels*. Filtering may be applied in all three dimensions of the frequency space. One-dimensional filters, operating along the horizontal axis, limit the baseband of the video spectrum. Such filters are of the continuous-frequency ("wave filter") type.

Two-dimensional filtering includes also the field-repetition axis, with a periodicity of 59.94 Hz (NTSC value). This is provided by a periodic filter of the line-comb type (Section 4.12). Three-dimensional filtering adds the third axis, the vertical dimension in cycles per picture height. This also employs a periodic filter that covers the picture scanning time, that is, a frame-store (or two field-stores having capacity for 262 and 263 lines, respectively).

The periodic filters most used for creating channels in the Nyquist space are of the nonrecursive type (Section 4.13) with taps taken at each junction between delay cells. Two- and three-dimensional filtering thus involves many interconnections among a large number of

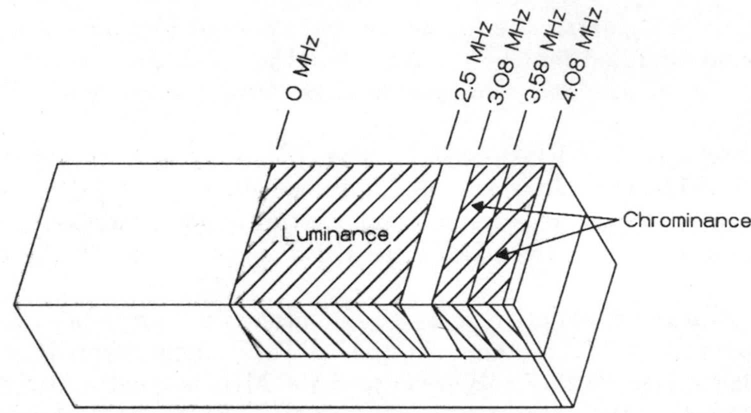

Figure 5.7 One-dimensional filtering between luminance and chrominance provided by a typical NTSC receiver.

cells. While integrated circuits can readily meet this requirement, multidimensional filtering must meet the demands of intricate design formulas, and many compromises are involved (e.g., between sharpness of frequency cutoff and the introduction of signal overshoot and ringing).

An example of one-dimensional filtering is provided by the typical receiver. No filtering between luminance and chrominance is provided at the transmitter, so crosstalk between them is inherent in the broadcast signal. To separate them at the receiver, the luminance response is cut off at about 2.5 MHz, and the chrominance sidebands are limited to ±0.5 MHz. The Nyquist frequency space of the received signal output from the second detector then appears as shown in Figure 5.7. The horizontal axis of the luminance space is cut off at 2.5 MHz and the chrominance sidebands extend on that axis from 3.08 to 4.08 MHz, beyond the luminance limit. As previously noted (Section 2.2), this cutoff of luminance bandwidth limits the horizontal resolution, but this loss has been accepted to obtain the reduction in crosstalk.

When a comb filter is introduced (Section 2.2), adjacent "wells" are introduced in the frequency space, into which the luminance and chrominance frequencies fall separately, appreciably reducing the crosstalk. These examples apply only to the receiver processing of the frequency space. A new approach, *cooperative processing,* involves filtering the space within the Nyquist volume at the transmitter (prefiltering) and the receiver (postfiltering). This arrangement (Section 2.6) deepens and sharpens the wells into which the

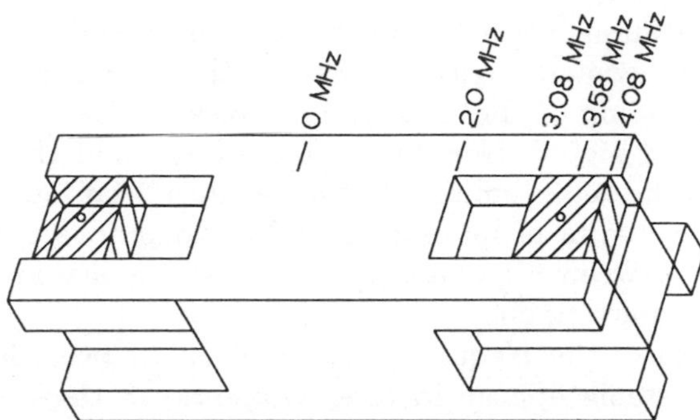

Figure 5.8 Three-dimensional filtering of the NTSC signal space to provide channels for auxiliary information. (Adapted from Reference 1.)

luminance and chrominance fit, making the crosstalk substantially invisible in stationary images.

5.8 Cooperative Three-Dimensional Filtering

An example of the extent to which prefiltering and postfiltering can open space for auxiliary information is offered by Isnardi.[1] It involves limiting the luminance bandwidth to 2.0 MHz, thus providing a channel 2.2 MHz wide for chrominance and auxiliary information. In the symmetry of the Nyquist volume, the channel appears under all four faces at both ends of the volume, as shown in Figure 5.8. The channels are produced by three-dimensional prefiltering from 2.2 to 4.2 MHz, ±14.985 Hz and ±121 c/ph. Within the channel, an SC-1 subcarrier with I and Q sidebands filtered to ±0.5 MHz occupies a central portion of the frequency space. An SC-2 subcarrier is also introduced into the channel, modulated with auxiliary information. The SC-1 and SC-2 signals occupy separate channels, as shown in the figure. In the channels containing the SC-1 subcarrier, on the lower side of the chrominance sidebands is an unoccupied space, 1.0 MHz wide, which is also available for auxiliary information.

The complexity of filtering required to introduce all these auxiliary channels is very great, perhaps beyond the present capabilities of economical design in the postfiltering at the receiver. The example serves, in any case, to show the clarity with which the options available in the frequency space can be visualized in the Nyquist volume.

When the NTSC system of compatible color television was developed in the early 1950s, all the mathematical and physical bases of the Nyquist volume were known to communications theorists, but not to many television engineers. Few of the NTSC members had a clear understanding of the unused resources in the spectrum of the compatible color signal. Nor, for that matter, could those engineers then envision the compromises that have since been found acceptable concerning the limits of luminance and chrominance spectra and of the trade-off between high resolution in a stationary image and low resolution in a moving one.

The specifics of the Nyquist-space analysis of proposed advanced and HDTV systems appear in later chapters, in those cases where the proponent has made use of its concepts in the system design.

References

1. Isnardi, M. A. Exploring and Exploiting Subchannels in the NTSC Spectrum. *SMPTE Journal* 97: 526–532 (July 1988).
2. Isnardi, M. A. Multidimensional Interpretation of NTSC Encoding and Decoding. *IEEE Transactions on Consumer Electronics* 34: 179–193 (February 1988).
3. Dubois, E. and W. F. Schreiber. Improvements to NTSC by Multidimensional Filtering. *SMPTE Journal* 97: 446–463 (July 1988).

Chapter

6

Compatibility in HDTV Systems

6.1 Compatibility Defined

Whenever a new system is proposed, the question of its impact on the utility and value of existing equipment and services is raised. As we have seen in the preceding chapters, the threat offered by new services to the NTSC, PAL, and SECAM systems has forced governing authorities and development organizations to face the issue of the *compatibility* of the proposed new systems with the existing ones.

A new system of television is defined as compatible with an old one when the receivers of the old system retain their utility when used with the new system. Few systems are totally compatible; some impairment in the performance of old receivers usually occurs because the new system is designed to meet different objectives.

A compatible system is fundamentally concerned with preserving the performance of receivers, that is, with the format of the transmitted signal. But there are many functions in the production and distribution of program material, prior to transmission, that contribute to, or are affected by, the compatibility requirement.

6.2 Protection of Existing Systems

There is general agreement that the old system should be protected while the new service establishes its technical validity and shows its economic strength in the marketplace. The alternative, a sudden change from the old to the new, would have economic and political consequences that no one cares to face. Cutting off NTSC service in the United States would render useless nearly 200 million receivers whose purchase value exceeds $35 billion, a loss that would cause the removal from office of any government body that authorized it.

Preserving the existing service has been, without exception, the procedure whenever new television services have been introduced. Moreover, shutting down the older service has often been delayed long after its value was too small to warrant the cost of continuing it.

An instructive example lies in the history of the television service in the British Isles. Public broadcasting began there in 1936, using the 405-line monochrome system. When the 625-line color system was introduced in Britain in 1967, it was incompatible with the 405-line service. To preserve the existing service, new means of distribution had to be found for the color transmissions. This was found in the then largely unused UHF channels. The color and monochrome systems continued as separate services operating in parallel.

The 405-line service continued for a life span of 50 years. When it was discontinued in 1986, so few 405-line receivers remained in use and the cost of serving them was so high that it was proposed that Parliament provide the 405-line audience with 625-line receivers free of charge. But this economy was deemed unacceptable politically.

A similar, if less extreme, situation occurred in France, where the 819-line monochrome service continued for 37 years after it was first introduced to the public in 1948. In both cases, the introduction of an incompatible color service depended entirely on the availability of the UHF channels.

6.3 Compromises in Compatible Systems

The impairment of the old service by a compatible system is well illustrated by the NTSC color system. When this system was introduced in 1954, many monochrome receivers (those responding to video frequencies at or above 3.58 MHz) could not discriminate against the color subcarrier, and this produced a pattern of fine dots, superimposed on the monochrome version of the color image. The NTSC designers attempted to minimize this effect by arranging for the subcarrier to have opposite phase on successive scans of each line, so that the dot pattern averaged out to zero during each frame scan. The eye was not completely fooled by this artifice and some vestige of the dots remained in the form of an upward crawl of the scanning lines. In the enthusiasm for the new color service, this effect was tolerated and in time the line crawl was eliminated by the decision of manufacturers to limit the bandwidth of color and monochrome receivers to less than 3.0 MHz, well below the 3.6 MHz of the subcarrier. This lowered the horizontal resolution to about 250 lines, 25 percent less than the 330 lines of the broadcast signal (Section 2.2).

The low definition was not a matter of wide concern until improved cameras and signal enhancement provided additional detail, hidden by the existing receivers. This suggested to receiver designers that a new market could be established by restoring the lost resolution. This led to the introduction of the comb filter (Section 2.2), which permitted, in the more expensive receivers, use of the full 4.2-MHz bandwidth in the broadcast signal. This is an example of how advances in technology can overcome problems once thought to be inherent in the system design.

6.4 Adapters for Compatible Service

A question concerning the utility of the service has risen with each new advance in television, notably the wide spectrum of protected channels provided by the cable service, the storage of programs on videotape, and the reception of programs from satellites. In each case, when the service was first offered, it was necessary for the viewer to purchase an adapter or recorder at a price often greater than that of the receiver to which it was attached. This requirement was thought to limit the market to enthusiasts who had the necessary resources. But this judgment was very wide of the mark.

Today in the United States, more than half the homes enjoying television have video recorders, and the same percentage applies to homes having cable service, for which fees are charged. This violation of the "free television" doctrine of the broadcasters testifies that a majority of viewers have found access to cable programs and the freedom to view them at a convenient time well worth the cost of the adapter, recorder, and fees. This fact was soon recognized by the manufacturers. Now even inexpensive receivers are "cable ready," and some receivers are now available with the tape recorder built into the cabinet. So the adapter problem has been put into a new perspective.

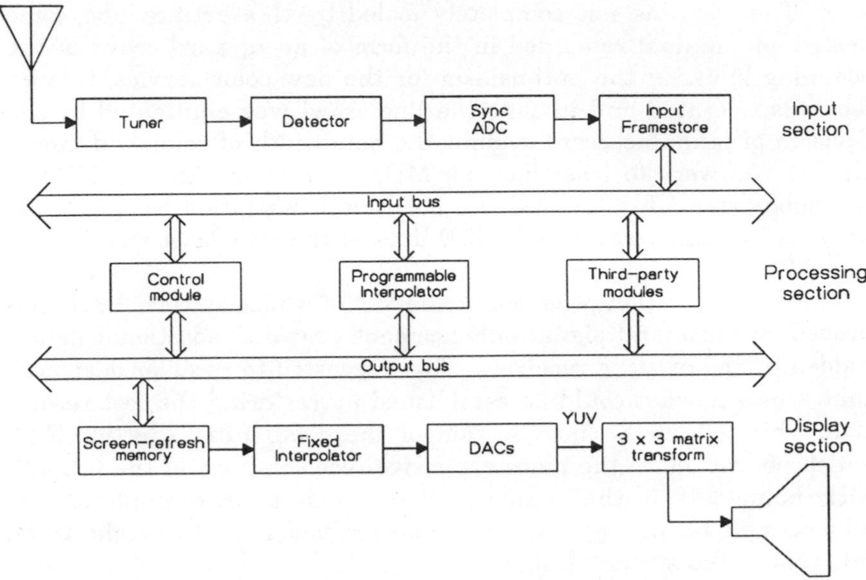

Figure 6.1 Open-architecture "smart" receiver proposed by W. F. Schreiber. The central section, organized like a personal computer, would receive new integrated-circuit modules as changes in the transmission system were introduced.

This change in outlook has encouraged those developing advanced and HDTV services to view the need for an adapter with equanimity, provided that its cost is not the major part of total receiver cost. The eventual inclusion of the adapter function in conventional receivers is taken for granted.

The time between inauguration of a new service and its inclusion in receivers has been greatly reduced by the use of integrated circuits. This trend has encouraged Professor W. F. Schreiber, Director of the MIT Advanced Television Research Program, to propose[1] that receivers be built to allow changes in their functions as the art advances. He describes this method as "open-architecture" receiver design. Such "smart" receivers (see Figure 6.1) would have a frame-store and signal busses that would allow new integrated circuits to be installed to take advantage of new signal inputs, without change in the expensive elements of the receiver, i.e., those associated with the display. Not surprisingly, Professor Schreiber's ideas have not been accepted by the television receiver industry, whose engineers are reluctant to provide for ill-defined eventualities or to add costs that competitors might decide to forgo.

6.5 Classes of Compatibility

In an excellent review of the current status of advanced and HDTV system development, in 1988 Robert Hopkins[2] suggested five levels of compatibility:

Level 0—Incompatible. The conventional receiver cannot display the HDTV image, and the HDTV receiver cannot display the conventional image.

Level 1—The HDTV signal can be converted by an expensive adapter to provide a display on the conventional receiver.

Level 2—Same as level 1, except that the adapter is inexpensive.

Level 3—The HDTV image is received and displayed by the conventional receiver, but at a loss of quality relative to the receiver's optimum performance.

Level 4—The same as level 3, except that the quality of the image is the highest of which the receiver is capable.

Level 5—The image displayed by the conventional receiver has all the quality of the HDTV image.

Most of the compatible systems described in this book fall into levels 1 to 3 inclusive, with level 4 applicable only to systems that use the conventional channel wholly unaltered.

Other terms applicable to compatible systems are "forward" versus "backward." In a forward system, the conventional receiver obtains service from the advanced or HDTV signal at levels 1 through 4. In the backward version, the HDTV or advanced receiver obtains service from the conventional signal. When the audience is served by substantial numbers of HDTV and conventional receivers, both types of compatibility are evidently highly desirable.

6.6 The Transcoding Functions

To convert otherwise incompatible signals to compatible ones, it is necessary to encode or decode ("transcode") them. Examples previously described are the encoders and decoders of the MUSE and MAC systems (Sections 2.7 and 2.8). Digital video transcoding methods have reached a point that, given unlimited resources, it should be possible to render a signal format into and out of any other, with little or no loss of signal quality. But resources are never unlimited, and the complications of transcoding are such that deterioration of signal quality occurs in the majority of practical applications.

An example is the transcoding of the 1125-line HDTV system into the 525-line NTSC system. The major stumbling block is the transfer from the 60-Hz HDTV field rate to the 59.94-Hz NTSC rate. The difficulty is such that, in common parlance, the 1125-line system is said to be "incompatible" with the NTSC system. The 1125-line-to-NTSC transcoding process illustrates many of the problems encountered in other systems.

The 1125-line system (for details see Chapter 7) is based on 1125 lines because that number contains, in common with the 525-line and 625-line systems, the factor 25. That is, 1125 = 25 × 45; 525 = 25 × 21; 625 = 25 × 25. Thus, integral counts of lines suffice for transcoding among all three systems. The ratios of lines 1125/525 = 45/21 and 1125/625 = 18/10 are also expressible in moderately sized integers, so the line-counting processes in 1125-line transcoding are straightforward.

A further requirement is that the content of the NTSC lines be compressed over a substantial number of the 1125 lines. Thus, 45 lines of the 1125-line image must be replaced by 21 lines in the transcoded NTSC image. Digital storage of the 45 lines must be followed by a readout, on a line-by-line basis, at a slower rate in the ratio 21/45. To preserve these integral relations, the digital sampling rate must be chosen at 45/21 times the line scanning rate of 15,734.25 kHz, or at some whole multiple of that rate. Fortunately, all these requirements are not difficult to meet in the design of the transcoder integrated circuits. What remains is the relationship between the content of a particular NTSC line and the 2.14 (= 45/21) lines it replaces. The term *interpolation* covers this process in line-content transcoding.

Here the complications, and sources of deterioration, arise. Each NTSC line must have the same general content, except for fine detail, as the 1125/60 lines it replaces, i.e., the geometry of the image must be preserved. This occurs only when 45 1126/60 lines are replaced by 21 NTSC lines. The time delay required to permit this interpolation is of no consequence in stationary parts of the image, but it may introduce errors when motion is present.

These errors occur when the field scanning rates of the two scanning patterns are the same. But another source of error occurs since they are not the same, namely 60 Hz for 1125/60 and 59.94 Hz for NTSC. The latter rate is rigidly tied to the color subcarrier frequency, with a tolerance of about ±3 parts per million, so no approximate values may be used. The integral ratio required for transcoding is then the precise ratio of the field rates, 6000/5994. The field stored in 6000 time units must be read out in 5994 units, while the geometry of the image is preserved.

As of 1989, transcoders capable of transforming the 1225/60 HDTV signal to 525/59.94 NTSC signal were available (see Section 7.11) but as of that date no organization had offered equipment capable of transcoding directly between the two principal proposed HDTV formats: 1125/60 and 1050/59.94 (see Table 6.1).

One method of deriving 59.94-Hz signals from the 1125/60 standard is to make a high-quality component-type 1125-line 60-Hz tape recording and to transform its content to the NTSC standard by changing the the playback rate from 60 to 59.94 Hz.

The Japan Broadcasting Corporation (NHK) was reported to have described to a CCIR Working Party in January 1989, a method of transcoding from the 1125/60 standard to the 1050/59.9 proposed standard in which both line-number and field-number conversions are achieved without image degradation.

Plans were announced to use tape-recorder transcoding in the subjective testing of the advanced and HDTV systems proposed to the FCC. But several proponent organizations offering systems using the 59.94-Hz field rate protested that such transcoding was not adequate for a fair presentation of their systems.

While direct transcoding from 60 to 59.94 Hz has proved feasible in professional equipment, it has been concluded that the process cannot be reduced to practice in receivers, not at least within a time span appropriate for system planning. Accordingly, an advanced or HDTV system based on the 60-Hz field rate is, for all practical purposes, incompatible with the NTSC service. This has led proponents of compatible systems to base their designs on the 59.94-Hz field rate.

A similar line of reasoning has led to the conclusion that the line rate of the HDTV system should be simply related to the NTSC rate. Thus several proposed systems employ a line rate of 1050 lines, twice the NTSC rate. In Europe similar arguments have led to proposals for 1250-line, 50-Hz systems. There are some exceptions, notably the Zenith system in the United States, which calls for 787.5 lines, 59.94 Hz (see Table 6.1).

Progressive scanning in the camera is called for by many proponents of advanced and HDTV systems, with the understanding that, in compatible systems, the transmitted signal would permit the conventional receiver to operate in interlaced fashion. This would double the bandwidth produced by the camera and required in subsequent circuit functions. While this requirement poses design problems, it is believed that they can be resolved by the time the HDTV service is standardized and offered to the viewing public.

Table 6.1 Proposed Advanced and HDTV Systems

Proponent/ name	Scanning (lines/field rate/ fields per frame)	System type	Aspect ratio NTSC	Aspect ratio ATV/HTDV
Single-Channel Compatible Systems (6-MHz Band)				
BTA/Japan	525/59.94/2:1	EDTV	4:3	Wide
Del Rey/Iredale	1125/59.94/2:1	NTSC HDTV	14:9	14:9
Faroudja	525/59.94/1:1 and 2:1 1050/59.94/2:1	NTSC	Not stated	Not stated
Hi. Res. Soc.	525/60.07/2:1	Modified NTSC	4:3	Not stated
MIT/Schreiber	525/59.94/2:1	NTSC R,G,B	Not stated	Not stated
NHK/MUSE-6	1125/60/2:1	NTSC EDTV	16:9	16:9
Production Services	1125/60/2:1	NTSC HDTV	4:3	16:9
Sarnoff/ACTV-I	525/59.94/1:1 1050/59.94/2:1	EDTV	4:3	16:9
Single-Channel Noncompatible Simulcasting Systems				
NHK/Narrow MUSE	1125/60/2:1.	Noncompatible	Not stated	16:9
Zenith	787.5/59.94/1:1	Noncompatible	Not stated	5:3
Dual-Channel Systems with 3-MHz Augmentation Channel				
NHK/MUSE-9	1125/60/2:1	NTSC	16:9	16:9
Philips/HDS-NA[a]	525/59.94/1:1 1050/59.94/2:1	NTSC HDTV	4:3	16:9

(Continued)

Table 6.1 ***(Continued)*** **Proposed Advanced and HDTV Systems**

Proponent/ name	Scanning (lines/field rate/ fields per frame)	System type	Aspect ratio NTSC	Aspect ratio ATV/HTDV
	Dual-Channel Systems with 6-MHz Augmentation Channel			
NYIT/Glenn[b]	1125/59.94/2:1	HDTV	5:3	5:3
Osborne	1125/60/2:1	NTSC HDTV	4:3	16:9
Philips HDS-NA[a]	525/59.94/1:1 1050/59.94/2:1	NTSC HDTV	4:3	16:9
Sarnoff ACTV-II	1050/59.94/2:1	NTSC HDTV	4:3	16:9
	Satellite Transmission Systems			
NHK/MUSE-E	1125/60/2:1	NTSC HDTV	16:9	16:9
Philips HDS-NA[a]	525/59.94/1:1 1050/59.94/2:1	NTSC HDTV	4:3	16:9
Scientific Atlanta	525/59.94/1:1 and 2:1 1125/59.94/2:1	NTSC HDTV	4:3	16:9

[a]The Philips' entries employ the same basic system.
[b]The NYIT system requires that the 6-MHz augmentation channel be time-shared by two broadcasters.

Source: Abstracted from information compiled by the Advanced Television Test Center from proponents' presentations to the FCC Advanced Television Advisory Committee. *TV Digest* 29(1): 6–7 (January 2, 1989).

6.7 Single- and Wide-Channel Compatibility

HDTV systems designed to be compatible with the conventional services fall into two broad classes. The single-channel systems use the conventional channel, to which auxiliary information is added. The wide-channel systems use additional spectrum space to carry auxiliary information. The extra channel space is intended only for the HDTV receiver, which also receives the conventional channel and the auxiliary information it may contain.

The first channel may remain free of any auxiliary information, and it then provides the conventional receiver with unaltered performance. In other cases, the first channel may be altered to operate cooperatively with the auxiliary channel. In that case the conventional receiver's performance may be impaired. An example is the ACTV-I system of the Sarnoff Research Center, described in Section 2.11. Means of inserting auxiliary information are also described in Sections 5.5, 5.6, and 5.7.

The wide-channel systems proposed to the FCC for HDTV service in the United States cover a wide range of techniques. A major approach is time compression, in which some of the picture element signals are spaced over one or more line times. The delay in displaying these elements causes impairment of the parts of the image in motion. An alternative approach avoids such signal delays and motion impairment, but at a sacrifice in the overall resolution of the image. Within these categories are many variations in transcoding, choice of scanning method, and related signal characteristics. Proposed wide-channel systems are listed in Table 6.1 and described in Chapter 9.

6.8 Simulcasting

An alternate approach to spectrum occupancy for HDTV systems is to separate the functions of the conventional and the HDTV channels, that is, to ignore the conventional system in designing the HDTV system. This was the approach taken in introducing the PAL and SECAM systems in Europe, when the existing monochrome services could be ignored (and continued in operation for many years) while the new services used different channels. If it is assumed that the NTSC, PAL, and SECAM systems will, at some time in the future, be discontinued as the 405-line and 819-line monochrome systems were, then it would be wise not to depend on them for the HDTV service.

The HDTV system based on this assumption is known as *simulcasting.* The HDTV channel (probably but not necessarily of the same width as the conventional channel) would have a Nyquist frequency space occupied by signals based only on the HDTV system requirements, wholly without reference to compatibility. Such an approach, fully exploiting the potential of the HDTV service, would require spectrum resources greater than those of a cooperative compatible system. But in the cable service, and in the design of tape and disk recorders, such demands could be met. Simulcasting has been advanced, on the assumption that the conventional services would be displaced on a short time scale, by Professor Schreiber and his associates at MIT. Their proposal[3] is listed in Table 6.1 and described in Chapter 8.

6.9 Compatibility in Production and Distribution

For many years, the television signal in production and distribution facilities has been handled using the standards of the transmitted signal, e.g., 525 lines, 59.94 Hz, two fields per frame. But the new opportunities and requirements presented by integrated circuits and digitization have encouraged designers to adopt different specifications for various items of production and distribution equipment.

One important example is the use of digitial signals in the distribution of program material within and among broadcast plants. The widely adopted CCIR Recommendation 601 for conventional distribution at 13.5 Ms/s for luminance and 6.75 Ms/s for chrominance is used in the compatible versions of HDTV signals. But these rates are insufficient for digital distribution of the HDTV signals themselves. Study groups are at work on an appropriate revision of this CCIR specification for wideband signals. CCIR Report 801 suggests that these rates be multiplied by 5.5 to 74.25 and 37.125 Ms/s, respectively. These rates have been used in the design of an 1125-line/60-Hz digital tape recorder (Section 7.11). Such high rates of sampling, applied to the HDTV signal, require that the overall sampling rate be 1.188 Gb/s, which requires that the recording drum employ eight heads, writing in parallel in strict synchronism. A commercial version of the 1.188 Gb/s recorder was anticipated in 1989.

The demands of compatibility extend to program origination and distribution, with specifications that exceed those of the broadcast signal. The need for excess bandwidth in studio equipment, for example, has been long established in conventional operations and is equally desired for advanced and HDTV systems. The majority of prime-time programs currently originate from film, in the 35-mm and 70-mm formats. Since 35-mm film as projected has definition equivalent to that of the HDTV image, excess definition is available only in 70-mm film. While several major film producers (MGM, Super-Panavision, and Todd-AO among others) offer 70-mm prints, the majority of producers use 35-mm film, and nearly all of the older films that are currently available are in that format. New film productions with the HDTV market in view are increasingly recorded on 70-mm prints.

Allied with the size of the film is its aspect ratio. The 16:9 ratio adopted by nearly all HDTV proponents (see Tables 6.1 and 9.1) covers only part of the wide-screen formats of motion pictures (see Figure 3.3). The compatibility issue arises when the HDTV image is displayed on the conventional 4:3 screen with the left and right portions of the HDTV image cut off. To maintain the center of interest on the 4:3 screen, the portion of the 16:9 image contained within the 4:3 limits must often be shifted laterally during scanning for the conventional audience. This lateral shift, known as "pan-and-scan" in telecine practice, must be available in compatible HDTV systems. But a difficulty is encountered in HDTV systems that achieve the 16:9 image by adding left and right portions to the 4:3 pattern. The "seams" between the center and outer portions must remain invisible during the pan-scan process. If there is any time delay between the signals representing the central and outer portions, the seam then becomes uneven. This problem has prompted one proponent (Philips HDS-NA system, see Table 6.1) to abandon time delay among signal components, even at some sacrifice in the overall definition of the system. This approach also avoids the loss of definition in moving objects inherent in time-delayed signals.

In addition to these complications in telecine operations, care must be taken in telecine and camera scanning to meet compatibility requirements. Nearly all HDTV proponents intend that progressive scanning be used in all program origination. Those offering compatible systems must also specify how the progressive scanning is to be altered to the interlaced form, including the four-field recurrence period, required by conventional receivers. We have already observed the complications involved in converting from interlaced to progressive scan in such receivers (Section 2.3). The inverse process is also complex but solved by straightforward digital processing.

6.10 Compatibility of Proposed Systems

Tables 6.1 and 9.1 list proposed advanced and HDTV systems, their proponents, scanning specifications, and aspect ratios. These lists were compiled by the Advanced Television Test Center (ATTC) from papers presented to the Advanced Television Advisory Committee of the Federal Communications Commission and were intended to indicate those systems that the ATTC expected to put to test in 1990. The tables offer a representative, if not exhaustive, cross section of advanced television development work in the United States.

The tables have 20 entries, three of which represent the application of one system to different environments. Of the 18 different proposals, ten are scanned at the NTSC-compatible field rate of 59.94 Hz and four are based on 1050-line images for the extended or high-definition version of the service. The 1125-line 60-Hz HDTV format has nine entries, four of which are different versions of the NHK MUSE family of systems. The greatest unanimity appears in the aspect ratio for the HDTV service; of the 18 different entries, all but four call for a ratio of 16:9, versus 15:9 and 14:9. It thus appears that if the stated intent of the FCC to combine the best features of the proposed systems in the ultimate HDTV standard is followed, there is ample room for choice. The systems listed in Tables 6.1 and 9.1 are described in detail in Chapters 8 and 9, respectively.

References

1. Schreiber, W. F., A. B. Lippman, A. N. Netravali, E. H. Adelson, and D. H. Staelin. Channel-Compatible 6-MHz HDTV Distribution Systems. *J. SMPTE* 98: 5–13 (January 1989).
2. Hopkins, R. Advanced Television Systems. *IEEE Transactions on Consumer Electronics* 34: 1–15 (February 1988).
3. Schreiber, W. F. and A. B. Lippman. Single-Channel HDTV Systems, Compatible and Noncompatible. Report ATRP-T-82, Advanced Television Research Program, MIT Media Laboratory, Cambridge, MA (March 1988).

Chapter

7

The 1125-Line HDTV System

7.1 Status of the 1125-Line System

The 1125-line HDTV system enjoys an outstanding place among advanced television systems. It has been under development for more than 20 years. During that time a full range of equipment for program origination, distribution, transmission, and display has been designed and offered commercially. A satellite-transmitted version, MUSE, broadcast experimentally on a daily basis in Japan since 1988, is scheduled to start as a full-scale public service in 1990.

A version of the 1125-line system has also been proposed as an international standard for program exchange. While there is currently disagreement on the form this international standard should take, there is wide acceptance that standardization is essential to the free interchange of program material among the many different operating and proposed public broadcast systems. If different studio standards are adopted, to conform to the differences in line and field rates of the conventional systems, it would be necessary to transcode among the studio standards prior to broadcast. To maintain "broadcast quality" during such transcoding is difficult and, many believe, not feasible in the present and foreseeable states of the art. So intense interest has been aroused by the potential of the 1125-line system for studio production standardization.

Early Development The first development of a high-definition television system began in 1968 at the Technical Research Laboratories of Nippon Hoso Kyokai (NHK) in Tokyo. In the initial work[1] film transparencies were projected to determine the levels of image quality associated with different numbers of lines, aspect ratios, interlace ratios, and image dimensions. Image diagonals of 5 to 8 feet, aspect ratios of 5:3 to 2:1, and interlace ratios from 1:1 to 5:1 were used. The conclusion reached in this study[2] was that a 1500-line image with 1.7:1 aspect ratio, and a display size of 6.5-foot diagonal, would have a subjective image quality improved by more than 3.5 grades (on the seven grade scale shown in Figure 7.1) compared to the conventional 525-line, 4:3 image on a 25-inch display.

To translate these findings to television specifications, the following parameters were adopted: 2-to-1 interlaced scanning at 60 fields per second, 0.7 Kell factor, 0.6–0.7 interlace factor, 5:3 aspect ratio, and a visual acuity of 9.3 cycles per degree of arc. Using these values, the preferred viewing distance for an 1125-line image was found to be 3.3 times the picture height. A display measuring 3 by 1.5 feet was constructed in 1972, using half mirrors to combine the images of three 26-inch color tubes. Using improved versions of wide-screen displays, the bandwidth required for luminance was established to be 20 MHz and, for wideband and narrowband chrominance, 7.0 and 5.5 MHz, respectively. The influence of viewing distance on sharpness was also investigated, with the results shown in Figure 7.1.

Tests were also conducted to determine the effect of a wide screen on the realism of the display. In an elaborate viewing apparatus (Figure 7.2), motion-picture film was projected on a spherical surface. As the aspect of the image was tilted, the viewers' reactions were noted. The results showed that the realism of the presentation increased when the viewing angle was greater than 20°. The horizontal viewing angles (Table 3.1) for NTSC and PAL/SECAM are 11° and 13°, respectively, while that of the NHK system is 30°.

7.2 Basis of the NHK 1125-Line System

The "provisional" standard for the NHK system, published in 1980, is shown in Table 7.1, together with the later versions for satellite transmission and studio operations. The NHK HDTV standards of 1125 lines and 60 Hz are incompatible with the conventional (NTSC) service used in Japan, so their adoption has raised questions now that the compatibility issue is prominent. No explanations appear in the literature, but justification can be found in the situation faced by

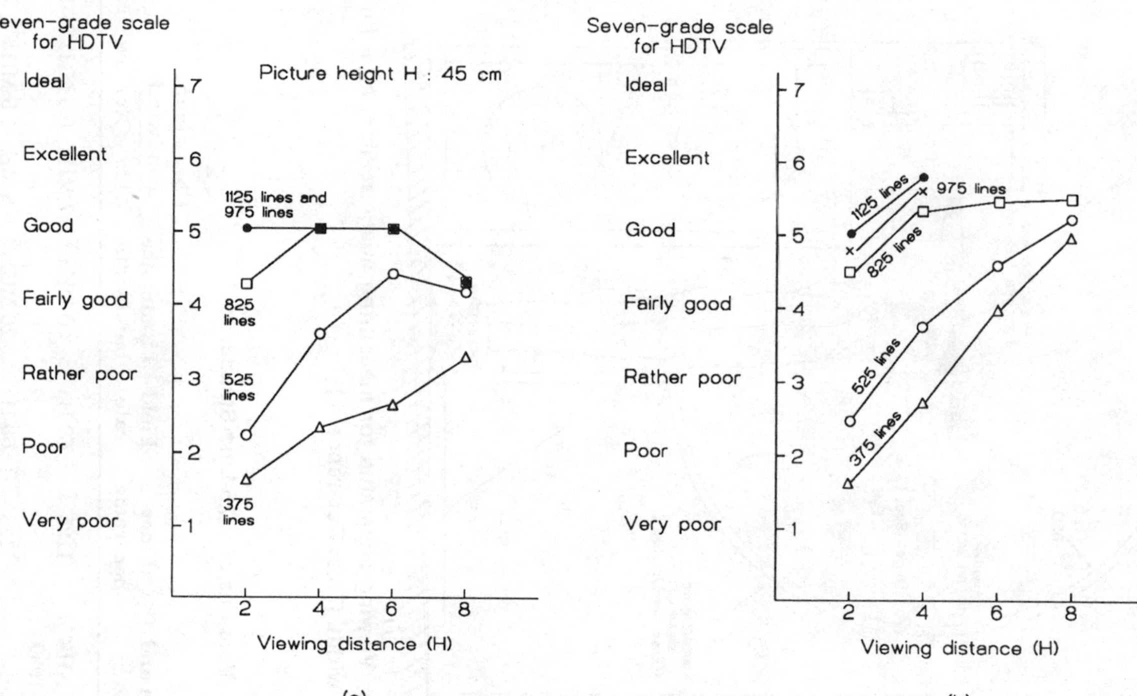

Figure 7.1 **Subjective evaluation of image quality (a) and sharpness (b) of images reproduced at line rates from 375 to 1125 lines per frame, at viewing ratios from two to eight times the picture height. (From Reference 3.)**

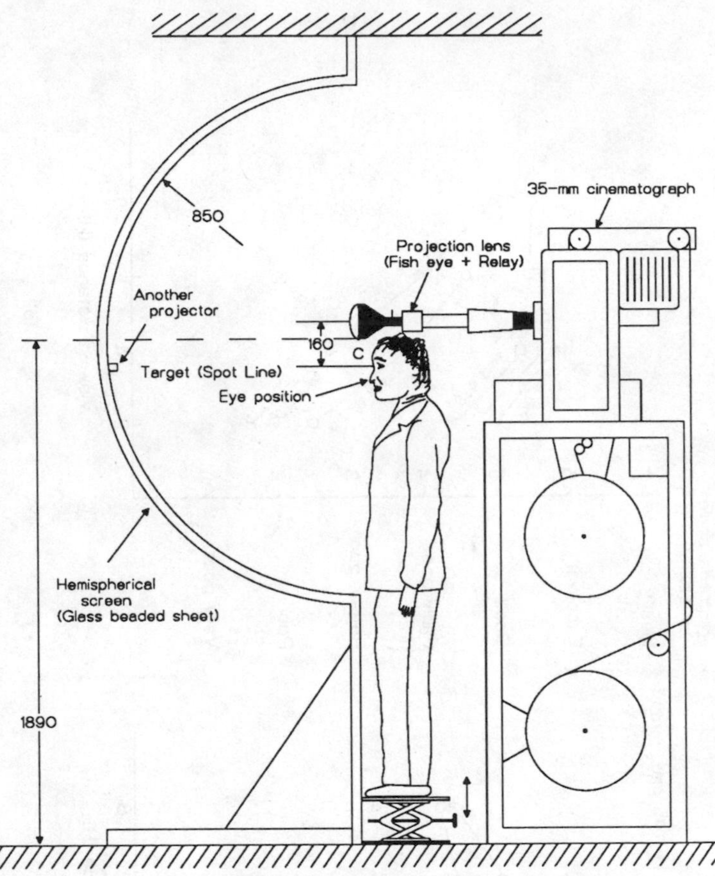

Figure 7.2 Viewing apparatus for measuring image realism as a function of picture width. (From Reference 4.)

Table 7.1 Versions of 1125-Line Standards

Version and date	Lines per frame	Field rate	Luminance bandwidth	Chrominance bandwidth: Wide	Chrominance bandwidth: Narrow	Aspect ratio
Provisional NHK Standard-1980	1125	60 Hz	20 MHz	7 MHz	5.5 MHz	5:3
Line-sequential MUSE-1986	1125	60 Hz	20 MHz	6.5 MHz	5.5 MHz	5:3
SMPTE Studio Standard-1987	1125	60 Hz	30 MHz	30 MHz	30 MHz	16:9

NHK prior to 1980. At that time there was widespread support for a single worldwide standard for HDTV service. If this were achieved, the NTSC and PAL/SECAM field rates of 59.94 and 50 Hz would prevent one (or all) of these systems from participation in such a standard. The 50-Hz rate was conceded to impose a severe limit on display brightness, and the 59.94-field rate posed more difficult problems in transcoding, versus 60 Hz. Thus, a 60-field rate was proposed for the world standard.

The choice of 1125 lines was also justified by the then-existing standards. The midpoint between 525 and 625 lines is 575 lines. Twice that number would correspond to 1150 lines for a "midpoint" HDTV system. This even number of lines could not produce alternate-line interlacing, then thought to be essential in any scanning standard. The nearest odd number, having a common factor with 525 and 625, was the NHK choice: 1125 lines. The common factor, 25, makes line-rate transcoding (Section 6.6) among the NHK, NTSC, and PAL/SECAM systems comparatively simple.

7.3 Equipment Development, 1970–1982[1,3–5]

The decade prior to the public demonstrations of the NHK system was a period of intense development of HDTV equipment in the NHK Laboratories. By 1980 the necessary camera tubes and cameras, picture tubes and projection displays, telecines, and video-to-tape recorders were available. Also the choices of transmission systems, signal formats, and modulation and demodulation parameters had been made. Work with digital transmission and fiber optics had begun, and a prototype tape recorder had been designed. Much HDTV equipment now on the market can be traced to these early developments. (See Reference 3 for descriptions of this NHK equipment.)

7.3.1 Cameras and Camera Tubes

The 1973 NHK cameras used three 1-1/2-inch vidicons, then commercially available. These lacked resolution and sensitivity. To replace them, NHK developed the return-beam Saticon (RBS), which had adequate resolution and sensitivity but about 30% lag. Cameras using three RBS tubes came into use in 1975. They were used for much of the development of the NHK system. By 1980 another tube, the diode-gun impregnated-cathode Saticon (DIS) tube, was ready for public demonstrations. This was a 1-inch tube, having a resolution of 1600 lines (1200 lines outside the 80 percent center circle), lag less than 1%, and 39 dB signal-to-noise ratio across a 30-MHz band. Misregistration among the primary color images, about 0.1 percent of the image dimensions in the earlier cameras, was reduced to 0.03 percent in the DIS cameras. When used for sporting events, the camera was fitted with a 14-times zoom lens of *f*/2.8 aperture. Its performance established the reputation of the NHK system among all viewers, including those from the motion-picture industry.

7.3.2 Displays

The major task of adapting the conventional display to high-definition began in 1973, when NHK developed a 22-inch picture tube with a shadow-mask hole pitch of 310 μm (compared with 660 μm for a standard tube) and an aspect ratio of 4:3. In 1978 a 30-inch tube of hole pitch 340 μm and 5:3 aspect ratio was produced. This tube had a peak brightness of 30 footlamberts (100 cd/m^2). This was followed in 1979 by a 26-inch tube with 370-μm hole pitch, 5:3 aspect ratio, and peak brightness of 45 footlamberts.

The resolution of these tubes (in terms of the percent modulation transfer) is shown in Figure 7.3a. The 26-inch tube offered the finest detail, about 50 percent transfer at 1000 lines. This figure illustrates the problem, not resolved in the late 1980s, of obtaining resolution in displays equal to that of the camera. The 26-inch tube was followed in 1983 by a 40-inch monitor of 5:3 aspect ratio and 450 μm hole pitch, developed with the assistance of several Japanese manufacturers.

The need for displays larger than those available in picture tubes led NHK to develop projection systems. A system using three CRTs with Schmidt-type focusing optics produced a 55-inch image (diagonal) on a curved screen with a peak brightness of about 30 footlamberts. A larger image (67 inch) was produced by a light-valve projector, employing Schlieren optics, with a peak brightness of 100 footlamberts at a screen gain of 10. The resolution of these projectors is shown in Figure 7.3b. The poor resolution of projection systems, compared with picture tubes, clearly shown in this figure, has persisted throughout the 1980s.

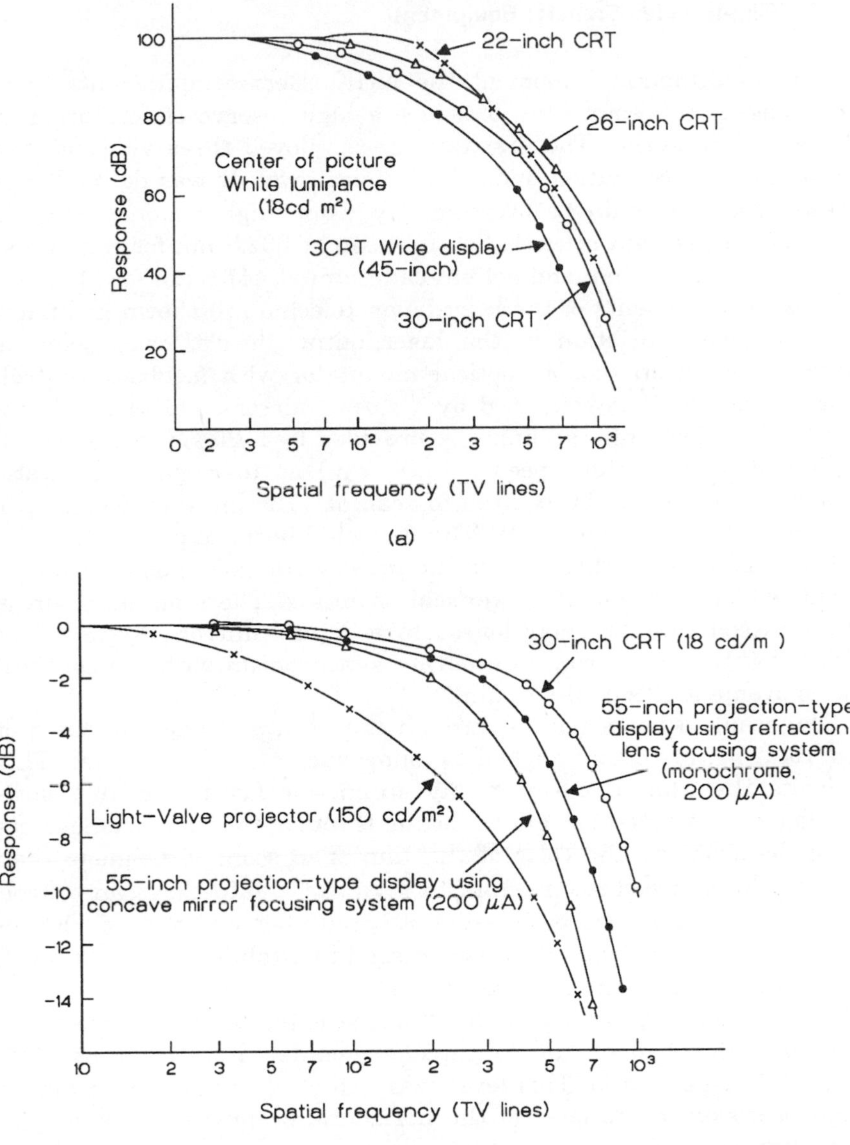

Figure 7.3 Percentage response of modulation transfer function for (a) cathode-ray picture tubes and (b) projection displays at resolutions up to 1000 lines. (From Reference 3.)

7.3.3 Film-to-Video Transfer Equipment

Throughout their development the NHK telecine equipments have been based on 70-mm film to assure a high reserve of resolution in the source material. The first telecine employed three vidicons, but these had low resolution and a high noise level. It was decided that these problems could be overcome by using highly monochromatic laser beams as light sources: helium-neon at 632.8 nm for red, argon at 514.5 nm for green, and helium-cadmium at 441.6 nm for blue.

The basic elements of the laser-beam telecine are shown in Figure 7.4. To avoid variation in the laser output levels, each beam is passed though an acoustic-optical modulator with feedback control. The beams are then combined by dichroic mirrors and scanned mechanically. Horizontal scanning is provided by a 25-sided mirror, rotating at such a high speed (81,000 r/m) as to require aerostatic bearings. This speed is required to scan at 1125 lines, 30 frames per second, with 25 mirror facets. The deflected beam is passed through relay lenses to another mirror mounted on a galvanometer movement, which introduces the vertical scanning. The scanned beam is then passed through relay lenses to a second mirror-polygon of 48 sides, rotating at 30 r/m, in accurate synchronism with the continuous movement of the 70-mm film.

The scanned beam passes through the film, and adjacent faces of the rotating mirror are aligned to follow successive film frames. This permits the film to move at any speed, for fast-frame and slow-motion effects. Passage of the beam through the film changes its color according to the color of the film. The scanned beam is then passed through dichroic mirrors to separate it into the primary colors and these are read by three photo-multiplier phototubes. The resulting R, G, and B signals are processed to establish the black level, gamma correction, and level correction.

The resolution provided by this telecine is limited by the mechanical scanning elements to 35 percent modulation transfer at 1000 lines, 2:1 aspect ratio. This level was achieved only by maintaining high precision in the faces of the horizontal scanning mirror and in the alignment of successive faces. To keep the film motion and the frame-synchronization mirror rotation in precise synchronism, an elaborate electronic feedback control was used between the respective motor drives. In other respects the performance was more than adequate. The signal-to-noise ratios of the output signals reached the high levels of 43, 44, and 41 dB for the red, green, and blue signals, respectively.

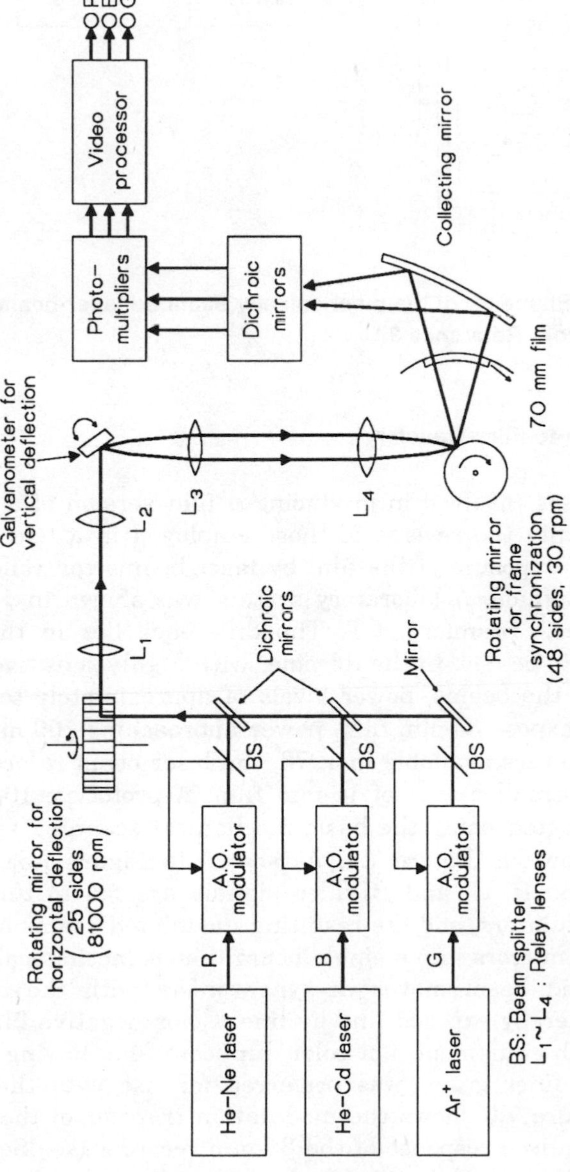

Figure 7.4 Elements of the mechanically scanned laser-beam telecine. (From Reference 3.)

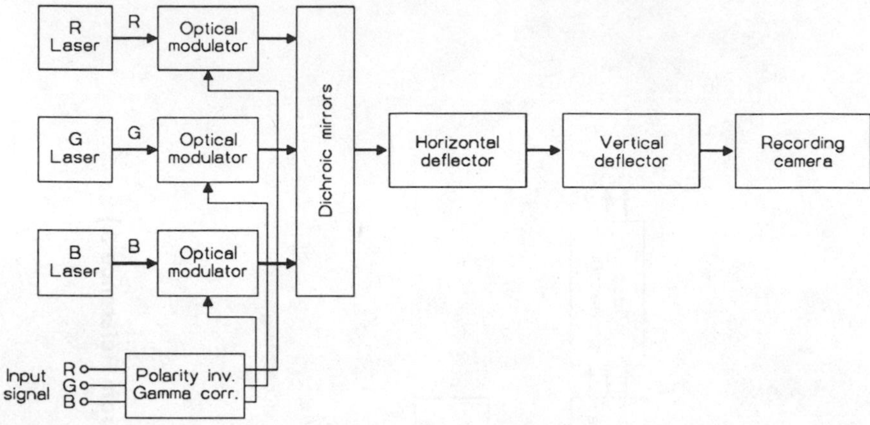

Figure 7.5 Elements of the mechanically scanned laser-beam video-to-film recorder. (From Reference 3.)

7.3.4 Video-to-Film Transfer

The processes involved in producing a film version of a video signal are essentially the reverse of those employed in a telecine, the end point being exposure of the film by laser beams controlled by the R, G, and B signals. A laboratory system was shown in 1971 by CBS Laboratories,[6] Stamford, CT. The difference lies in the power required in the beams. In the telecine, with highly sensitive phototubes reacting to the beams, power levels of approximately ten milliwatts suffice. To expose 35-mm film, power approaching 100 mw is needed. With the powers available in 1978, the laser-beam recorder was limited to the smaller area of 16-mm film. A prototype 16-mm version was constructed using the basic mechanical scanning system of the telecine shown in Figure 7.4. As shown in Figure 7.5, in the laser recorder, the R, G, and B video signals are fed to three acoustic-optical modulators, and the resulting modulated beams are combined by dichroic mirrors into a single beam that is mechanically deflected. The scanned beam moves in synchronism with the moving film, which is thereby exposed line by line. Color negative film is usually used in such equipment, but color duplicate film, having higher resolution and finer grain, was preferred for use with the 16-mm recorder. Figure 7.6 shows the modulation transfer of the film proper and the required response of the 35-mm recorder (see Section 13.21). Since that time the laser system has been discarded for the electron-beam (EBR) system discussed in Section 13.2.2.

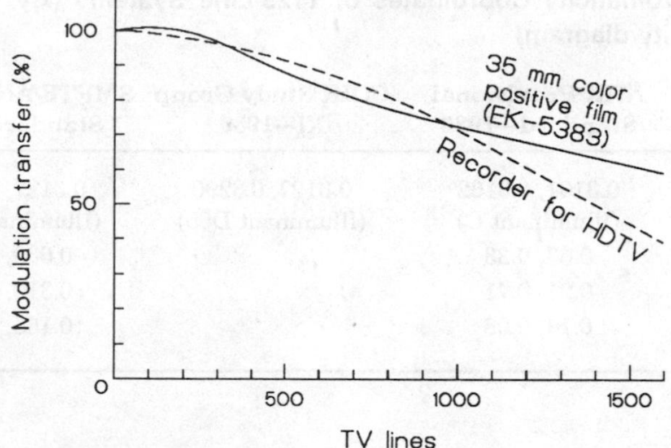

Figure 7.6 Modulation transfer percentage for typical color-positive film and specifications for video to 35-mm film recorders. (From Reference 3.)

7.4 Colorimetric Standards

The colorimetric standards of the initial NHK system were based on the NTSC primary colors and Standard Illuminant C. The *x,y* coordinates on the CIE diagram are shown, together with those used in other 1125-line systems, in Table 7.2.

Dr. Fujio and his associates at NHK Laboratories decided that the *I* and *Q* chrominance axes of the NTSC system were not the optimum choice. Based on tests[5] at NHK Laboratories, they determined that the chrominance axes for the NHK system should be those shown in Figure 7.7a, positioned with respect to the color gamut as in Figure 7.7b. These are shown on the uniform chromaticity scale (UCS) version of the CIE diagram, selected because the chromatic intervals are equally spaced along the chrominance axes.

The luminance and chrominance signals, in terms of the *R, G, B* primaries, for the provisional NHK standards are:

Luminance: $Y = 0.30R + 0.59G + 0.11B$

Wideband chrominance: $C_W = 0.63R - 0.47G - 0.16B$

Narrowband chrominance: $C_N = -0.03R - 0.38G + 0.41B$

These signals have bandwidths of 20.0, 7.0, and 5.5 MHz, respectively. The corresponding definitions of luminance and chrominance

Table 7.2 Chromaticity Coordinates of 1125-Line Systems (*x,y* values on CIE chromaticity diagram)

Color quantity	NHK Provisional Standard—1980	CCIR Study Group XI—1986	SMPTE/ANSI Studio Standard—1987
White	0.3101, 0.3162 (Illuminant C)	0.3127, 0.3290 (Illuminant D65)	0.3127, 0.3290 (Illuminant D65)
Red primary	0.67, 0.33		0.630, 0.340
Green primary	0.21, 0.71		0.310, 0.595
Blue primary	0.14, 0.08		0.155, 0.070

for the SMPTE studio standard (Section 7.13) differ, since the reference white and primaries are not the same.

7.5 Transmission and Modulation Methods

Among the NHK engineers, conventional vestigial-sideband broadcasting was not considered feasible for the wide bandwidths of the 1125-line system. Fujio and his co-authors[1] were explicit in 1980 when they wrote, "The VHF and UHF bands are unusable for the transmission of the broadband signals of high-definition television" (page 582). The alternative then seen was satellite broadcasting using frequency modulation. This choice was influenced by the fact that the Japanese Islands are readily covered by satellite transmissions.

In frequency modulation the noise in the reproduced image increases with the baseband frequency. In the composite signal the chrominance signal occupies the upper range of the baseband, so chrominance noise determines the transponder power required to obtain an acceptably low noise level. In typical cases, an 1125-line signal with 30-MHz base bandwidth, occupying an fm radiofrequency bandwidth of 100 MHz, feeding a receiving antenna with a diameter of 5 feet, would require a transponder power of 3.2 kw in the 22-GHz satellite band for a signal-to-noise ratio of 41 dB. Since such a power level was unobtainable in the satellite service, composite transmission was ruled out.

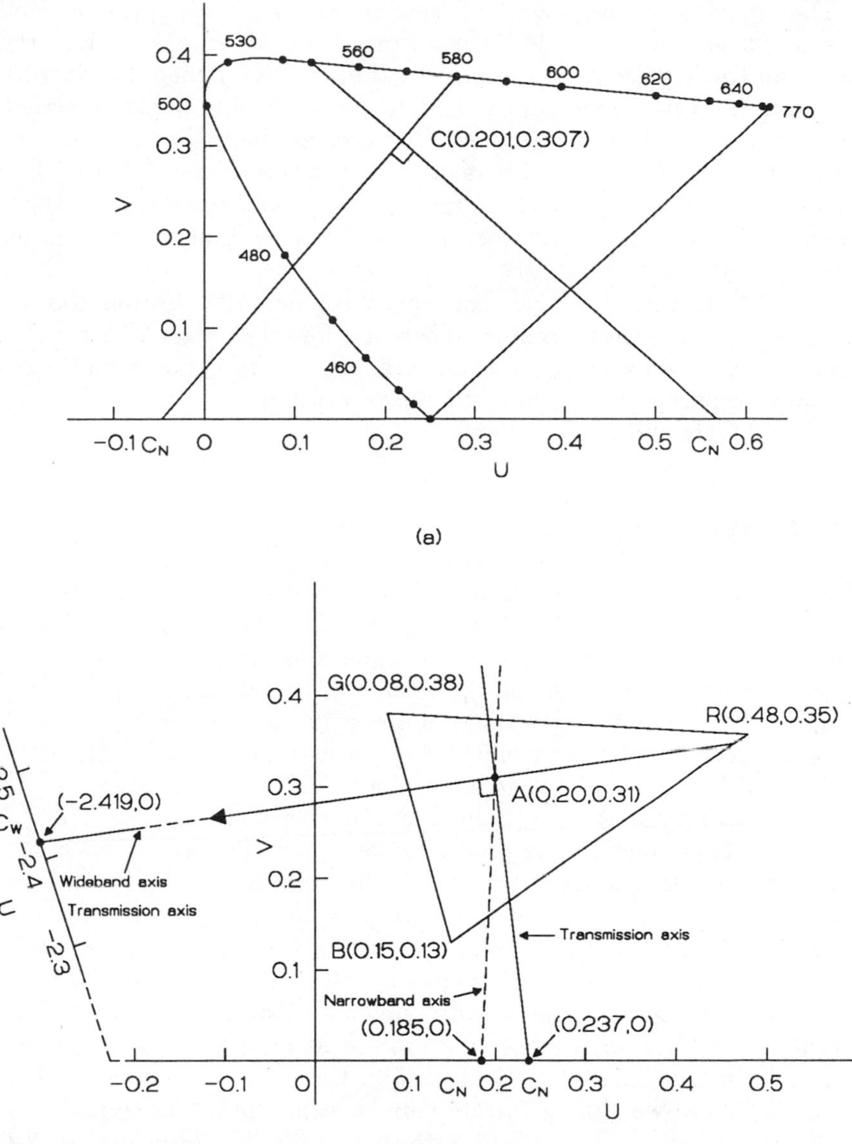

Figure 7.7 Chrominance axes chosen for the NHK system displayed on the uniform chromaticity scale version of the CIE chromaticity diagram: (a) relationship to the monochromatic locus, (b) relationship to the color gamut. (From Reference 7.)

The alternative was to use separate transmission of luminance and chrominance,[1] the *separate Y-C* system. This confines the high signal levels to the low-frequency region of the frequency modulation with 20 MHz for luminance and 6.5 MHz for each chrominance signal, and 75 and 25 MHz, respectively, for the fm channel widths on the 22-GHz band. The required transponder power with the 5-foot receiving antenna, would be reduced from 3.2 kw to 570 watts (360 watts for luminance, 210 watts for chrominance). Signal-to-noise ratios would be greater than 42 dB.

The Y-C transmission system would require 12.6 kw on the 42-GHz band, to permit greater allowance for rain attenuation. This finding prompted the NHK engineers to conclude that satisfactory satellite broadcasting of the 1125-line service could be provided only on the 12-GHz and 22-GHz bands.

7.6 Noise in 1125-Line MUSE Service

The limited power of the transponders, with the high loss due to rain attenuation on the higher-frequency satellite bands, required that close attention be paid to the permissible noise level in televised images. In 1980 Fujio[8] reported the results of an extensive study of the sources of noise and the means of improving signal-to-noise ratios.

The threshold of detectability of noise was taken at 53 dB below the signal level. This level is far lower than the 40–45 dB generally accepted as excellent performance in cameras and displays. The difference is explained by the properties of vision. The eye is most sensitive to low-frequency noise in both the spatial and temporal domains. It can be considered to act as a low-pass filter whose response is the inverse of $1 + (f/f_o)^2$, where f is the baseband video frequency and f_o is a cutoff frequency determined by the scanning constants, the aspect ratio, and viewing distance. Typically, the cutoff is 3.2 MHz in the 525-line system at an aspect ratio 4:3 and 6.9 MHz in the 1125-line system at 5:3. Using these filter characteristics, a "noise-weighting factor" can be computed that adjusts the visible noise to a value smaller than the 53-dB limit. In the Y-C system the weighting factor was found to be 13.4 dB for luminance and 9.5 dB for chrominance. These values permit signal-to-noise ratios at the display of the order of 40–45 dB.

The low visual response to high-frequency noise also allows a reduction in the transponder power if the higher video baseband frequencies are emphasized (increased in amplitude) prior to modulation at the transmitter and correspondingly deemphasized after demodulation at the receiver. This emphasis/deemphasis technique was adopted for the satellite system, with crossover frequencies, above which the emphasis was applied, taken at 5.2 MHz for luminance and 1.6 MHz for chrominance. In one example given by Dr. Fujio, the total transponder power was stated to be 260 watts (190 for luminance, 69 for chrominance) with the noise-weighting factors and emphasis/deemphasis values given above.

7.7 Signal Compression in the MUSE System

The separate transmission of luminance and chrominance in the Y-C system requires either that three channels be used or that the signals be compressed sufficiently to fit into a single channel. This was an easy choice: spectrum economy demanded the compression approach. The technique adopted is known as time-compression integration (TCI). Figure 7.8 shows three versions of TCI signals, in which the luminance and chrominance signals are sent sequentially during the scanning of one line (Figure 7.8a and b) or two lines (Figure 7.8c). These patterns are similar to that adopted in the MAC family of DBS systems (Section 2.7). They require complex signal processing in the receiver, so NHK decided not to pursue them and concentrated on devising a system that could take advantage of the advanced techniques then employed in the 12-GHz satellite band.

Thus was born the present NHK system of satellite broadcasting, the MUSE.[9] This system, described in greater detail in Section 2.8, was announced in 1984. It involves digital sampling of the Y and C components at a high frequency, followed by subsampling at a rate lower than the Nyquist limit (Section 4.6). As originally proposed, the sampling rate was 64.8 Ms/s, followed by subsampling at 16.2 Ms/s. As outlined in Section 2.8, this resulted in the sequential transmission of signals representing every fourth picture element, so that the full detail was filled in only in four successive fields. The picture elements of an object in motion were thereby displaced from their proper positions, causing a loss of horizontal resolution. Stationary parts of the image were displayed in the full 1125-line resolution of the NHK system. Fujio reported in 1985[4] that the base bandwidths for luminance and chrominance were 20–22 and 7.0 MHz for stationary images, respectively, 12.5 and 3.5 MHz for objects in motion. It is thus implied that the level of the horizontal resolution for objects in motion is about 60 percent of that for stationary objects. The audio and control information is digitally sampled and multiplexed in the vertical blanking interval.

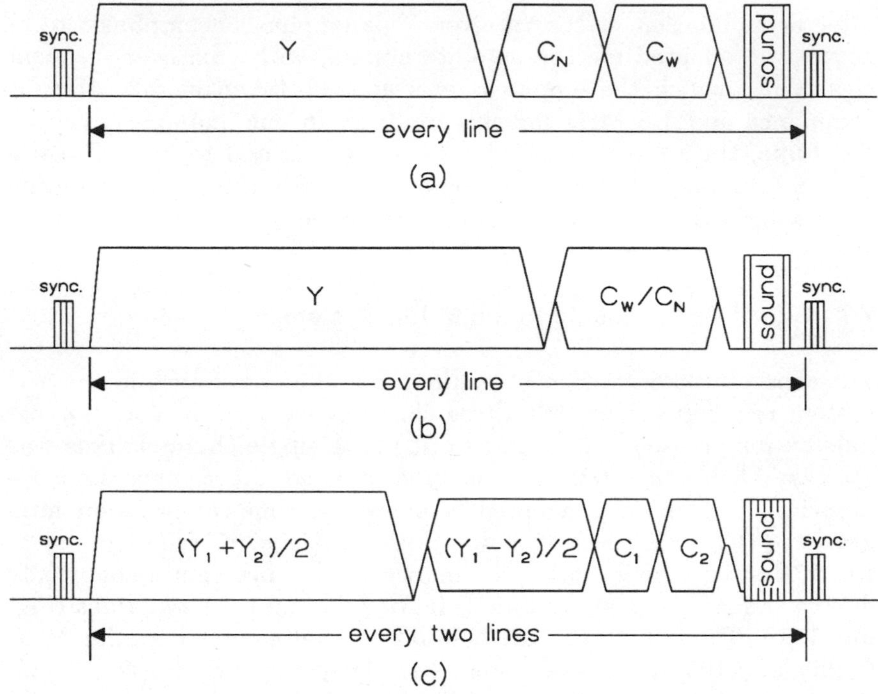

Figure 7.8 Three variations of time-compressed integration signals with interspersed luminance and chrominance segments. (From Reference 4.)

The version of MUSE on which the Japanese public DBS service will be offered employs an initial sampling rate of 48.6 Ms/s (rather than 64.8 Ms/s) while retaining the 16.2 Ms/s subsampling rate. Thus, the picture elements are spread over only three fields, and the blurring of moving objects is reduced accordingly. The encoding and decoding apparatus is complicated (see Figure 2.7) and the converter circuits for reception employ complex integrated circuits. (See Reference 9 for complete details of the video and audio processing required.) The current version of the MUSE system will be the first to offer an HDTV service to the public in 1990. The response of the market, i.e., sales of receivers capable of making full use of the system, will be watched closely by the proponents of HDTV systems in other countries. The wide expanse of services planned for the MUSE system is illustrated in Figure 7.9.

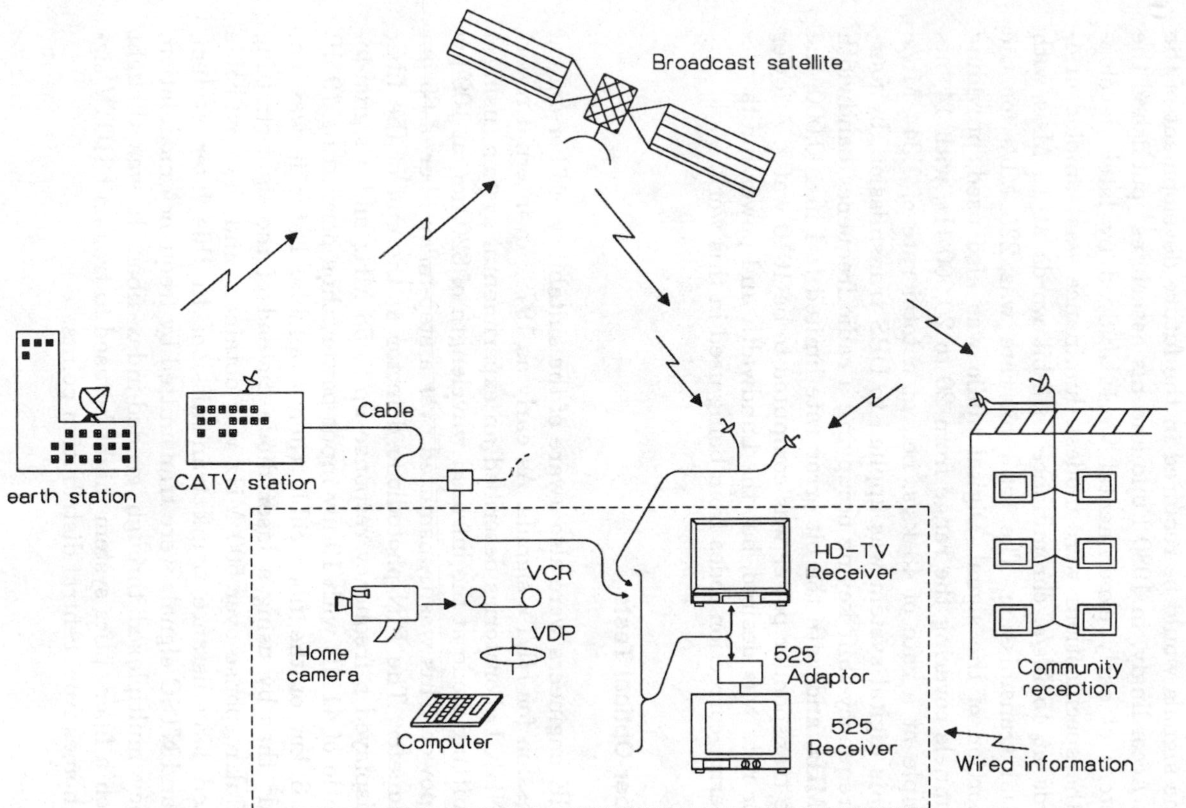

Figure 7.9 The MUSE DBS service is intended to serve the cable, terrestrial broadcast, and tape recorder audiences, as well as the individual viewer. (From Reference 9.)

7.8 Digitization of 1125-Line Signals

Prior to 1980 the NHK engineers realized that digital versions of the 1125-line signals would be required in the future development of the system. Accordingly, in 1980 Fujio and his associates[1] published the parameters of early experimental work. Judged by later developments the specifications were modest: luminance was sampled in 4-bit words at 45 Ms/s, chrominance in 3-bit words at 15 Ms/s with sequential transmission. The total bit rate was 225 Mb/s for the video portion of the signal. Digital audio was also used, providing two channels covering the range from 20 to 20,000 Hz, with 14 bits per sample at a rate of 48 ks/s, i.e., at a total rate of 1.344 Mb/s. When this digital system was applied to DBS transmission, by four-phase frequency-shift keying occupying a radio frequency bandwidth of 150 MHz, and with the bit error rate limited to 1 in 100,000, the satellite transponder power was computed to be 1010 watts. A lower bit error rate was desired, but the bandwidth and power limits prevented error-correction codes from being used in this work.

7.9 Fiber-Optical Tests

The NHK engineers were also aware of the suitability of fiber-optical transmission for HDTV signals. As early as 1978, an internal report of the NHK Laboratories described an experimental system, using a light-emitting diode at the infrared wavelength of 820 nm, at 109 μw output power. This was transmitted over a step-index fiber of 65 μm core diameter. The PIN photodiode detector at the end of the fiber cable displayed a frequency response up to 58 MHz and a signal-to-noise ratio of 41 dB with 10 μw input power, thus permitting 10 dB loss in 3 km of the fiber cable then available. Later it was announced[1] that by using a laser diode instead of the light-emitting diode, flat response over 500 MHz was obtained, and the new cable permitted long-distance trunk transmission. In this later system, HDTV and NTSC signals were transmitted by frequency modulation, frequency multiplexed through graded-index fiber. It was thought that such a fiber-optics system might be used to transmit HDTV signals to homes from central distribution points.

7.10 Tape Recording of 1125-Line Signals

The 30-MHz bandwidth of the 1125-line system can be recorded on magnetic tape only by pushing the state of the art to its outer limits.

After ten years of study and experimentation, the NHK engineers were able to report[3] in 1982 that a protoype recorder had been built, but that its performance was limited by the types of magnetic tape available and by the difficulty of building recording and playback heads capable of withstanding the centrifugal force produced by high-speed rotation. The basic elements of the prototype are shown in Figure 7.10. The HDTV signal modulates an fm modulator, as is customary in conventional recorders. The recording head contains a wideband transformer to produce the magnetizing current. The velocity of the recording head relative to the tape surface must be of the order of 50 m/s to record the fm signals to an upper frequency range of the order of 50 MHz and 75 m/s is needed to reach 70 MHz. The quality of reproduction depends on the coercive force and retentivity of the tape used; the metal-coated tape then available offered substantially higher values than the iron-oxide tapes used in conventional recorders. The 1982 NHK report gives no detailed specifications of the performance of the prototype; it stated that much further development was needed.

Ten years later a commercial version[10] of an 1125-line videotape recording system showed the extent of that further development. Two equipments are involved: the HDV-1000 recorder and the HDT-1000 signal processor. The recorder uses reels of 1-inch tape similar to the Type-C format tape commonly used in conventional studio practice. But the tape speed is nearly twice as great, 48.3 cm/s versus 25 cm/s. This higher speed limits the recording time to 63 minutes with an 11.75-inch reel.

The 1125-line input may be separate RGB or luminance-chrominance components. These are recorded on high-density tape by frequency modulation, with a bandwidth of 20 MHz (+0.7 dB, –3.0 dB) for luminance, 10 MHz for chrominance. The three RGB components are kept in synchronism within ±5 ns. Up to three sound tracks can be accommodated with a range of 50 Hz to 15 kHz (+1.5 dB, –3.0 dB).

The recording drum has a writing speed, relative to the tape surface, of 25.9 m/s (approximately the same as in the conventional Type-C recorder). To write at 20 MHz, versus the 5 MHz of NTSC recording, four video tracks are written in parallel by the drum. To bring these signal segments into precise alignment during playback requires elaborate signal processing, which is carried out in the separate signal processor mentioned above. This involves digital processing with 8-bit words at a sampling rate of 69.1 Ms/s, with a residual timing error of ±3 ns. The signal-to-noise ratio in playback is better than 41 dB over the 20-MHz band (measured in the green channel).

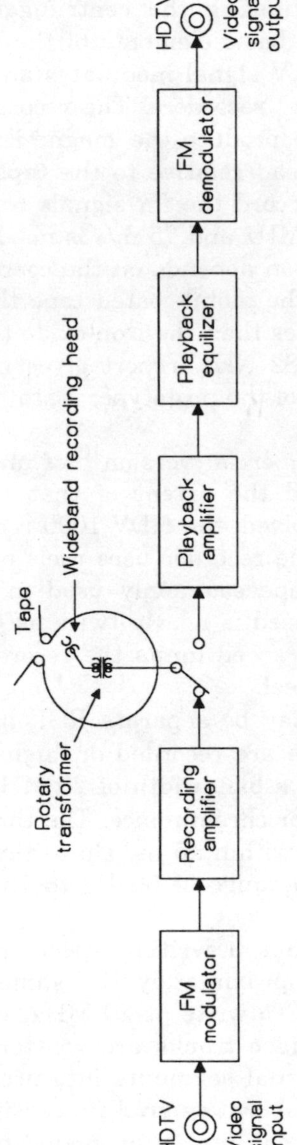

Figure 7.10 Elements of the HDTV tape recorder. (From Reference 3.)

This is complex, expensive, and heavy equipment. The combined weight of the recorder and signal processor is 470 pounds, and their cost was approximately $200,000 in 1989. Their performance, as indicated by the specifications given above, is adequate in relation to other elements of the system, notably the displays (Section 7.11). But it suffers from a major limitation common to all analog signal recorders, i.e., deterioration of quality in successive generations of duplication. In the distribution of motion-picture programs produced by HDTV, such duplication is an essential production process.

The success of digital tape recording in allowing multiple duplication in conventional broadcasting has encouraged Japanese engineers to convert HDTV recorders to the digital domain. Two recent papers describe their efforts. In one experimental model,[11] the general principles of the commercially available recorder described above are retained, namely, multiple-track recording by frequency modulation, based on the Type C 1-in tape format. The recording drum (Figure 7.11) has no fewer than 18 heads: 8 for recording, 8 for playback, and 2 for erasure. The drum rotates at 7200 r/m, and the head-to-tape velocity is 51.5 m/s, nearly twice that of the analog/composite HDTV recorder.

The sampling rates follow the suggested values of CCIR Report 801: 74.25 Ms/s for luminance, 37.125 Ms/s for the chrominance signals, 8 bits per sample. These values (5.5 times those of the CCIR Recommendation 601 for conventional digital operations) require an overall bit rate of $(74.25 + 37.125 + 37.125) \times 8 = 1.188$ Gb/s. After dividing the signal among eight heads, the recording rate of each head is 148.5 Mb/s. At the head velocity of 51.5 m/s, the minimum recording wavelength is 0.69 μm, and the same small dimensions apply to the magnetic tracks, whose pitch and width are 37 and 34 μm, respectively. At these rates and dimensions, the number of 8-bit samples in the complete line time is 2200, of which 1920 are active. The quantization noise (Section 4.5) is 58.8 dB below the signal.

Each of the eight recording heads must record only one-eighth of the picture content. After the input signal is sampled at 74.25 Ms/s, the three image areas in luminance and chrominance are each divided into four adjacent vertical segments, and each of these is digitally stored and read out at the subsampling rate of 18.5625 Ms/s. The expanded luminance and chrominance areas of each segment are then multiplexed, with every other chrominance segment removed. This results in a total of eight channels, with the 8-bit picture-element words rigidly synchronized.

Before these eight signals are applied to the respective recording heads, each is coded for error correction based on a Reed-Solomon

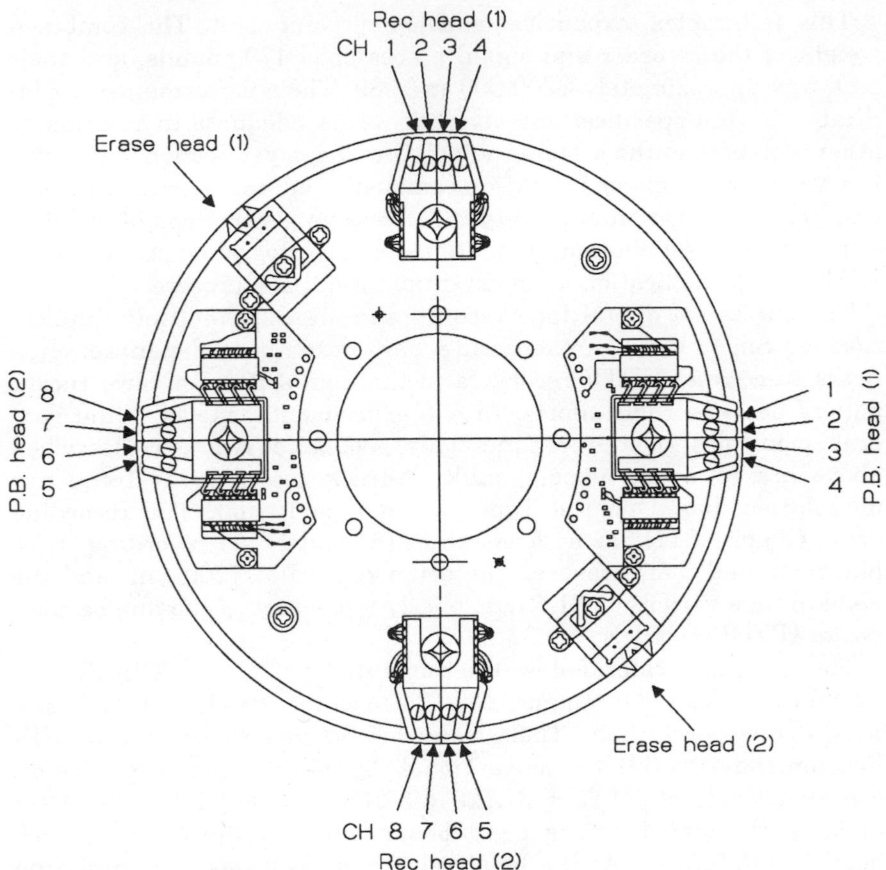

Figure 7.11 Recording, playback, and erase heads of the 1.188 Gb/s digital HDTV tape recorder. (From Reference 11.)

product code and then arranged in serial form. The bit error rate, about 10^{-5} before correction, is negligible after correction. This permits as many as 20 generations of duplicate tapes to be made without degradation. Metal-particle tape, running at 805 mm/s, allows one hour of recording on an 11.75-inch reel.

Eight channels of audio are available, sampled at 48 ks/s in 16-bit words. With parity and redundancy codes, the total recording rate is 1.152 Mb/s. The sampling rate and word length are the equal of the compact-disk audio standards (Section 4.15). A commercial version of this recorder has been announced by the Sony Corporation, whose engineers developed it.

The second experimental HDTV digital recorder[12] is intended for use in an environment of analog equipment. It is also based on the Type-C 1-inch tape format. Eight heads are used on the drum. For use with analog equipment, a lower bit rate, 648 Mb/s, is the basis of the design. The luminance bandwidth is 21 MHz, chrominance 9 MHz. The tape speed is 0.511 m/s, head-to-tape velocity 45.9 m/s, and the minimum wavelength on the tape is 0.85 μm. Luminance is sampled at 54 Ms/s, chrominance at 27 Ms/s. The error rate is sufficiently low to allow ten generations of duplicates to be made without degradation. Playing time is 90 minutes with a 14-inch reel.

7.11 Equipment Available for the 1125-Line Service

A wide variety of equipment for the 1125-line system is commercially available from Japanese manufacturers, notably Hitachi, Ikegami, Matsushita, Sony, and Toshiba. Space is available here to mention only a few items. They have been selected from the equipments offered by the Sony Corporation. Detailed discussion may be found in subsequent chapters.

7.11.1 Camera and Accessories

The Model HDC-300 camera employs three 1-inch Saticon tubes, offering resolution of 1200 lines at the center of the image. The registration error is 0.05 percent or better. The sensitivity is 2000 lux at *f* /4.5. Three viewfinders with 16:9 aspect ratio are available with diagonals of 1.5, 3, and 7 inches, with resolutions of 350, 450, and 1000 lines, respectively. Control of the camera is provided from the console by the HDC0-300 Camera Control Unit. Monitoring of the image and its waveform are provided by the HDM-90 9-inch monitor and 1730HD waveform monitor. Lenses for the camera are available from Nikon and Fujinon. Fixed-focus lenses of *f*/1.2 aperture are available in focal lengths from 9.5 to 50 mm. Zoom lenses with ratios of 5 to 22, at apertures from *f*/1.2 to *f*/2.2, cover a focal length range from 12 to 400 mm. The 1989 price for the complete camera system with a 15- to 180-mm *f*/1.8 zoom lens was approximately $300,000.

7.11.2 Displays

Four-color monitors are available, with diagonals of 12, 18, 28, and 38 inches, of which only the last size has an aspect ratio of 16:9 (the smaller sizes are offered with the 5:3 ratio). The center horizontal resolutions are 600, 760, 1000, and 1000 television lines, respectively, and the vertical resolution at the center is 750 television lines in each monitor. No data are offered on image brightness.

Several projector displays are offered. One uses a concave 120-inch screen with a light-gain of 13 and an aspect ratio of 5:3. The front-projection system provides resolution of 1000 lines at the screen center, at a brightness of 50 footlamberts. A second front-projection system can be adjusted for a flat screen at diagonal sizes of 60 to 240 inches, with 1000-line resolution. No brightness levels are given, since they depend on the screen size.

7.11.3 Tape Recording

See Section 7.10.

7.11.4 Videodisk Recording

Model HDL-2000 has a luminance bandwidth of 20 MHz, chrominance 6 MHz. The signal-to-noise ratio is 43 dB for luminance. The audio covers 20 Hz to 20,000 kHz at a dynamic range of 90 dB, harmonic distortion less than 0.05 percent, crosstalk below 80 dB, and no measurable wow or flutter.

7.11.5 Tape-to-Film Recording

The limited performance of the laser-beam recorder developed by NHK (Section 7.3) has led to a totally new approach for this equipment requirement using an EBR (Electron-Beam Recording) system. The input signal is derived from the HDV-1000 tape recorder in the playback mode, with signal processing performed by the HDT-1000 time base corrector/signal processor. The 1125-line, 60-field RGB outputs of that processor are first passed through analog-to-digital converters and stored in three frame-stores. These are read out as three 2250-line progressively scanned signals, at the standard film-advance rate of 24 frames per second. The parallel signals are realigned in sequential form, and gamma correction is performed with the aid of a microcomputer. The gamma-corrected signals, after digital-to-analog conversion, are fed to a light-emitting diode that produces a light signal corresponding to the sequence of RGB video.

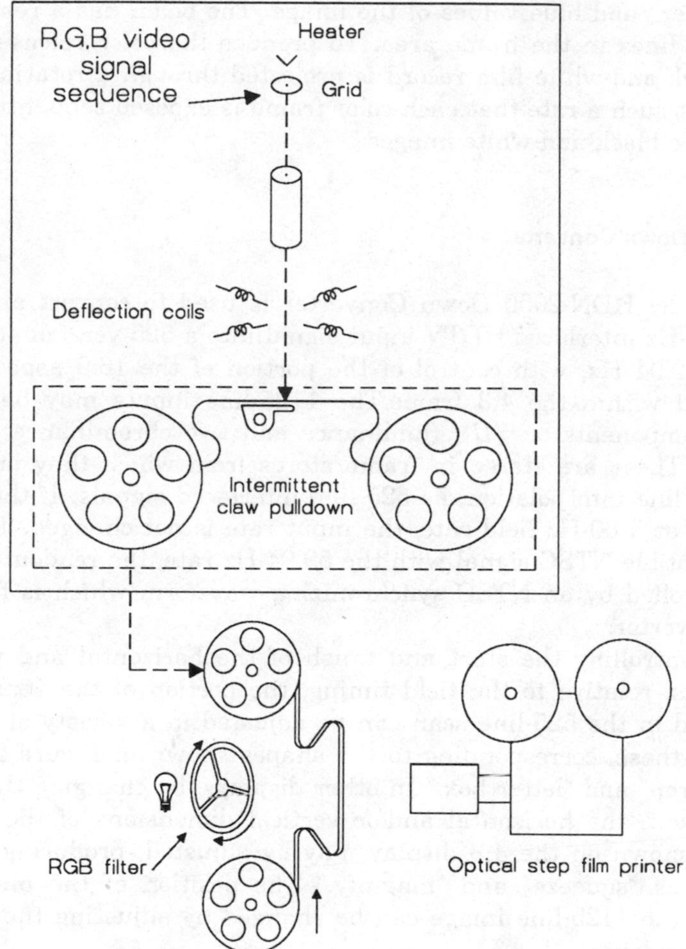

Figure 7.12 Exposure system of the Sony Electron-Beam Recorder.

The light is passed through fiber optics to a photodiode in the electron-beam recorder (Figure 7.12). The video signal thus recovered is applied to control the beam of the electron gun of the recorder. The beam lands directly on the surface of the fine-grain black-and-white film beneath it.

The beam is scanned at 24 frames per second, matching the film motion in the claw pulldown mechanism beneath the gun. The film frames are exposed by the electrons of the beam, producing a sequence of three black-and-white frames representing the respective red, green, and blue values of the image. The beam has a resolution of 2090 lines in the frame area. To produce the color release print, the black-and-white film record is projected through a rotating color wheel at such a rate that each color frame is exposed sequentially by the three black-and-white images.

7.11.6 Down Converter

The model HDN-2000 Down Converter is used to convert an 1125-line, 60-Hz interlaced HDTV input signal into a 525 version at either 60 or 59.94 Hz, with control of the portion of the 16:9 aspect ratio included within the 4:3 frame.The 1125-line inputs may be either RGB components or *YUV* (luminance and two chrominance components). These are stored in frame stores from which they are read out, by line interpolation, as 525-line interlaced signals. If these are desired at a 60-Hz field rate, the input rate is not changed. To form a compatible NTSC signal with the 59.94-Hz rate, the readout timing is controlled by an NTSC synchronizing waveform which is fed into the converter.

By controlling the start and finish of the horizontal and vertical readouts, relative to the field timing, the portion of the 16:9 frame included in the 525-line scan can be adjusted in a variety of modes. Two of these, corresponding to the shapes shown in Figure 2.2, are "edge crop" and "letter box." In other displays, by changing the rates of readout, the horizontal and/or vertical dimensions of the HDTV image shown on the 4:3 display may by adjusted, producing effects known as "squeeze" and "magnify." The position of the magnified part of the 1125-line image can be changed by adjusting the timing of the readouts.

The 525-line outputs are available in either RGB or *YUV* component form. Freeze-field images may be produced, and a full range of image-enhancement functions is provided. An NTSC color-bar generator is included for calibrating color balance levels after conversion.

7.11.7 HDTV Switcher

An essential item of equipment in production and postproduction operations is a means of switching among program sources and of creating associated effects such as "dissolves" and "wipes." The Model HDS-1000T Switcher is such a device. It can accept seven RGB component color input signals and two monochrome inputs for titles. The bandwidth accorded to each component is 30 MHz (+0.3 dB, –3.0 dB), and crosstalk between channels is –40 dB at 30 MHz. There are four RGB outputs: two for program, one for review, and one for return video. Chroma-key facilities and a color-bar generator are included. Inputs are provided to assure synchronism ("genlock") among the signals.

7.12 HDTV Demonstrations in North America

In the fall of 1987 a comprehensive demonstration[13] of the 1125-line system was held under Canadian auspices. Nearly 7000 observers were shown comparisons of the MUSE and the NTSC systems in Ottawa, Toronto, Seattle, and Danbury, CT. Two types of sites were used: shopping centers where a large sample of the general public was available, and rooms set up to resemble living-room conditions.

The NTSC images were deliberately chosen to be of studio quality, reproduced on 26-inch displays from Type-C 1-inch tape recorders. The 1125-line images, transmitted by satellite via the MUSE-E system, were displayed on 28- and 30-inch tubes, and in one location an additional 50-inch rear-projection display was available. Seating arrangements for viewing distances of three to nine times the picture height were provided in the living-room settings, with a minimum of four times for the NTSC displays.

Participation in the living-room tests was sought by radio and newspaper advertisements, and by random selection among the participants in the shopping centers. The test results were consistent among those participating in Ottawa, Montreal, and Seattle. In Danbury, tapes of the MUSE reception were used and the quality was lower than at the three direct-reception sites. The typical viewer at the living-room sites was a 41-year-old male with a college degree and a yearly income of $40,000 (U.S. dollars) who owned two TV receivers and a tape recorder.

Among the observers in the living-room situations, the strong preference was for the HDTV displays, based on color quality, sense of depth, sharpness, brightness, and the shape of the screen, and to a lesser degree on size of the display and quality of the display of motion. These observers also indicated that they would be willing to pay higher rates for pay-channel movies when available on HDTV.

Answers to questions on the acceptable cost of HDTV equipment were less encouraging. When asked to estimate how much more an HDTV receiver might cost than a high-quality NTSC set, the observers, on average, thought $300 to $400. When informed that the estimated price would be in the range of $1500 to $2500 (U.S.), only 34 percent in shopping centers and 24 percent in living rooms expressed interest in purchasing at those prices. The $1500 price was of interest to only 26 percent if the service were available only on cable (pay television) and to 10 percent if only tape rentals were available. The sharply lower interest in the HDTV demonstrations at Danbury, where the HDTV quality was poor, indicated that signal quality will be an important requirement in the HDTV service. The report on these tests emphasizes that the NTSC displays were of exceptionally high quality, thus underscoring the competitive aspect of improvements in the NTSC service.

7.13 Abstract of the SMPTE 1125-Line Studio Standard

The SMPTE 240M Standard[14] for the 1125-line production system (see Figure 10.3) is based on the following parameters, abstracted from the complete text:

Scanning: 1125 lines per frame; 1035 active lines per frame; interlaced 2:1; aspect ratio 16:9; 60.00-Hz field repetition rate; 33,750 lines per second.

Colorimetry: (1931 CIE chromaticity coordinates, x,y)

Reference primaries: green 0.310, 0.595; blue 0.155, 0.070; red 0.630, 0340.

Reference white: D_{65} 0.3127, 0.3291.

Transfer Characteristic of the Reference Camera:

$$V_c = 1.1115 \times L_c^{(0.45)} - 0.1115 \quad (\text{for } L_c \geq 0.0228)$$
$$V_c = 4.0 \times L_c \quad (\text{for } L_c < 0.0228)$$

where V_c is the video signal output of the reference camera, normalized to the system reference white and L_c is the light input to the reference camera, normalized to the system reference white.

Transfer Characteristic of the Reference Reproducer:

$$L_r = \left(\frac{V_r + 0.1115}{1.1115}\right)^{\left(\frac{1}{0.45}\right)} \text{ for } V_r \geq 0.0913$$

$$L_r = \frac{V_r}{4.0} \text{ for } V_r < 0.0913$$

where V_r is the video signal level applied to the reference reproducer, normalized to the system reference white and L_r is the light output from the reference reproducer normalized to the system reference white.

Reference Clock: Signal durations are specified both in microseconds and in reference clock periods; there are 2200 reference clock periods in the total line; the reference clock frequency is 74.25 MHz.

Color Receivers: The image is represented by three parallel, time-coincident video signals. The "color set" is E_G' green, E_B' blue, and E_R' red; the "color different set" is E_Y' luminance, E_{PB}' blue color difference, E_{PR}' red color difference. The color sets are the signals appropriate to directly drive the reference reproducer, normalized to the system reference white. The color difference set is derived from the color set through the linear matrix given below.

Luminance and Chrominance Signals:

$$E_Y' = (0.701 \times E_G') + (0.087 \times E_B') + (0.212 \times E_R')$$

$$E_{PB}' = \frac{E_B' - E_Y'}{1.826}$$

$$E_{PR}' = \frac{E_R' - E_Y'}{1.576}$$

Transformation Equations:

$$\begin{bmatrix} E_G \\ E_B \\ E_R \end{bmatrix} = \begin{bmatrix} 1.000 & -0.227 & -0.477 \\ 1.000 & 1.826 & 0.000 \\ 1.000 & 0.000 & 1.576 \end{bmatrix} \begin{bmatrix} E_Y \\ E_{PB} \\ E_{PR} \end{bmatrix}$$

$$\begin{bmatrix} E_Y \\ E_{PB} \\ E_{PR} \end{bmatrix} = \begin{bmatrix} 0.701 & 0.087 & 0.212 \\ -0.384 & 0.500 & -0.116 \\ -0.445 & -0.055 & 0.500 \end{bmatrix} \begin{bmatrix} E_G \\ E_B \\ E_R \end{bmatrix}$$

Video and Synchronizing Signal Waveforms:

See Figure 7.13.

The video signal shown is of the form E_Y', E_G', E_B', or E_R'. The details of the synchronizing signals are identical for the E_{PB}' and E_{PR}' color-difference signals.

Timing of Events in a Video Line: The reference clock periods of a video line are:

Rising edge of sync (timing difference)	0
Trailing edge of sync	44
Start of active video	192
End of active video	2112
Leading edge of sync	2156

Durations of Video and Sync Waveforms: The time durations of the waveforms in Figure 7.13 are:

	Clock periods	Time in μs
a	44	0.593
b	88	1.185
c	44	0.593
d	132	1.778
e	192	2.586
f (sync rise time)	4	0.054
Total line	2200	29.63
Active line	1920	25.86

Bandwidth: The bandwidth of each signal in the color set is nominally 30 MHz. The bandwidth of each signal in the color-difference set is nominally 30 MHz for analog-originating equipment. In digital-originating equipment, the nominal bandwidth of the E_{PB}' and E_{PR}' signals is 15 MHz.

Signal Levels in Analog Representation:

E_Y', E_G', E_B', and E_R' Signals

Reference black level (mv)	0
Reference white level (mv)	700
Synchronizing level (mv)	±300

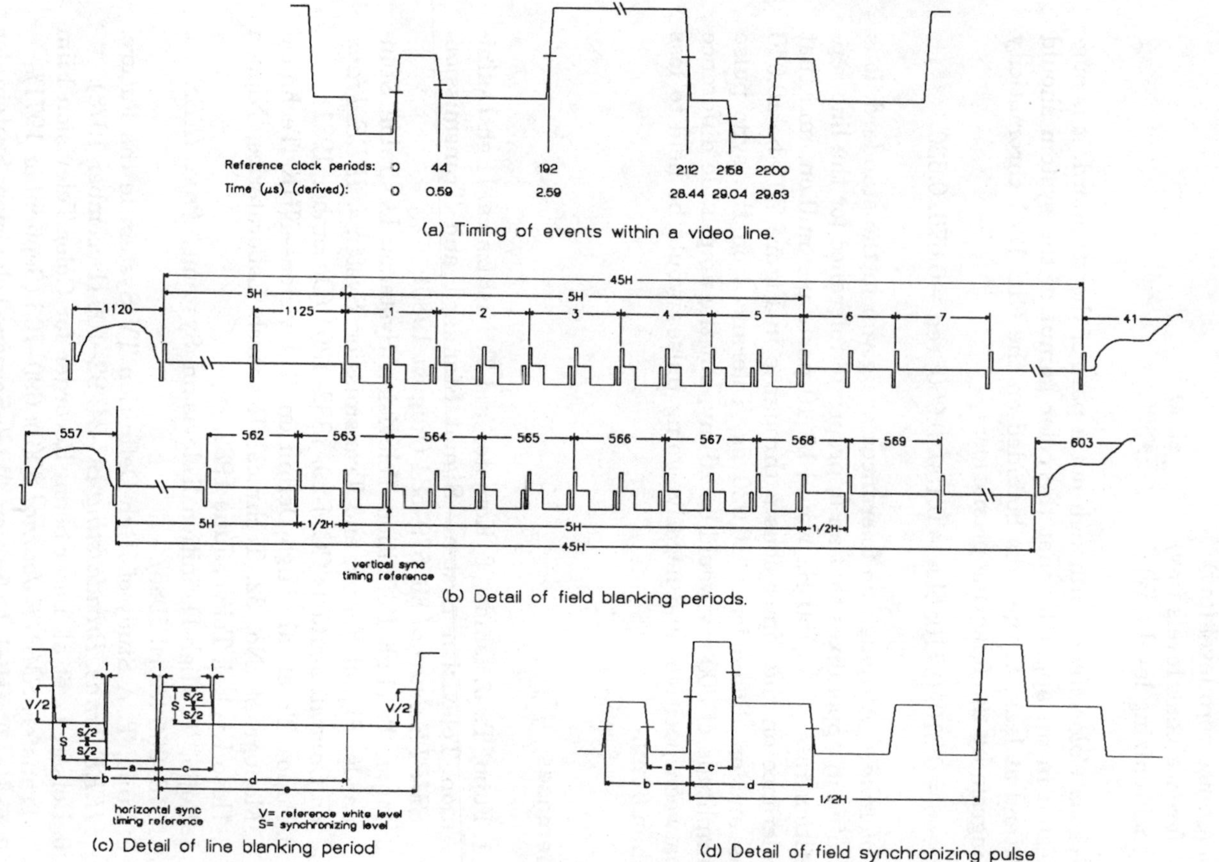

Figure 7.13 Synchronization waveform of the SMPTE 1125-line Standard 240M.

E_{PB}' and E_{PR}' Signals

Reference zero level (mv)	0
Reference peak levels (mv)	±350
Synchronizing level (mv)	±300

System Colorimetry: Although not a part of the standard, it is suggested in an appendix that the color gamut of the system should extend at least to the area bounded on the CIE 1931 chromaticity diagram by the following primaries:

green = 0.210, 0.710; blue = 0.150, 0.060; red = 0.670, 0.330.

Tolerances: Although no tolerances are given in the standard, it is stated in appendixes that an appropriate tolerance for the line repetition rate of 33,750 Hz would be 10 parts per million, and that tolerance on the dimensions a through e in Figure 7.13 be ±0.040 μs and on dimension f, ±0.020 μs. Tolerances on the sync pulse amplitude of 300 mv would be ±6 mv, but the amplitude difference between positive- and negative-going pulses would be held to less than 6 mv.

References

1. Fujio, T., J. Ishida, T. Komoto, and T. Nishizawa. High Definition Television System—Signal Standard and Transmission. *SMPTE Journal* 89: 579–584 (August 1980).
2. Fujio, T. High Definition Television Systems: Desirable Standards, Signal Forms, and Transmission Systems. *IEEE Trans. on Communications* COM-29: 1882–1891 (December 1981).
3. Fujio, T. et al. High-Definition Television—NHK Technical Monograph No. 32. Technical Research Laboratories, Nippon Hoso Kyokai, Tokyo, June 1982.
4. Fujio, T. High-Definition Television Systems. *Proc. IEEE* 73: 646–655 (April 1985).
5. Fujio, T. A Study of High-Definition TV System in the Future. *IEEE Trans. Broadcasting* BC-24: 92–100 (December 1978).
6. Beiser, L. et al. Laser-beam Recorder for Color Television Film Transfer. *SMPTE Journal* 80(9): 699–703 (September 1971).
7. Fujio, T. High-Definition Wide-Screen Television System for the Future. *IEEE Trans. Broadcasting* BC-26: 113–124 (December 1980).

8. Fujio, T. A Universal Weighting Function of Television Noise and Its Application to High-Definition TV System Design. *IEEE Trans. Broadcasting* BC-26: 39–48 (June 1980).
9. Ninomiya, Y., Y. Ohutsuka, Y. Izumi, S. Gohshi, and I. Iwadate. An HDTV Broadcasting System Utilizing a Bandwidth Compression Technique-MUSE. *IEEE Trans. Broadcasting* BC-33: 130–160 (December 1987).
10. VTR System: HDV-1000 Video Tape Recorder and HDT-1000 Time-Base Corrector and Signal Processor. SONY Communications Products Co., Tokyo, and Teaneck, NJ 07666.
11. Hashimoto, Y., H. Nakaya, and T. Yoshinaka. An Experimental HDTV Digital VTR with a Bit Rate of 1.188 Gbps. *IEEE Trans. on Broadcasting* BC-33: 203–209 (December 1987).
12. Umemoto, M., Y. Eto, K. Takeshita, and N. Ohwada. An Experimental 648 Mbit/s HDTV Digital VTR. *IEEE Trans. on Broadcasting* BC-33: 210–213 (December 1987).
13. The North American Public Demonstrations of High Definition Television. Available from the Engineering Department, Canadian Broadcasting Corporation, Montreal, April 1988.
14. SMPTE 240M Standard for Signal Parameters—1125/60 High-Definition Production System. Society of Motion Picture and Television Engineers, White Plains, NY, April 1987.

Chapter

8

Proposed Single-Channel Advanced Television Systems

8.1 Proposed Advanced Systems

The advanced television systems described in this and the following chapter are listed in Tables 8.1 and 9.1. They have been presented by their proponents for consideration and test by the Advisory Committee on Advanced Television Service (ACATS) of the Federal Communications Commission. The ACATS has set up an elaborate investigation procedure to be carried out by its Planning Subcommittee through seven Working Parties: WP-1, system attributes and assessment; WP-2, testing and evaluation specifications; WP-3, spectrum utilization and alternatives; WP-4, alternate media technology and broadcast interface; WP-5, economic factors and market penetration; WP-6, subjective assessment; and WP-7, audience research.

The information on the systems listed in Tables 8.1 and 9.1 has been turned over to the Advanced Television Test Center, which has set up the equipment necessary to test each system in three areas: propagation of the channel or channels used on the VHF, UHF, and 2.5- and 12-GHz bands; laboratory testing for comparison of system performance on baseband and modulated r-f conditions; and over-the-air testing of the proposed systems on the bands listed. During these

Table 8.1 Proposed Single-Channel Advanced and HDTV Systems

Proponent name	Scanning (lines/field rate/ fields per frame)	System type	Aspect ratio NTSC	Aspect ratio ATV/HTDV
Single-Channel Compatible Systems (6-MHz Band)				
Brdcst.Tech Assn./Japan	525/59.94/2:1	NTSC EDTV	4:3	16:9
Del Rey/ Iredale	1125/59.94/2:1	NTSC HDTV	14:9	14:9
Faroudja SuperNTSC	525/59.94/1:1 1050/59.94/2:1 or 2:1	NTSC EDTV	4:3	4:3 or 29:18
High Resol. Sciences	525/60.07/2:1	Modified NTSC	4:3	Not stated
MIT-RC Schreiber	525/59.94/2:1	NTSC RGB	16:9	16:9
NHK/MUSE-6	1125/60/2:1	NTSC EDTV	16:9	16:9
Production Services	1125/60/2:1	NTSC HDTV	4:3	16:9
Sarnoff/ ACTV-I	525/59.94/1:1 1050/59.94/2:1	EDTV	4:3	16:9
Single-Channel Noncompatible Simulcasting Systems				
MIT-CC Schreiber	1125/60/2:1	HDTV	Noncompatible	16:9
NHK/ Narrow MUSE	1125/60/2:1	HDTV	Noncompatible	16:9
Zenith	787.5/59.94/1:1	HDTV	Noncompatible	16:9

Source: Abstracted from information compiled by the Advanced Television Test Center from proponents' presentations to the FCC Advanced Television Advisory Committee. *TV Digest* 29(1): 6–7 (January 2, 1989).

tests, WP-6 on subjective testing is to conduct viewer evaluations of the images produced by each system. The Test Center was scheduled to have its equipment ready for these tests by 1991.

Each proponent was asked to list the attributes of the proposed system under the following headings, which reveal the depth of the proposed study:

I. General Description:

1. *Compatibility:* with NTSC receivers, videotape recorders, channel allocations, and other advanced systems.
2. *Transmission Scenario:* numbers of channels and channel widths; whether contiguous or noncontiguous.
3. *Terrestrial Implementation:* requirements for terrestrial broadcast of the system.
4. *Display:* intended size and viewing angle.

II. System Attributes:

1. *Image Issues:* luminance and chrominance spatiotemporal resolutions; chromaticity and colorimetry; artifacts; transient response; aspect ratio; baseband video bandwidth; subjective assessment of picture quality.
2. *Compatibility Issues:* same as image issues, above, plus sync/blanking/subcarrier modifications; use of overscan/underscan.
3. *Audio for Advanced Systems:* number of channels; modulation scheme; signal-to-noise ratio; nonlinear distortion; channel crosstalk; visible delay (lip-sync); and dynamic range.
4. *Degradation of NTSC Audio:* audio intercarrier.
5. *Ancillary signals.*
6. *Terrestrial Transmission Issues:* compatibility characterization; susceptibility to noise, multipath or echo, interference, group delay error, and nonlinear distortion; transmitter/antenna requirements; bandwidth requirements; field testing; coverage relative to NTSC; gracefulness of degradation.
7. *Suitability for Alternative Media Distribution:* interference; adjacent channel, to/from other services; effect of micro-reflections; intermodulation distortion; channel loading; suitability for satellite distribution and for other distribution systems (microwave, fm links, multichannel distribution, fiber optics, telephone lines, video cassette and disk recorders); security considerations.
8. *Consumer Equipment Issues:* receiver complexity; ATV input/output characteristics; baseband video and audio interfaces; ancillary signal interfaces; compatibility with NTSC equipment; baseband video and audio compatibility.

9. *Other Considerations:* near-term implementation; long-term viability and obsolesence; upgradability/extendability; studio/plant compatibility.

The responses of the proponents to this extensive inquiry were by no means uniform, and in several cases information was withheld because patent applications were pending. But in most cases, a full description of the proposed system was offered. Except as noted in the references, the descriptions given below are based on the material submitted to Working Party 1. The order of presentation is alphabetical, within each of the system classifications in Tables 8.1 and 9.1.

8.2 Compatible Systems

8.2.1 BTA Clearvision System

The Clearvision system has been proposed by the Broadcast Technology Association, an organization of Japanese manufacturers and research workers. It is an NTSC-compatible system: 6-MHz channel, 525-line scanning, 59.94 fields per second, interlace 2:1. Five changes from conventional NTSC practice are proposed: progressive scan in the receiver display, separate luminance-chrominance processing in the receiver, compensation of detail rendition in highly saturated color images, adaptive emphasis of the high-frequency components at low levels of the luminance signal, and higher-resolution signal sources.

The BTA Receiver The presentation[1] by BTA states that, while the receiver improvements it proposes are new to NTSC practice, they have in fact recently been introduced commercially. An example has already been described, the Philips IDTV receiver (Section 2.3). To transpose from the interlaced signal of the compatible broadcast, congruent field scans are stored and their picture elements are compared with those of the preceding and following lines. The median of the three values is chosen for the picture element to be imposed on the extra line required to convert from the interlaced to the progressive scan. This aspect of the BTA system has thus been reduced to practice, but only by the use of elaborate digital storage and processing, which makes the receiver expensive.

High-Resolution Signal Sources To provide the full resolution that the BTA receiver can display, the BTA proposes that the camera also employ progressive scanning at 525 lines. The report points out that

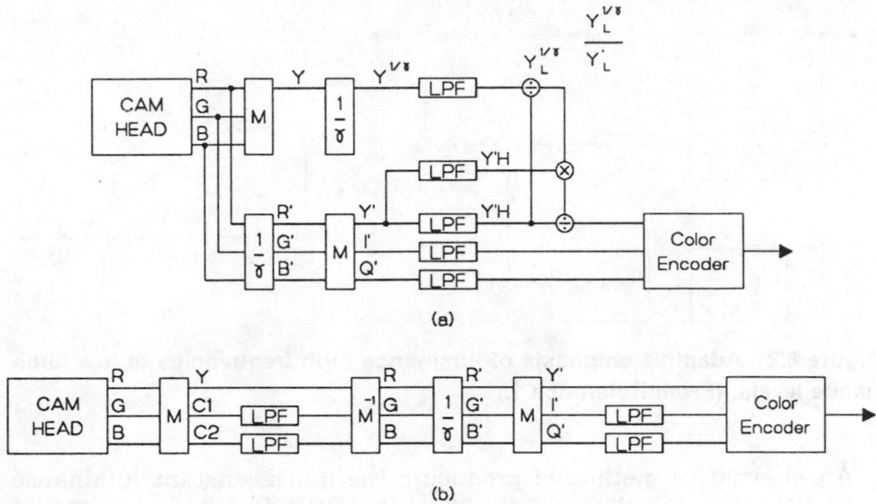

Figure 8.1 Quasi-constant luminance processing of the BTA proposal: (a) by passage of *Y* signal through a high-pass filter; (b) by inverse matrixing prior to gamma correction. (From Reference 1.)

HDTV cameras now available (e.g., the BTS Model KCH-1000 Multistandard Camera [Section 1.11] and the Sony Model Model HDC-300 [Section 7.11]) can provide the higher resolution needed.

Compensation of Detail in High Saturation The BTA report points out that the constant-luminance objective of the NTSC standards is not met when encoding occurs after the RGB signals are gamma corrected, because part of the luminance signal is then carried by the *I* and *Q* signals. Since these chrominance signals have limited bandwidth, the luminance detail in saturated color portions of the image is thereby restricted.

Accordingly, in the BTA presentation it is proposed to preemphasize the luminance signal, prior to encoding, by such amounts as to offset the loss to *I* and *Q*. The methods of so doing, illustrated in Figure 8.1, are named "quasi-constant luminance processing." In Figure 8.1a, two versions of the *Y* signal are produced. The upper version, after gamma correction, is passed through a low-pass filter. The lower version, matrixed after gamma correction, is passed through two filters, high-pass and low-pass. The filtered outputs are combined with the upper version, and the combined *Y* signal is encoded as shown. The result is that the high-frequency region of the luminance is emphasized. BTA points out that the improved resolution is displayed by conventional receivers.

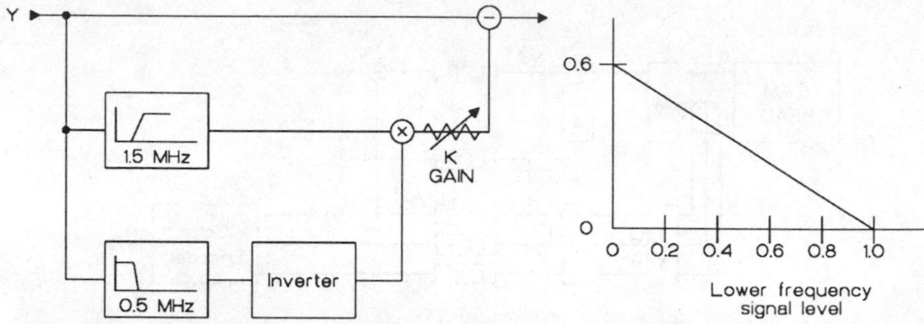

Figure 8.2 Adaptive emphasis of luminance high frequencies at low luminance levels. (From Reference 1.)

An alternative method of producing the quasi-constant luminance signal is shown in Figure 8.1b. Here the RGB signals are matrixed and the two color-difference signals are low-pass filtered. An inverse matrix then recovers the RGB signals, which are gamma corrected and then matrixed again to form the Y′, I′, and Q′ signals. The color-difference signals are then again low-pass filtered and encoded with the luminance in the conventional manner.

Adaptive Low-Level Emphasis A further improvement over conventional NTSC practice is suggested in the BTA report: emphasis of detail in the dark portions of the image. It is pointed out that when displays are viewed in well-lighted rooms the detail in dark portions of the image tends to be washed out. A compensation method is shown in Figure 8.2. The luminance signal is passed through two filters, high-pass and low-pass. The low-pass output is inverted and combined with the highpass output and the combined signal is added back to the luminance through a variable gain amplifier. The gain is controlled by the level of the low-frequency portion of the luminance signal, as shown in the figure. The net result is an increase in the high-frequency luminance signal level as the image darkens. While this compensation system is intended for the transmitter, it is pointed out that, when used in the receiver, an improvement in signal-to-noise ratio can be obtained.

The BTA system was tested in Japan in 1988. The results showed that the improved resolution permitted a preferred viewing distance of four times the picture height, and provided a quality improvement of 1.5 steps on the CCIR seven-step scale. A display size greater than a 30-inch diagonal was found desirable.

Further Development Within the compatibility limits of the BTA system, the report lists a number of techniques that are under develop-

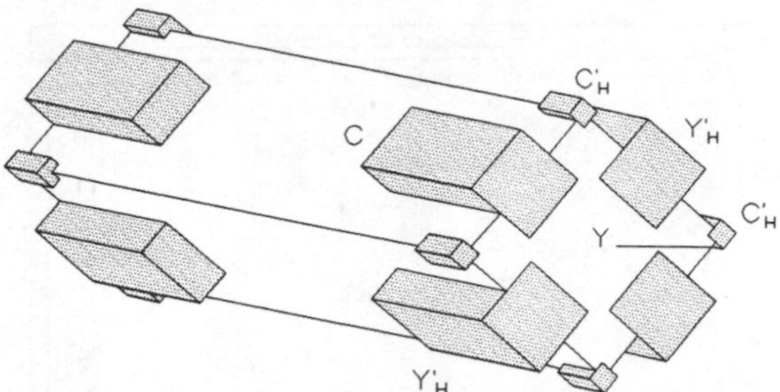

Figure 8.3 Nyquist volume for advanced BTA system, showing holes in frequency space filled by extended-bandwidth Q chrominance (corners of volume). (From Reference 1.)

ment for further improvement, although these were not offered as part of the system currently proposed. One such technique is the transmission of the luminance frequency range from 4.2 to 6 MHz in unoccupied frequency space (Fujinuki hole) in the Nyquist volume (Section 5.3). Another is extending the frequency range of the I chrominance signal beyond 0.5 MHz, by the same method, i.e., by inserting an ancillary signal at the corners of the Nyquist volume. The combined additions to the frequency space would then appear as shown in Figure 8.3. Still another proposed improvement would be the transmission of side-panel information, at 16:9 aspect ratio, by quadrature modulation. The report states that the degradation of conventional NTSC reception by the presence of these signal components, while visible, is so small as to be "negligible."

8.2.2 Del Rey HD-NTSC System

The HD-NTSC System[2] proposed by the Del Rey Group is an NTSC-compatible system, operating on the 6-MHz channel at 59.94 fields per second. It differs from conventional practice in two major respects: an aspect ratio of 14:9 and double-trace scanning at the camera and the HD-NTSC receiver, which spreads the scanning sequence over three frames.

Choice of 14:9 Aspect Ratio The Del Rey presentation[2] states that the advanced television receiver, to attract attention, should have a different appearance from the conventional set even when not operating. Noting the preference of viewers for a wider screen, the choice of 14:9 was made, rather than 16:9, because the 14:9 image would more nearly fill the 4:3 screen of the conventional receiver. The number of

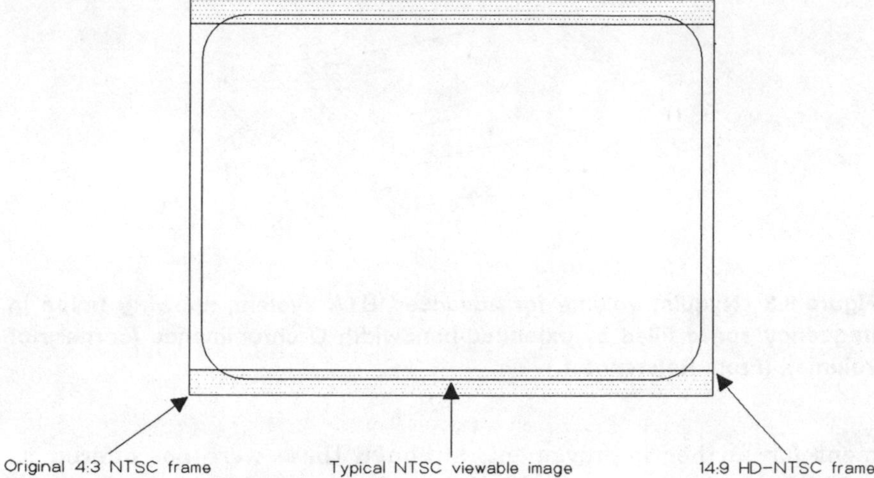

Figure 8.4 Display of the Del Rey HD-NTSC system. The number of active lines is reduced to 414, producing 70 additional blank lines, 35 above and below the 14:9 display. (From Reference 2.)

active lines is reduced from the NTSC value of 484 to 414, and the 70 additional blank lines are disposed above and below the display as shown in Figure 8.4. When the display is overscanned, only a portion of these lines is visible. It is noted that means must be taken to avoid distracting the viewer by the appearance, in the visible blank lines, of the synchronization and data signals present during the vertical retrace interval, but no implementation is described.

Digital Audio During Vertical Blanking Interval The additional vertical blanking time provides a channel for digital transmission of stereo sound. The Del Rey proposal is for adaptive delta modulation at 600 kb/s, intended to provide a frequency response from 20 Hz to 16 kHz, with a dynamic range of 85 dB, 60-dB channel separation, and distortion less than 0.2 percent. An additional signal would also be transmitted during the vertical interval reference (VIR) time for scanning synchronization as described below.

Double-Trace Scanning In the Del Rey system, the camera and display scan at 1050 lines, of which 828 are active, at 14:9 aspect ratio. The camera must also be capable of producing 1320 picture elements per line, three times the NTSC value of 440. The camera signal thus represents 828 × 1320 = 1,092,960 picture elements per frame,

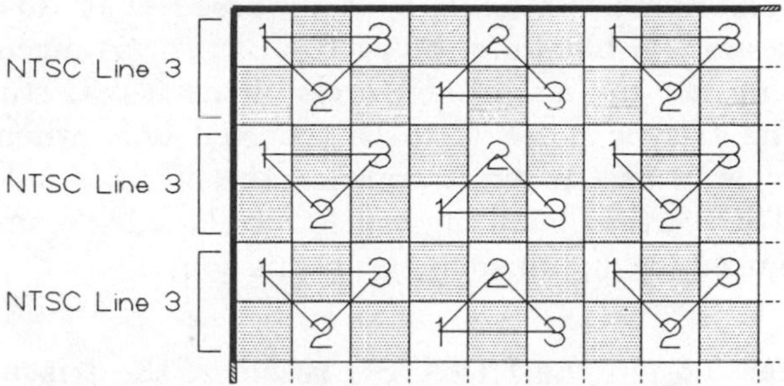

Figure 8.5 Array of "subpixel" triplets on camera faceplate in HD-NTSC system. Each class of subpixel (e.g., "1") is transduced by the camera at intervals of six picture-element spaces. The complete scan occupies three frames. (From Reference 2.)

progressively scanned. This signal is stored and read out as pairs of lines, which are combined. This signal is then low-pass filtered to 4.2 MHz and transmitted as the NTSC-compatible 525-line, 59.94-Hz field, 2-to-1 interlaced signal. The 4.2-MHz limit is made possible by the distribution of the picture elements along each of the 828 active scanning lines in the camera.

As shown in Figure 8.5, the picture elements on the camera faceplate are arranged in triplets ("TriScan" concept), shown numbered 1, 2, and 3. When two successive camera lines are combined to form one NTSC line, the NTSC line contains all the picture elements in the order 1, 2, 3. In the camera, however, the 1 elements are transduced at intervals of six elements, and similarly for the 2 and 3 elements. This is the equivalent of scanning three times the number of picture elements as are present in the standard NTSC image, but doing so over the scanning time of three NTSC frames. The rate of scanning elements thus remains unchanged and can be accommodated in the 4.2-MHz NTSC baseband. It will be noted that this temporal subsampling of picture elements out of their normal order is similar to that of the MUSE system (Section 2.8), but that the Del Rey system uses a quite different technique.

Receiver Synchronization At the HD-NTSC receiver the individual picture elements shown in Figure 8.5 are presented on a 1050-line progressively scanned display having 828 active lines and 1320 picture elements per line. The recovery of the individual elements must be synchronized with the selection of picture elements at the camera. This is arranged by a pulse (the "trisync pulse") inserted every sixth field on an odd-numbered field. It is proposed that this pulse be placed in the vertical blanking interval (VBI), so arranged that neither the timing nor the amplitude levels of the NTSC standard are disturbed. The trisync pulse is to be inserted only when the HD-NTSC system is in use. When it appears, the 35 lines following and preceding NTSC vertical blanking are set at black level, unless those intervals are used for digital sound transmission.

Receiver Processing In the HD-NTSC receiver, the trisync pulse is detected and brings into operation two processing circuits. One of these accepts the digital sound signal during the extra blank line periods during vertical retrace, decodes the bit stream, and converts it to analog form for the loudspeakers. The other accepts the 4.2-MHz analog video luminance signal (after passing a comb filter), encodes it digitally, and passes it through a dual resolution processor. This processor derives from each NTSC line the individual picture elements shown in Figure 8.5, stores them, and reads them out on two lines (corresponding to those from which the NTSC line was formed at the transmitter). Also performed in this processor, by methods not described, are conversions of parts of the image in motion to low spatial resolution. The output of the processor is converted to analog form and fed to the display, which presents the 1050-line image, progressively scanned.

The Del Rey Group is explicit in its recommendation for a large receiver display. A typical display shown measures 49 by 29 inches (54-inch diagonal), at a recommended viewing distance of 10 feet (approximately four times the picture height).

Effects of Scanning Time Delay The long delay between the successive appearances of a particular picture element, 3/29.97 second, introduces low-frequency flicker and noticeable displacement of the detail of objects in motion. These artifacts are described in the Del Rey paper as having a noticeable effect on image quality. The 1/10 second flicker introduces a shimmering effect or "scintillation" that is most evident at the edges of extended stationary dark objects against a light background. It is suggested that storage of the picture element signals in a display buffer, to prolong their duration over the 3/29.97 second interval, may be necessary.

When motion is present, the displacement of the position of a given picture element between successive scans introduces a loss of resolution. The report gives an example in which both vertical and horizontal smear occurs (Figure 8.6). At the right is shown the repro-

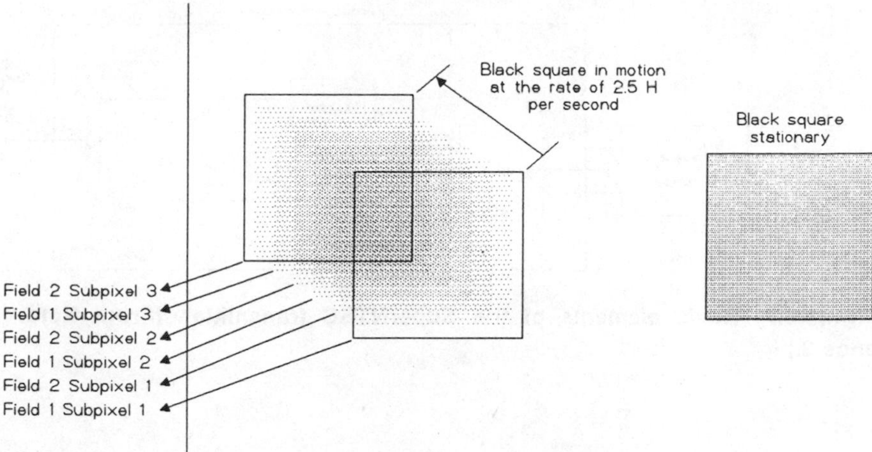

Figure 8.6 Vertical and horizontal smearing caused by six-field time delay between start and completion of the HD-NTSC scanning sequence. (From Reference 2.)

duction of a stationary black square. When this square moves diagonally downward, the "subpixels" in Figure 8.5 appear in six different positions in six successive fields, as shown at the left. The black area is thereby much reduced in size, and it is surrounded by a larger area shading from dark gray to white, vertically as well as horizontally. To mitigate this smearing effect, the previously mentioned dual resolution processor arranges to lower the temporal resolution in areas of high spatial resolution, and vice versa.

8.2.3 The Faroudja SuperNTSC System

The SuperNTSC system[3] is an NTSC-compatible system operating on a 6-MHz channel with a 525-line, 59.94-fields-per-second, 2-to-1 interlaced signal. The aspect ratio is 4:3 initially, with 29:18 (= 1.61) as a later objective.

Scanning Progressive scanning is used in the camera and advanced receiver display. Camera scanning is at 525 lines for initial low-cost production, with a subsequent preferred value of 1050 lines, 59.94 frames per second. The progressive scan is converted to interlaced

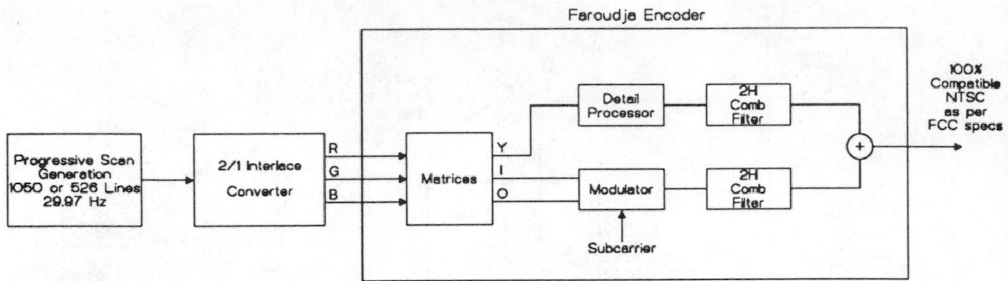

Figure 8.7 Basic elements of the SuperNTSC transmitter. (From Reference 3.)

for compatible service in a scan converter that stores the progressive lines and reads them out at the interlaced intervals. Included in the converter are Nyquist prefilters that reduce aliasing from vertical, temporal, and interlaced sampling.

Transmitter Encoding The Faroudja system uses the prefiltering techniques described in Section 2.6. As shown in Figure 2.5, the *Y*, *I*, and *Q* signals are passed through 2-H comb filters that are adaptively controlled by the image content. This reduces the cross-color and cross-luminance artifacts to negligible levels. The encoding of the R, G, B signals to *Y*, *I*, *Q* is conventional except that the *Q*-signal bandwidth is extended to match that of the *I* signal, 1.3 MHz. The primaries and reference white have the NTSC chromaticity coordinates. The elements of the SuperNTSC transmitter are shown in Figure 8.7.

Luminance Detail Processing Based on the finding that, at low-level luminance transitions, the eye is not sensitive to preshoot or overshoot oscillations, the SuperNTSC transmitter sharpens the rise-time of such small transitions by selective aperture correction ("boost"), while remaining within the 4.2-MHz limit of the NTSC standard. The amount of boost is controlled by five factors: the luminance transition level, the gray level, the hue and saturation levels, and the chrominance-to-luminance ratio. This shortening of the low-level luminance rise-times does not produce modulation beyond the limits of the baseband, so it produces a sharper image on any conventional receiver equipped with a comb filter.

SuperNTSC Receiver The display of the Faroudja advanced receiver is scanned at 1051 lines, 2-to-1 interlaced at 59.94 fields per second. The horizontal resolution, enhanced by the luminance spectral expansion technique described below, is stated to be 500 lines per picture height. The vertical resolution, reduced by blanking and interlace from the 1051-line scan, is 600 lines per picture height. At 4:3 aspect ratio, these values produce an image of 400,000 picture elements per frame, approximately three times that of the conventional display.

The interlaced signal is transformed to progressive scan by a line doubler comprising a field store, a 2-line store, circuits for detecting motion, and a time-compression device in the ratio of 2.0019/1. In the absence of motion, the output of the line doubler produces interlaced fields of 525.5 lines. The line content is provided by two successive fields of the broadcast signal. When motion is present, the lines are generated from a single input field, and line doubling is performed by interpolation within the field. The motion-detection logic continuously makes the choice between the interfield and intrafield modes. The receiver also employs comb filters that are complementary to those used at the transmitter, thus preserving the wide separation between the luminance and chrominance components.

The receiver also contains circuitry which complements the chrominance transition function at the transmitter (Section 2.6). Normally the luminance and chrominance transitions are congruent in the scene, but the narrow bandwidths of the I and Q components cause chrominance transitions to be smeared and laterally displaced. The transmitter chrominance circuits are arranged to extend the bandwidth of the filters governing the I and Q signals, when a luminance transition occurs. A complementary function, known as "chrominance bandwidth expansion," is performed in the receiver filters.

A final receiver process is used to extend the luminance bandwidth to a stated limit of 15 MHz, from which the 500-line horizontal resolution is obtained. As noted above, small luminance transitions (less than 20 percent of the range) are sharpened at the transmitter. At the receiver, transitions greater than 20 percent are likewise sharpened by a similar selective aperture correction. Prior to the display, the whole range of a transition is sharpened by a factor of 2, and the horizontal resolution correspondingly increased. The basic elements of the SuperNTSC receiver are shown in Figure 8.8.

The SuperNTSC system has been reduced to practice and has been exhibited in the United States in 1988 and in Japan in 1989. Faroudja line-doubling equipment has been purchased by the Sarnoff Research Center for use in its work with the ACTV systems.

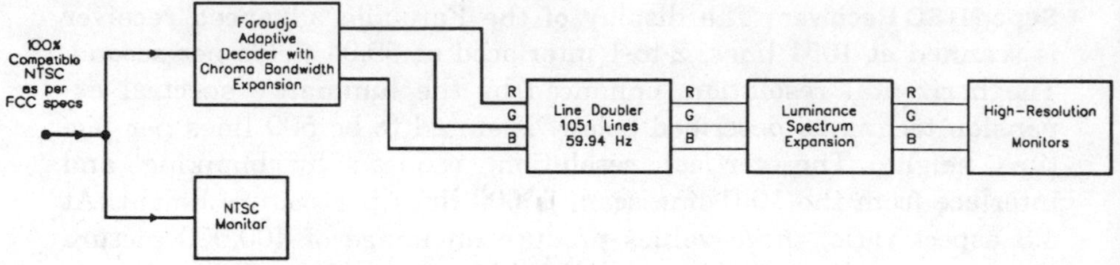

Figure 8.8 Basic elements of the SuperNTSC receiver. (From Reference 3.)

8.2.4 High-Resolution Sciences CCF Proposal

The standards proposed[4] by High Resolution Sciences, Inc. for the "Chroma Crawl Free" (CCF) system are identical to the NTSC standards in all respects except the vertical and horizontal timings. The 59.94-Hz NTSC field rate is raised 0.21 percent to 60.07 Hz, with corresponding increases in the frame rate and the line-scanning frequency. Table 8.2, taken from the CCF report, compares the NTSC and CCF scanning parameters. It will be noted that the color subcarrier frequency remains unchanged; the frequency difference between the picture and sound carriers is also retained at the NTSC standard value, 4.5 MHz.

Purpose of 60.07-Hz Field Rate The purpose of the change in the field rate is to eliminate the upward motion ("color-dot-crawl") of the colored edges in the cross-color artifact of the NTSC system. This crawl

Table 8.2 Scanning Parameters of the NTSC and CCF Systems

Parameter	NTSC	CCF	Difference
Scanning lines	525	525	None
Horizontal frequency	15,734.26 Hz	15,768.92 Hz	+0.21%
Vertical frequency	59.94 Hz	60.07 Hz	+0.21%
Color subcarrier	3,579,545 Hz	3,579,545 Hz	None
Subcarrier cycles/line	227.5	227	-0.21%
Chroma crawl	Yes	No	Improved
Cross-color	Yes	Reduced	Improved

is caused by the 180° phase difference between color subcarrier cycles on successive lines of each field. This effect arises from the choice of the subcarrier frequency as an odd multiple, 455, of half the line-scanning rate. This choice causes the luminance and chrominance components to be interleaved in frequency in the video baseband. The CCF system is based on an even multiple, 456, which produces the 15,768.92-Hz line rate shown in Table 8.2. The frequency interleaving of luminance and chrominance is thus sacrificed, but the 180° shift between subcarrier cycles on successive field lines is reduced to zero. What remains is an 180° shift between adjacent lines in each frame. This shift is integrated by the eye, and no vertical motion appears. A reduction in cross-color artifacts is also thereby obtained. The CCF presentation states that the overall effect of the change to the 60.07-Hz field rate provides a "definite improvement" in the quality of the image.

Compatibility of 60.07-Hz Scanning To test the effect of the change in field rate on conventional domestic receivers and tape recorders, High Resolution Science arranged in July 1988 for the CCF system to be carried for many days over a major cable network to 700 cable systems serving 10 million subscribers. No reports of malfunction of receivers or recorders were received. It thus appears that equipment in the home can accommodate the 0.21 percent change in the field rate. Professional equipment, whose timing is directly derived from the NTSC subcarrier frequency, cannot make this adjustment. In such cases, the change in rate must be introduced after the composite NTSC signal is produced, prior to broadcast. For this purpose, an NTSC-to-CCF converter, comprising a decoder and encoder, is employed. A frame store and frame comb in the decoder stores the NTSC signal digitally, from which the RBG or *Y-IQ* signals are read out at the CCF field rate. The encoder portion of the converter then produces the composite version of the CCF signal. Time control of the converter is provided by a synchronization generator operating at the standard NTSC subcarrier frequency, with the line rate derived from it by the 456 multiple mentioned above. An incidental effect of the even-number multiple is that the frequency interval between the color subcarrier and the sound carrier, while maintaining a value identical to the NTSC standard, undergoes a 180° phase shift at the end of each line scan. This cancels the effect of the aural-subcarrier sound beat. The report states that the CFF system shows no sound-to-picture or picture-to-sound intermodulation effects.

8.2.5 MIT Receiver-Compatible System

The MIT-RC (Receiver Compatible) system[5,8] is NTSC-compatible, occupying the standard 6-MHz channel, with a transmitted signal of 525 lines, 59.94-Hz field rate, interlaced 2-to-1 with an aspect ratio of 16:9. The display on a conventional receiver has 60 blank lines above and below the 16:9 outline, minus those hidden by overscan. The chrominance information is transmitted at 14.985 frames per second, half the NTSC rate. The extra time of the blank lines and the frames free of chrominance are used to carry ancillary information sufficent to produce a 1050-line luminance display with a resolution of 535 lines horizontally and 600 lines vertically for a stationary image. This is reduced to 315 and 360 lines, respectively, for objects in motion. The chrominance resolutions are 180(I), 130(Q), lines (stationary), and 180(I), 65(Q), lines (in motion), respectively. The RC system has been designed to be used cooperatively with a simulcast channel-compatible (CC) system[6] described in Section 8.3.1.

Studio Origination and Encoding The preferred studio origination would use a progressively scanned 1050-line camera at 59.94 frames per second. An encoder would transform this signal to the NTSC 525-line interlaced standard, but with only the central 360 lines actively scanned with the camera signal, low-pass filtered to 4.2 MHz. This signal would serve conventional receivers with a 16:9 display centered on the 4:3 frame.

The encoder would occupy the time of the blank lines with the higher-frequency content of the camera output. The RC HDTV receiver, scanned at 1050 lines, would use both the low-resolution (4.2-MHz) information, available for 75 percent of the frame time, and the higher-frequency information carried during the blank lines of the remaining 25 percent of the time.

The reduction of the chrominance scan rate to 14.985 frames per second is allowable, according to the report, because the visual response to color-difference information does not require the 29.97-per-second rate of the NSTC standard. (The report states that "this is the only respect in which the NTSC system is over-designed.") To reduce the 29.97-per-second frame rate to half its value, successive chrominance line scans may be stored, averaged, and read out every other frame. One method of inserting information in the empty frame times is to form an ancillary signal which has the same value on two successive frames. This signal is alternatively added to and subtracted from the chrominance signal and can be separated from

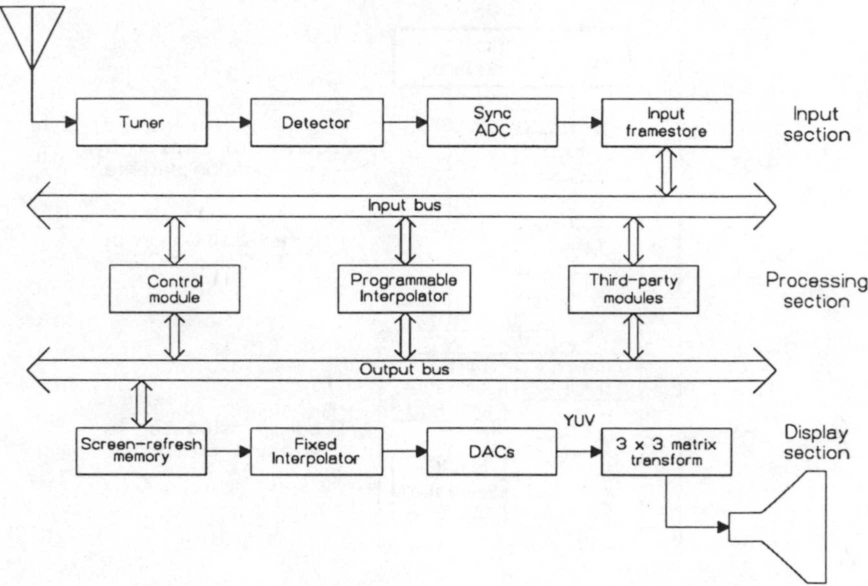

Figure 8.9 Open-architecture receiver proposed for the MIT-RC and CC systems. The central section, organized like a personal computer, would receive different integrated circuit modules as changes in the transmission system were introduced. (From Reference 6.)

the chrominance by the inverse subtraction and addition in the receiver. An alternative method of filling the open times, by subsampling as in the MUSE and Del Rey systems, is also mentioned.

Open-Architecture Receiver The details of the methods of encoding and modulating the ancillary information are not specified in the MIT-RC presentation. Rather, it is proposed that these processes be changed as the state of the art progresses. To permit the RC system receiver to accommodate such changes, it would be designed with digital elements that could be replaced as changes are introduced at the transmitter. This "open architecture" design[7,8] is confined to the digital processing functions in the receiver, as shown in Figure 8.9. This approach is taken from the current practice in "upgrading" personal computers by insertion or replacement of integrated circuits. This flexibility of receiver processing is also used in the simulcast (MIT-CC) system described in Section 8.3.1.

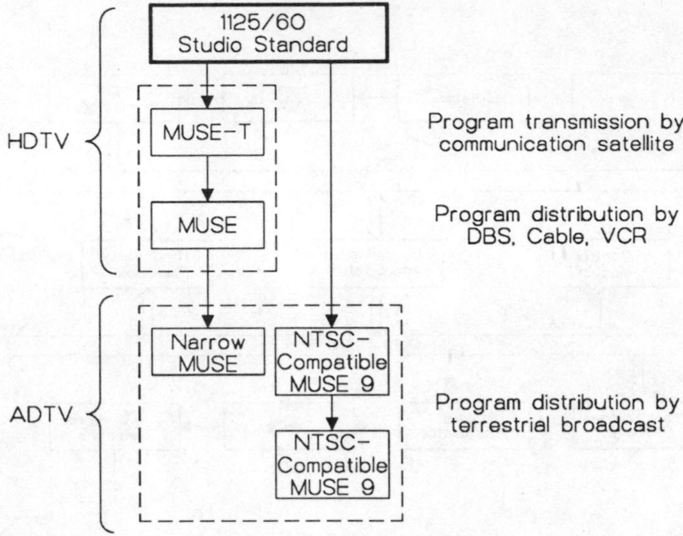

Figure 8.10 The family of MUSE systems. The "advanced definition television" (ADTV) systems were devised by NHK at the suggestion of American terrestrial broadcast organizations to fit in one 6-MHz channel or in a 9-MHz dual-channel assignment. (From Reference 9.)

8.2.6 NHK MUSE-6 System

The NHK MUSE-6 system is one of two 6-MHz channel MUSE systems presented to Working Party 1, listed in Table 8.1.[9] Known as "advanced definition television" (ADTV) systems, they were devised by NHK at the suggestion of two American television broadcast organizations, the National Association of Broadcasters (NAB) and the Association of Maximum Service Telecasters (AMST), to provide terrestrial broadcast service. The family of MUSE and ADTV systems is shown in Figure 8.10. At the time of their presentation to Working Party 1, the ADTV systems were described as being tested by computer simulation and under hardware development.

Scanning and Resolution The MUSE-6 signal originates at the NHK standard of 1125 lines, interlaced 2-to-1 at 60 fields. For NTSC compatibility, the 60-Hz rate is transcoded to 59.94 Hz, and this rate is maintained throughout the remainder of the MUSE-6 system to the display.

The parent MUSE system (Section 2.8) uses time compression over three fields to reduce the baseband to 8.15 MHz. Since this exceeds the bandwidth available in the NTSC channel, an intermediate change of scanning from the total 1125 lines to 750 lines is encoded at the transmitter and reverse decoded at the receiver, as described below. The display on conventional receivers offers a 16:9 aspect ratio centered on the 4:3 frame, with 345 active lines. The advanced MUSE-6 receiver offers the resolutions shown in Figure 8.11. The loss of resolution for moving objects is caused by the time compression of the signal. The chrominance resolutions are one-half or less of the luminance resolution.

Motion Detection As in the parent system, the time compression used in MUSE-6 introduces a 2-to-1 loss in the horizontal resolution of moving objects. To control the transcoding when motion is present, a motion detector based on comparison of three successive fields generates a signal that disables the high-frequency processing of moving objects. The motion detector is also used to control the 1125- to 750-line converter and the 60- to 59.94-Hz field converter.

Transmitter Transcoding To reduce the 30-MHz bandwidth of the HDTV camera signal to the 4.2-MHz NTSC baseband requires elaborate transcoding by both time and frequency multiplexing. Figure 8.12 shows the elements of the encoder. The camera signal is sampled at 31.4 Mb/s for luminance, 15.7 Mb/s for chrominance. After 1125- to 750-line and 60- to 59.94-Hz field conversions, the signals are processed for constant luminance and matrixed to 750-line Y, I, and Q signals. At this point, additional vertical resolution, required because the active portion of the 16:9 display contains only 345 lines, is introduced by the mask processor. This increases the vertical detail to 690 lines by time-domain multiplexing in the upper and lower blanked regions of the 4:3 frame. The horizontal resolution limit is extended to 7.7 MHz by the frequency-multiplex method shown in Figure 8.13. Two subcarriers, separated by one-half and one-quarter the frame rate, carry sidebands that are interleaved in the baseband range from 1.9 to 3.9 MHz. This interleaving function is under the control of the motion detector and is disabled during motion in the scene.

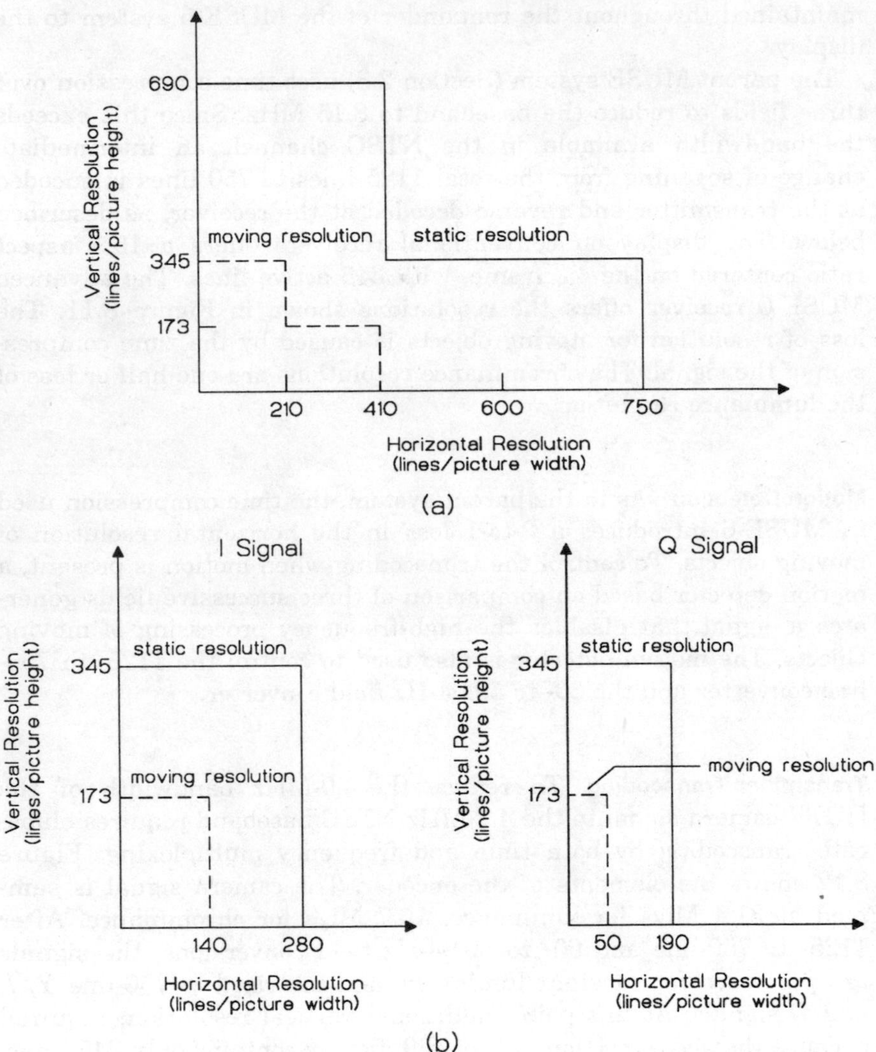

Figure 8.11 Spatial resolutions of the MUSE-6 system: (a) luminance; (b) chrominance. (From Reference 9.)

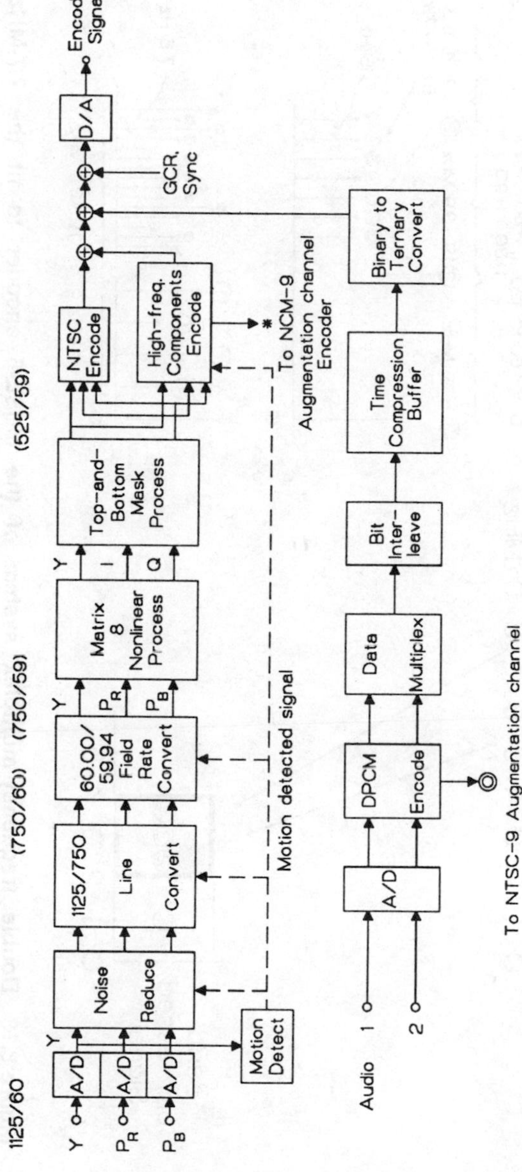

Figure 8.12 Elements of the MUSE-6 transmitter encoder. (From Reference 9.)

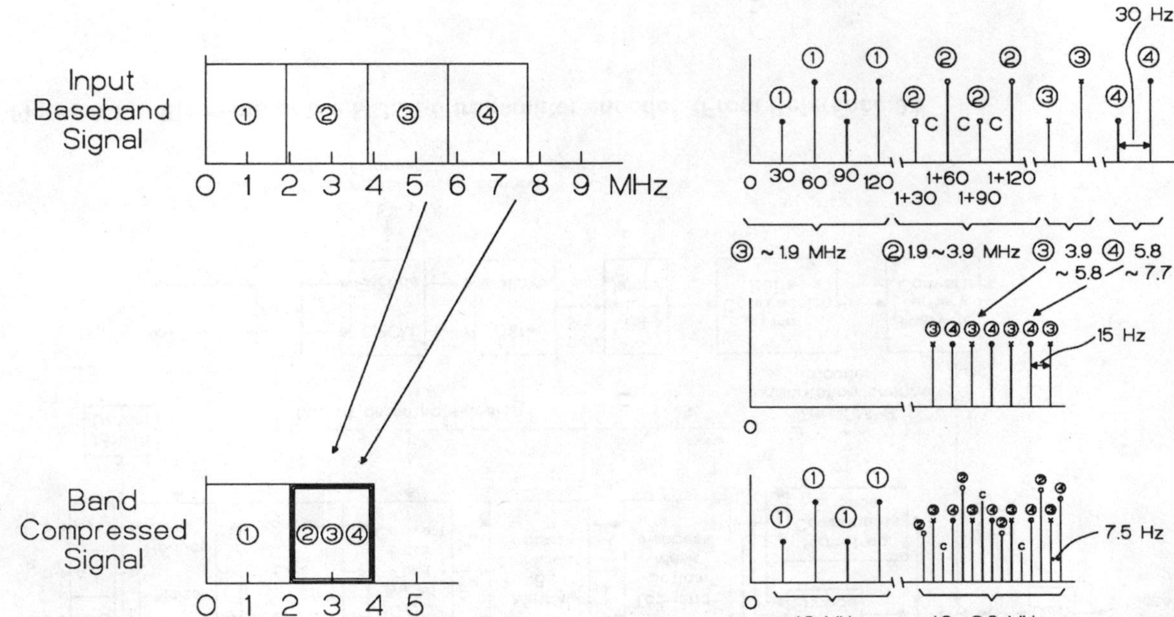

Figure 8.13 Double frequency-multiplex system of the MUSE-6 encoder to fit the 7.7-MHz baseband into the 4.2-MHz NTSC channel. The band, from 2 to 4 MHz, contains three sets of interlaced signals, arranged as shown at the right. (From Reference 9.)

The sound signal is digitized at 32 kb/s in 15-bit words, with redundant bits to bring the data rate to 585 kb/s. After compression, the digital audio is inserted in the video waveform after the color burst. Overscan in conventional receivers is assumed to hide any evidence of the sound bursts during active scanning. One stereo-pair sound channel is offered at 15 kHz upper limit, with a dynamic range exceeding 90 dB.

Receiver Decoding The receiver decoding, shown in Figure 8.14, is the inverse of the transmitter encoding, with the exception of the field-rate conversion, which is omitted since the MUSE-6 display operates at 59.94 fields. The display therefore offers only 999 frames for each 1000 frames scanned by the camera, and one frame in 1000 must be discarded in the transmitter processing. This "frame-cut" operation takes place under the control of the motion detector when there is a switch between cameras or when the scene is stationary or slowly moving. The omission of the frame is then not evident to the viewer.

In Figures 8.12 and 8.14, entry points for the MUSE-9 augmentation channel are marked. The MUSE-6 system has been designed to operate as a part of the MUSE-9 system (Section 9.2).

8.2.7 Production Services Genesys System

The submission[10] of the proposal for the Genesys Transmission system by Production Services, Inc., to Working Party 1 is limited to a portion of the encoding and decoding equipment, following the establishment of the program-source line and field rates and prior to processing of the Y, I, and Q signals at the receiver, respectively. Since patent applications had not been filed at the time of the submission, no technical details were presented. While the encoder will handle the program origination at any line and field rate, the report recommends the 1125-line, 60-field, 2-to-1 interlaced 16:9 format. The encoder accepts the HDTV signal and transforms it to modulation on a 70-MHz carrier. This is intended for the transmitter visual exciter, which must be of the intermediate-frequency type. It is assumed that the HDTV signal is also transcoded, by equipment not part of the Genesys system, to the 525-line, 59.94-field signal required for NTSC compatibility and that this signal is also applied to the if-type exciter.

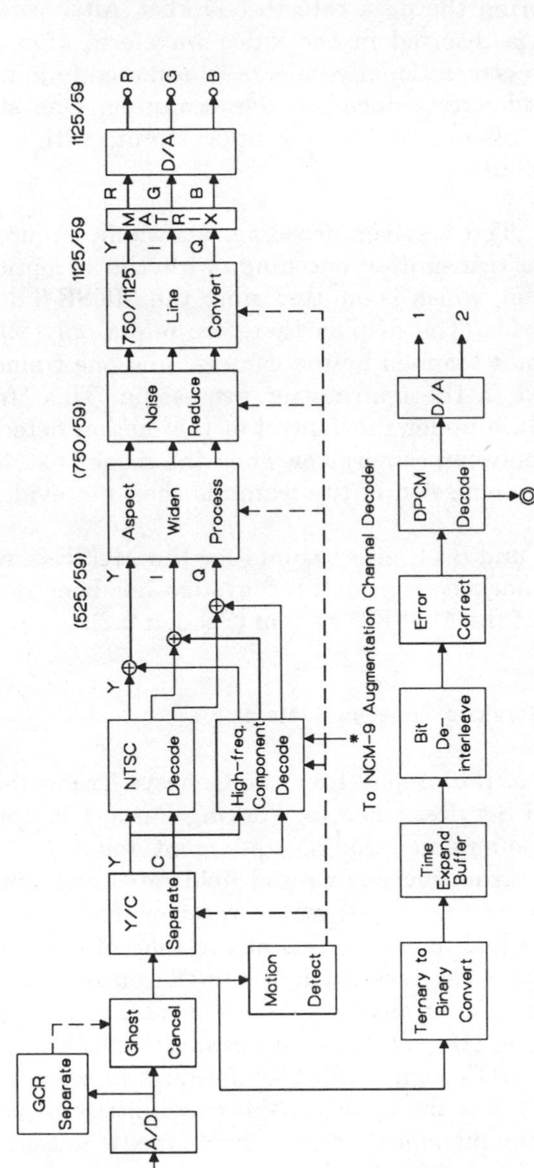

Figure 8.14 Elements of the MUSE-6 receiver decoder. (From Reference 9.)

At the receiver, a set of integrated circuits under development by Production Services would act as a decoder to introduce the inverse of the transmitter encoding, recovering the HDTV line and field rates at video baseband. The baseband signals would be processed conventionally for the display and sound systems.

8.2.8 Sarnoff Research Center ACTV-I System

The Advanced Compatible Television (ACTV) systems[11,12] of the Sarnoff Research Center comprise two cooperative systems. The first, ACTV-I, is intended to provide NTSC-compatible service on a 6-MHz channel. The second, ACTV-II, is an evolutionary extension of ACTV-I, occupying a 6-MHz augmentation channel, not necessarily contiguous with that occupied by ACTV-I.

The ACTV-I system is intended to serve while the limited performance of HDTV displays prevents full utilization of the HDTV service, and until the spectrum needed for augmentation channels in the terrestrial service is available. When and as these limitations are removed, the ACTV-II service would be introduced.

Camera Scanning The ACTV camera produces a widescreen signal of 16:9 aspect ratio, the duration of each line being 52 μs. The camera scanning could be either 1125 lines or 1050 lines, progressively scanned, at 59.94 frames per second. Since cameras capable of this performance are not yet available, the scanning initially recommended would be 1050/2:1 lines interlaced or 525/1:1 progressive scan. The source scanning on which the proposed system is based is 525/1:1, transcoded to 525/2:1 for the compatible NTSC transmissions.

The ACTV-I service is based on the component signals and processing illustrated in Figure 8.15. There are four components, the first of which serves conventional receivers, and the remaining three are transmitted in portions of the NTSC Nyquist signal space to which conventional receivers have minimal response.

ACTV-I Component 1 As shown in Figure 8.15, component 1 is the standard NTSC signal, including the center portion of the widescreen display. The left and right edges of the display, each 1 μs wide, are reserved for the low-frequency portions of the sidepanels of the widescreen display. The contents of these portions of the image, as

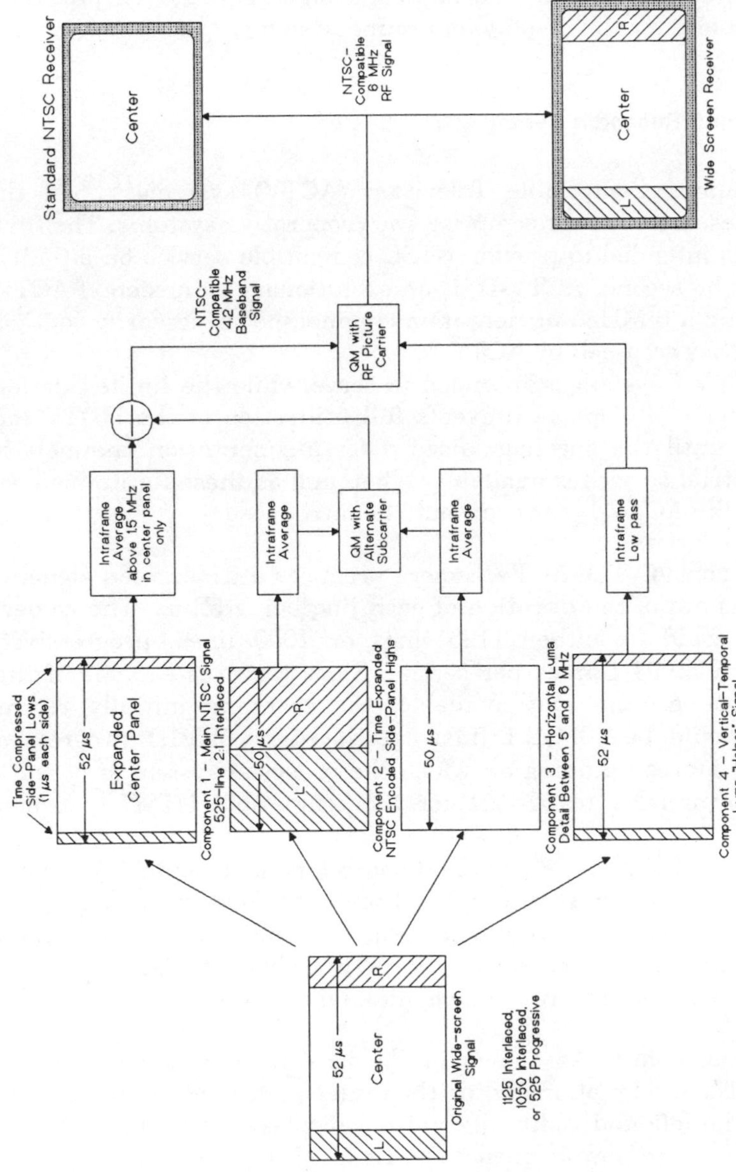

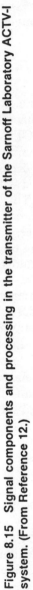

Figure 8.15 Signal components and processing in the transmitter of the Sarnoff Laboratory ACTV-I system. (From Reference 12.)

displayed on conventional receivers, are intended to be hidden by the overscan of such receivers. Component 1 is transmitted with 4.2-MHz video baseband in accordance with the NTSC standards; it is the only component to which conventional receivers are intended to respond.

ACTV-I Component 2 The second component, as shown in Figure 8.15, provides the high-frequency portions of the sidepanels. To fit within the 4.2-MHz channel, the sidepanel signals are limited to 700 kHz for luminance and 83 kHz for chrominance. The "seams" between the center and sidepanel portions are minimized by the low-frequency content of the sidepanels, which is sent with the NTSC signal in component 1. At the encoder producing the four components, redundant information is provided in the high-frequency portions of the sidepanels, which produces "feathered" boundaries at the seams.

Components 1 and 2 carry the horizontal luminance information of the 525/2:1 widescreen transmitted signal, up to 5 MHz.

ACTV-I Component 3 Component 3 carries the luminance signal detail between 5 and 6 MHz. This signal is encoded by transposing the 5-MHz signal to 0 MHz, and filtering to eliminate the upper sideband. The remaining signal, representing 1 MHz of additional horizontal detail, is time compressed and transmitted during the 50-μs active line scan of component 1.

ACTV-I Component 4 This component constitutes a "helper" signal to transmit additional information in the vertical and temporal domains. The derivation and use of the helper signal is illustrated in Figure 8.16. The progressive scan from the camera is lowpass filtered in a five-tap vertical-temporal filter. The black samples A and B are transmitted by component 1 and thus are available at the receiver. Each white sample is derived as the average (A + B)/2 of two black samples, as in the figure. The white samples are the helper signal. The purpose of this signal is to restore detail in vertical motion, lost during transcoding from the progressively scanned source, 525/1:1, to the transmitted compatible form, 525/2:1. The helper signal is of low amplitude, since it consists only of the differences representing the lost detail. The helper signal is encoded to 4:3 aspect ratio, and low-pass filtered to 750 kHz, then transmitted by quadrature modulation of the rf picture carrier, as shown in Figure 8.15.

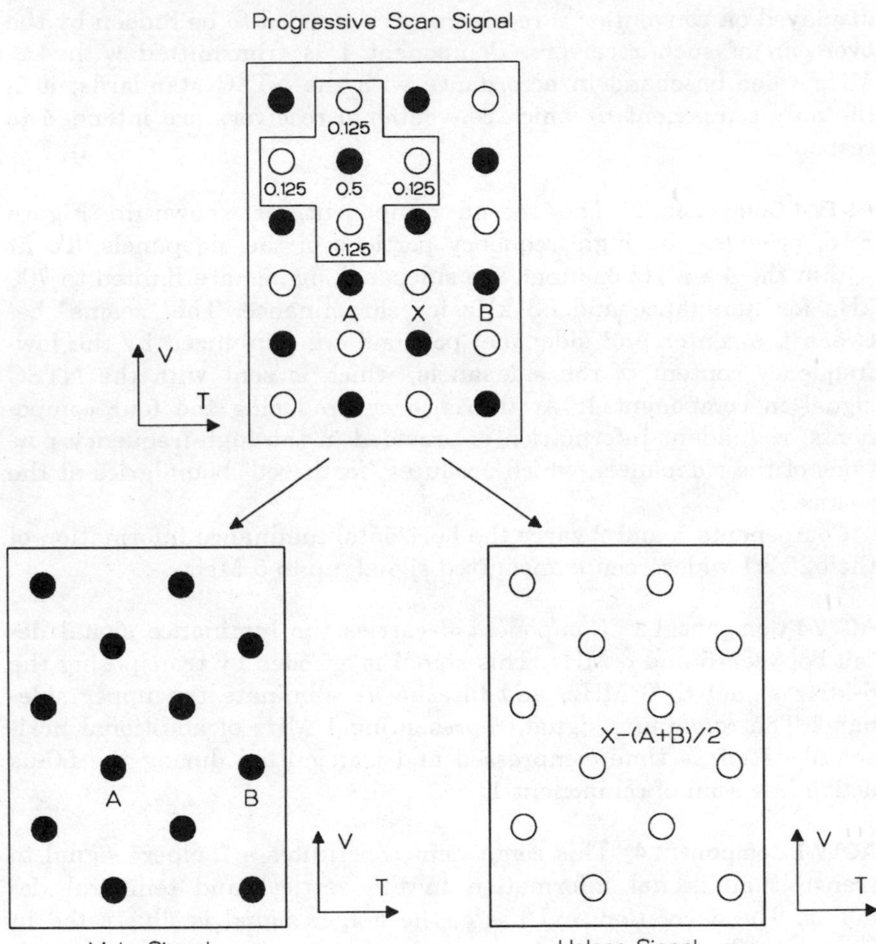

Figure 8.16 Derivation of the ACTV-I helper signal. (From Reference 12.)

Intraframe Processing Components 1, 2, and 3 are processed by intraframe averaging as shown in Figure 8.17. Pairs of picture elements within each frame, separated by 262-line periods, are averaged. In component 1 the luminance and color difference signals are added as shown and the intraframe average is taken for video frequencies above 1.5 MHz. In component 2, the sidepanel high-frequency signals are averaged and the averages are substituted for the picture-element pairs. In component 3, pairs of elements in the high-

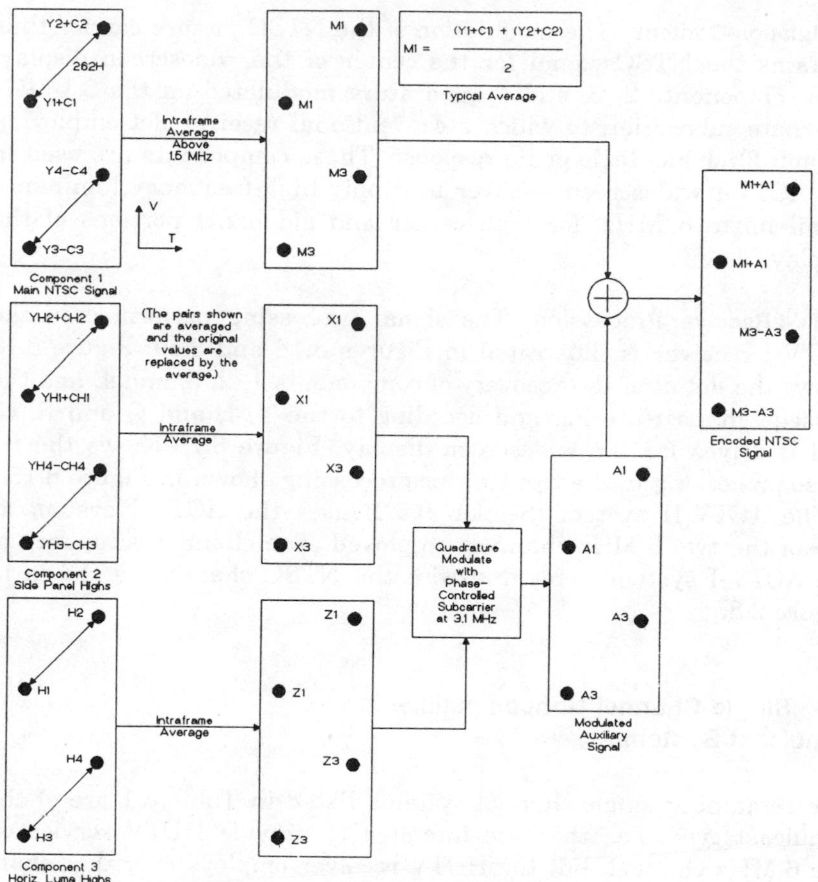

Figure 8.17 Intraframe averaging of components 1, 2, 3, and 4 in the ACTV-I encoding of the transmitted signal. (From Reference 12.)

frequency horizontal luminance signal are averaged and the averages substituted for the pairs. The averaged signals of components 2 and 3 are then applied as quadrature modulation of an alternate subcarrier, taken as the 395th multiple of half the line frequency, i.e., 3.1075171 MHz. This modulated subcarrier is added to the intraframe-averaged component 1, producing an NTSC-compatible baseband video signal, which is quadrature modulated with the NTSC picture carrier. The other quadrature modulation signal is component 4, after passage through a 750-kHz low-pass filter.

Modulation Content The modulation of the NTSC picture carrier thus contains the NTSC signal for the center of the widescreen display, plus components 2, 3, and 4 quadrature modulated on the 3.1-MHz alternate subcarrier, to which a conventional receiver not employing a comb filter has little or no response. These components are used in the ACTV-I widescreen receiver to supply high-frequency luminance detail up to 6 MHz, for both center and sidepanel portions of the display.

ACTV-I Receiver Processing The signal processing performed in the ACTV-I receiver is illustrated in Figures 8.18 and 8.19. Figure 8.18 shows the detail of the recovery of components 1, 2, 3, and 4, and the subsequent postfiltering and decoding to the Y, I, and Q and R, G, and B drives for the widescreen display. Figure 8.19 shows the inverse processing to the transmitter processing shown in Figure 8.15.

The ACTV-II system (Section 9.5.2) uses the ACTV-I system as one of the two 6-MHz channels employed. The channel occupancy of the ACTV-I system, compared with the NTSC channel, is shown in Figure 2.8.

8.3 Single-Channel Noncompatible Simulcast Systems

The remaining single-channel systems listed in Table 8.1 are of the simulcast type, i.e., they are intended to provide HDTV service on one 6-MHz channel, but the HDTV receiver employs only that channel and ignores the NTSC-standard channel which is simulcast separately to provide compatible service. In this approach, the HDTV service is designed to utilize the full resources of the channel, free of the limitations imposed by compatible use of the channel.

8.3.1 MIT Compatible-Channel System

The MIT-CC simulcast system[8,13] employs double-sideband quadrature-modulation of a carrier in the center of the 6-MHz channel. This signal is not intended for NTSC receivers, i.e., the system is noncompatible. The preferred originating signal is produced by a camera operating at 16:9 aspect ratio and 1200 lines, progressively scanned. The preferred frame rates would be either 60 or 72 frames per

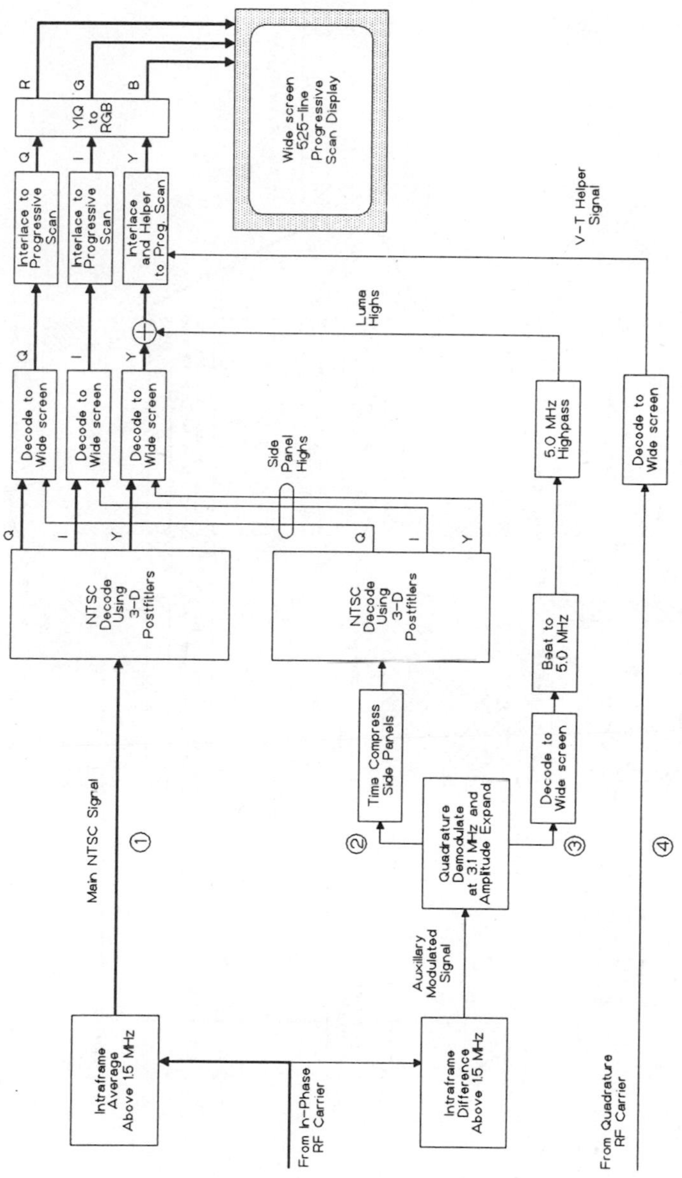

Figure 8.18 Decoding in the ACTV-I receiver. (From Reference 12.)

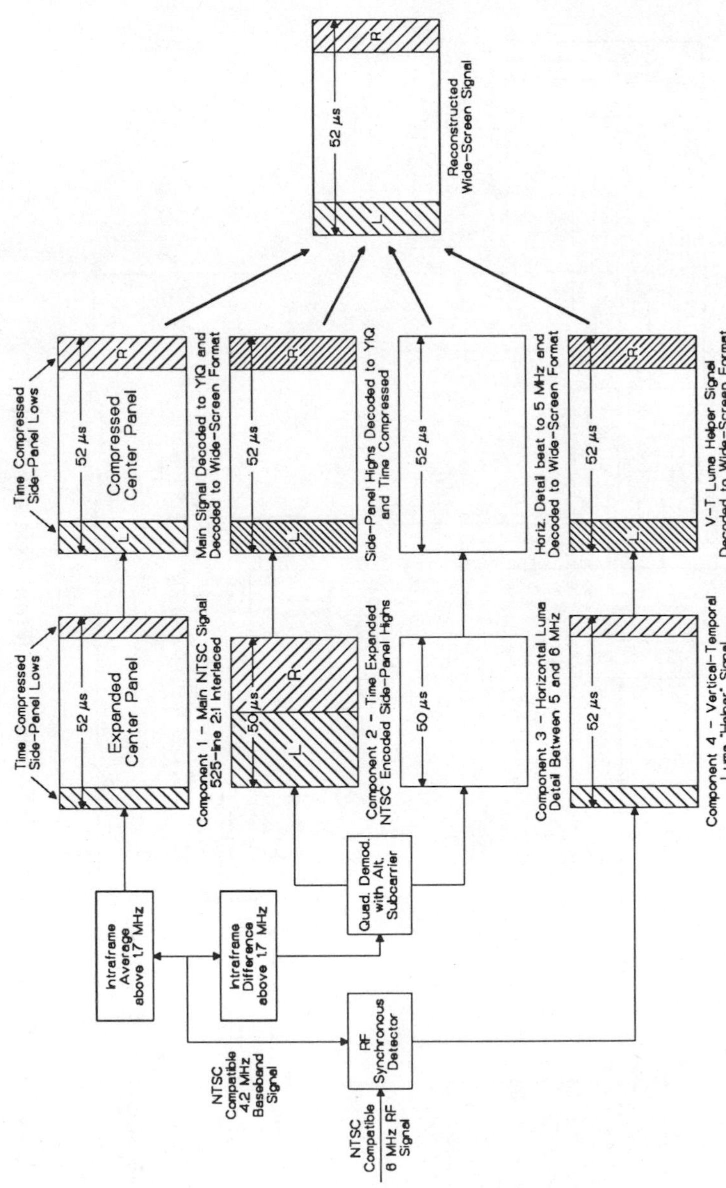

Figure 8.19 ACTV-I receiver processing, the inverse of the transmitter processing shown in Figure 8.15. (From Reference 12.)

second, these being multiples of a basic picture repetition rate of 12 per second. This signal source, feeding the NTSC-compatible simulcast channel, would produce a display of the "letter-box" type, with blank bars above and below the 16:9 area displayed on the 4:3 conventional display. The RGB output of the camera is scanned at resolutions of 240 picture elements per picture height, 400 picture elements per picture width, scanned progressively at 12 frames per second. The luminance resolution for stationary portions of the images is 762 × 1200 at 12 frames per second and 508 × 800 at 36 frames per second. The chrominance resolutions for static portions are 400 picture elements per picture width, 254 elements for portions in motion.

The transmitter signal processing is shown in Figure 8.20. The analog output of the camera is digitized in a converter operating at 8.4 or 12 Mbytes/s. The digitized signal is passed through a quadrature mirror filter from which are produced a number of components, of which nine are selected with frame rates for different portions of the image, in accordance with the rate of motion present in each. The digitizing clock rate is precisely 6 MHz. Each frame of digital data occupies 1/12 second during which time nine successive signal segments of 127 lines each are sent in sequence, followed by 105 lines for audio and data components. The digital components are stored in a frame-store.

The digital components corresponding to odd and even lines of the image are converted to analog form by the two digital-to-analog converters shown in Figure 8.20. These analog signals are lowpass filtered at 3 MHz, and are quadrature-modulated on the picture carrier at the center of the channel. For fm transmission on a direct-broadcast satellite channel, an alternate digital-to-analog converter delivers a 6-MHz channel suitable for transponders.

Figure 8.21 shows the vertical and horizontal resolutions of the MIT-RC system (Section 8.2.5) and the MIT-CC system compared with the Narrow-MUSE system (Section 8.3.2).[14]

In the MIT-CC receiver the inverse of the processing shown in Figure 8.20 is performed. This results in a display of 1200 lines, 16:9 aspect ratio, with the resolutions of the several modes employed at the transmitter.

The MIT-CC system was initially designed for distribution over 6-MHz cable channels, but it can be transmitted on the DBS service, and on terrestrial 6-MHz channels. It is intended that the MIT-CC

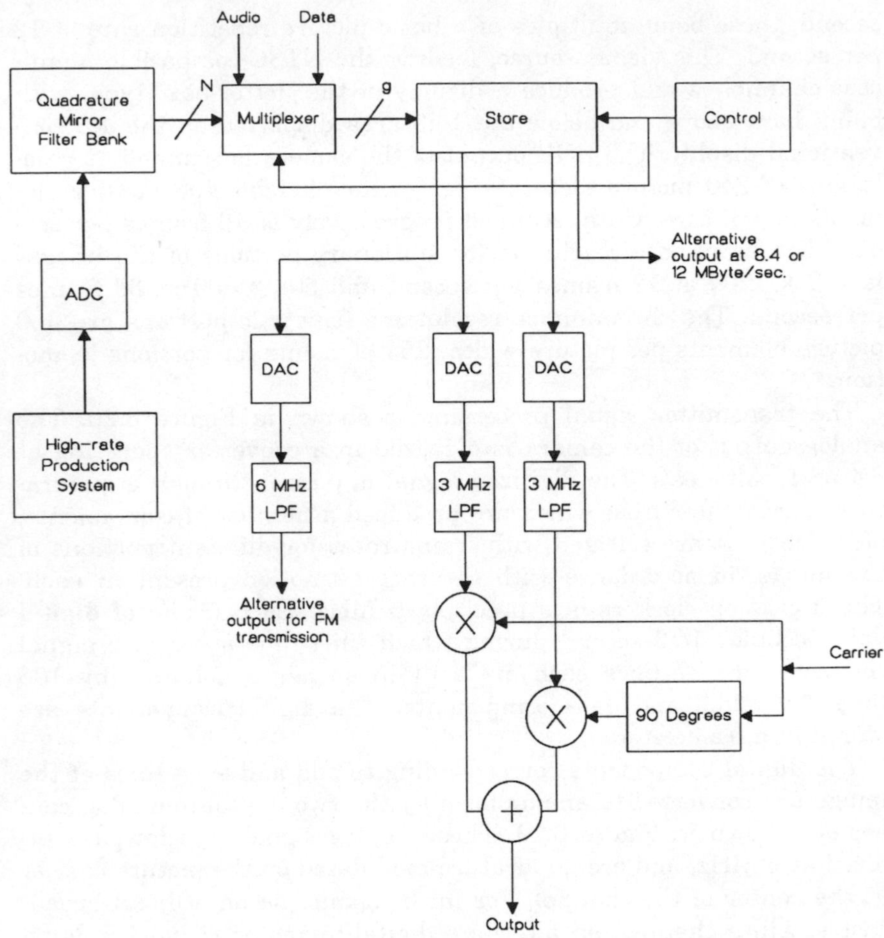

Figure 8.20 Encoder elements of the MIT-CC simulcast system. (From Reference 13.)

receiver employ an adaptive channel equalizer (not detailed in the submission document). The bandwidth available for audio and data is 0.5 MHz, corresponding to a bit rate of 100 Mb/s. This is sufficient for digital transmission of stereo sound channels having specifications that approach compact-disk audio quality.

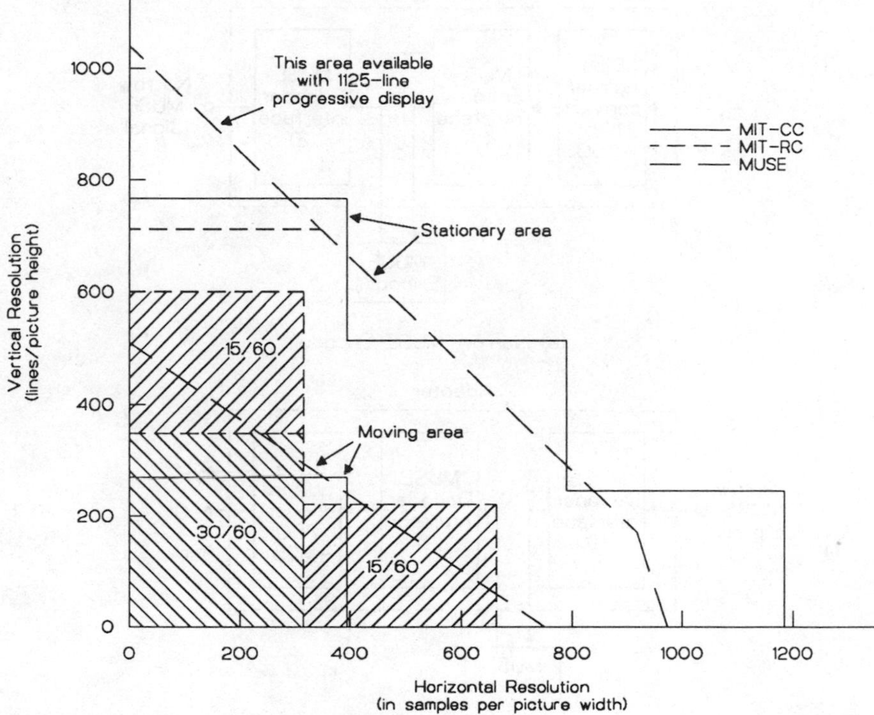

Figure 8.21 Vertical and horizontal resolutions of the MIT-RC, MIT-CC, and Narrow-Muse systems, for stationary and moving areas. (From Reference 13.)

8.3.2 NHK Narrow-MUSE System

The NHK MUSE family of systems[9] (see Figure 8.10) contains two proposals intended for 6-MHz channels. The MUSE-6 system (Section 8.2.6) is an NTSC-compatible system that meets the requirements of the compatible service. The NarrowMUSE system, being a simulcast noncompatible proposal, uses more fully the resources of the 6-MHz HDTV channel. It was designed primarily for the terrestrial service, simulcast with a standard NTSC signal.

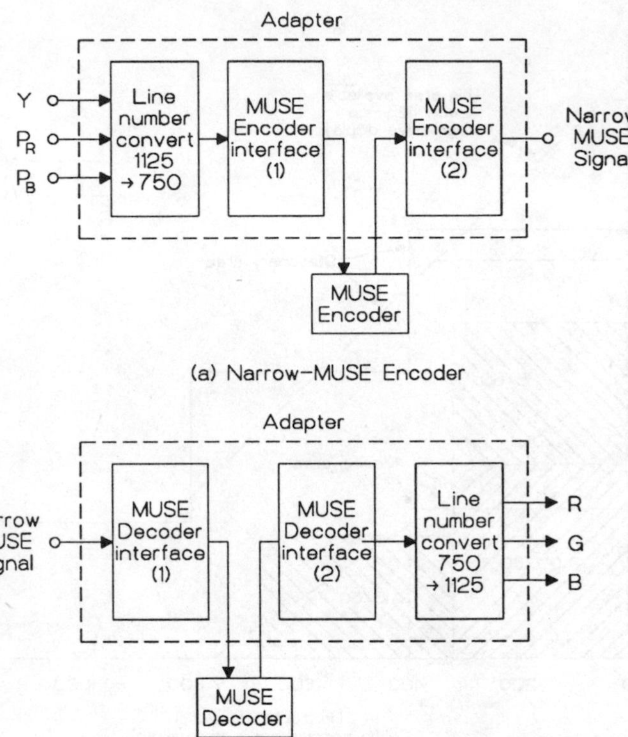

Figure 8.22 Encoder and decoder elements of Narrow-MUSE. (From Reference 14.)

Relation to Basic MUSE System The basic MUSE system (Section 2.8) originates with an 1125/60/2:1 16:9 camera signal, which is encoded by the processing shown in Figure 2.7b. The processed signal is further encoded, as shown in Figure 8.22a, to 750 lines. The processing from 1125 to 750 lines is shown in Figure 8.23. The 750-line video base bandwidth is 4.86 MHz, which fits within the 6-MHz channel. The MUSE subsampling permits recovery of additional detail to 15 MHz.

Display The displayed luminance vertical resolution (Figure 8.24a), is 650 lines per picture height for stationary parts of the image, 325 lines for moving portions. The chrominance resolutions (Figure

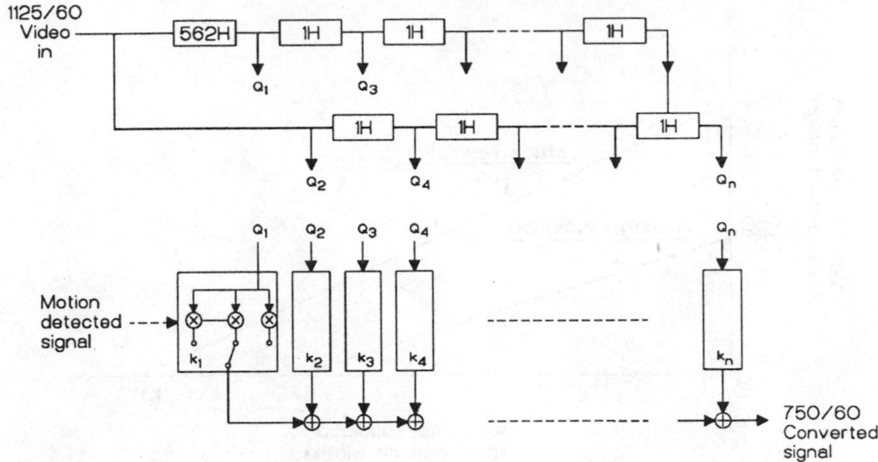

Figure 8.23 Narrow-MUSE line converter from 1125 to 750 lines. (From Reference 14.)

8.24b) are half the luminance values. The image displays 650,000 luminance picture elements.

Channel Content Figure 8.25 shows that the picture carrier (the same frequency as that employed in the NTSC simulcast channel) is modulated by three signals, and fills 5.946 MHz of the 6-MHz channel.

Receiver At the receiver, inverse decoding occurs as in Figure 2.7b and the 750-line signal is decoded to 1125 lines, 60 fields, interlaced 2-to-1. The audio signal, digitized and multiplexed in the vertical blanking interval, may have either of two formats. Format A provides four channels of 15-kHz bandwidth, 90-dB dynamic range. Format B provides two channels of 20-kHz bandwith, 96-dB dynamic range.

8.3.3 Zenith Spectrum-Compatible HDTV System[15–18]

The SC-HDTV system is a single-channel simulcast system specifically designed to coexist with NTSC in the existing TV bands of terrestrial broadcasting. Sharing of the existing TV bands requires the use of the so-called "Taboo" channels. These channels cannot now be

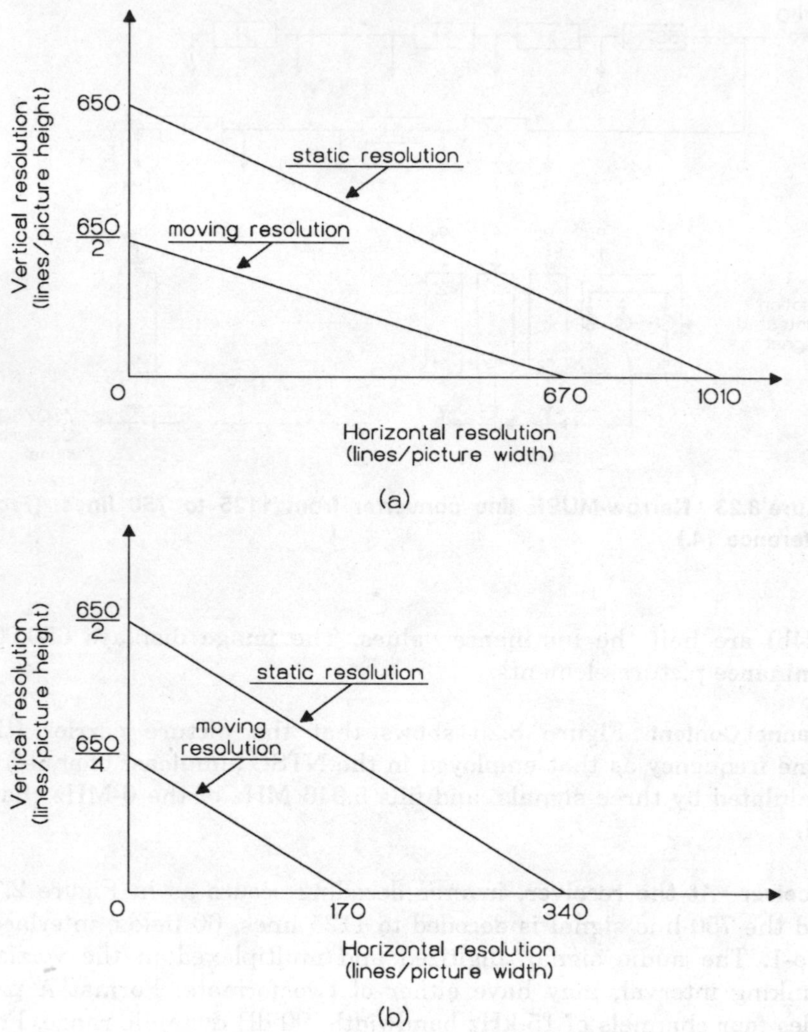

Figure 8.24 Luminance (a) and chrominance (b) resolutions of Narrow-MUSE. (From Reference 14.)

used, mainly because of excess interference. A primary design feature of the SC-HDTV transmission signal is reduced and more evenly distributed power when compared with NTSC. This reduces cochannel interference into NTSC. In addition, the SC-HDTV transmission signal horizontal and vertical timing are made equal to those of NTSC. This allows the use of precision carrier frequency

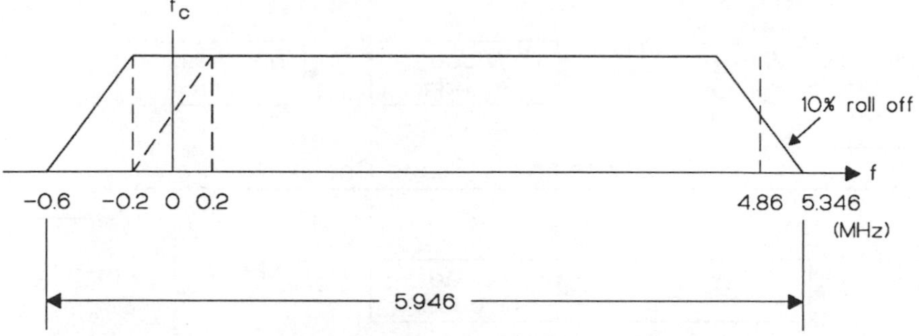

Figure 8.25 Spectrum elements of Narrow-MUSE 6-MHz channel. (From Reference 14.)

offset which reduces the visibility of cochannel interference in both directions. It will be applied in extreme situations of interference.

Figure 8.26 shows the source-to-transmission encoding and the emission-to-display decoding of the system.

Source Scanning The camera of the SC-HDTV system progressively scans 787.5 lines, at 59.94 frames per second for a horizontal deflection frequency of 47.203 MHz, exactly three times the NTSC rate. The R, G, and B bandwidths extend over 37 MHz. The aspect ratio is 16:9. The number of active lines per frame is 720 and there are 1280 active pixels per line.

The video encoder (Figure 8.27) processes the source signal into two simultaneous components, I and Q, of 3 MHz bandwidth each. These I and Q components are not to be confused with the *I* and *Q* signals of the NTSC standard.

Video Processing The objective of video encoding is redundancy reduction in the spatial as well as in the temporal domain. The R, G, and B camera signals are A/D converted and digitally matrixed into a luminance component, *Y*, and SC-HDTV chroma difference components C_1 and C_2, as shown in Figure 8.27. The video encoder applies adaptive transform/sub-band coding. Potentially, a full picture's content is retained but in flat areas with little detail and little variation in the image little information is transmitted. This makes time available for the transmission of more information needed for areas of greater detail. Adaptation and control information are transmitted separately during the vertical blank interval as part of the Digital Data signal as will be explained below.

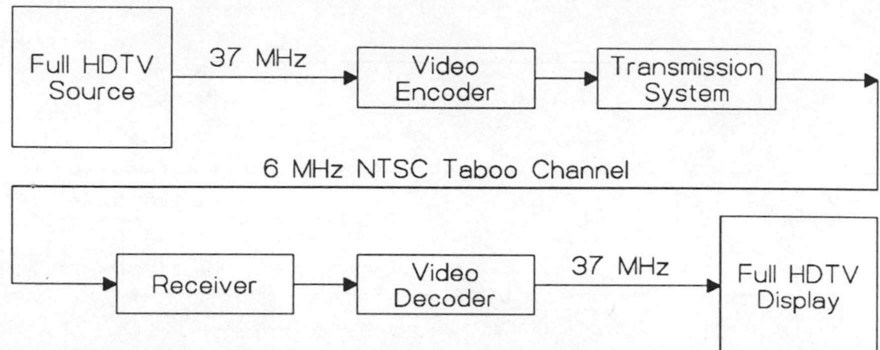

Figure 8.26 Source-to-transmission encoding and emission-to-display decoding of the Spectrum-Compatible system.

The Y signal is transformed into 64 spatial sub-bands represented by the 64 rectangles in Figure 8.28. In any given frame (frame and field are equivalent in progressive scan), on the average only a fraction of the 64 sub-band coefficients need to be transmitted, the others are zero or negligibly small. Each of the C_1 and C_2 signals is transformed maximally into 32 spatial sub-bands every other frame. On the average, only a similar fraction of the 32 coefficients per

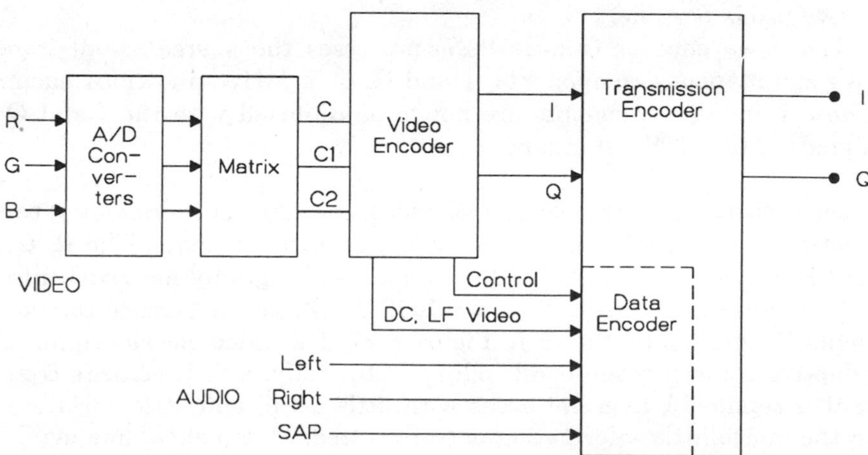

Figure 8.27 Baseband encoding of R, G, B video signals to S-C HDTV *I* and *Q* signals (different from NTSC *I* and *Q* signals) of equal bandwidth for transmission.

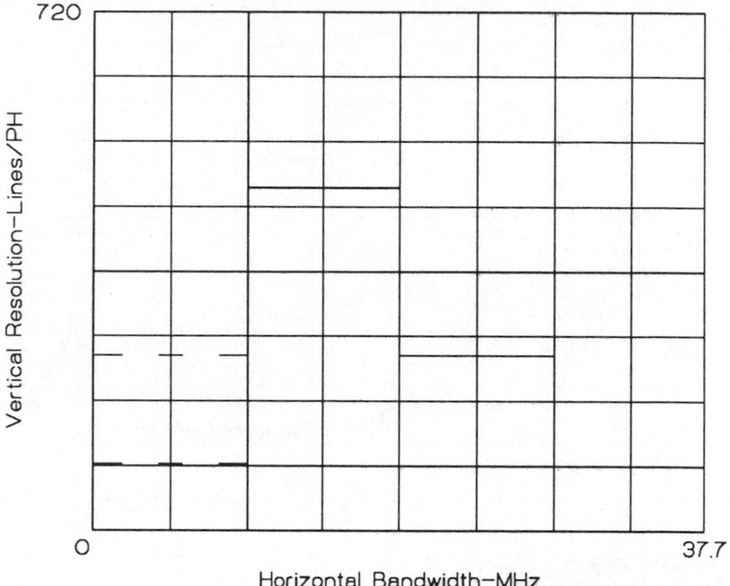

Figure 8.28 Horizontal bandwidth and vertical resolution of the 64 Y-signal spatial sub-bands in the Spectrum-Compatible system.

frame need to be transmitted. The chroma difference coefficients are time-multiplexed with the luminance coefficients. Many of the higher-order coefficients are low in amplitude, which allows compression. This improves the signal-to-noise ratio. The coefficients are D/A converted and transmitted in analog form after transmission encoding.

Transmission Encoding The video processing as described changes the spectral distribution of the source signal to a certain degree but the weighting towards DC and low frequencies, as illustrated for a typical NTSC program signal in Figure 8.29, is retained. The curve shows amplitudes averaged over a one-hour period. The minor peak near 3.6 MHz is due to the NTSC chroma signal and is eliminated from the SC-HDTV transmission signal due to the time-multiplexing of the chroma difference coefficients, as mentioned.

The substantial peak near DC and low frequencies is eliminated from the SC-HDTV transmission signal as well. This is accomplished by filtering off the DC and the first approximately 200-kHz components from the coeeficient signal and transmitting these components as part of the Digital Data signal.

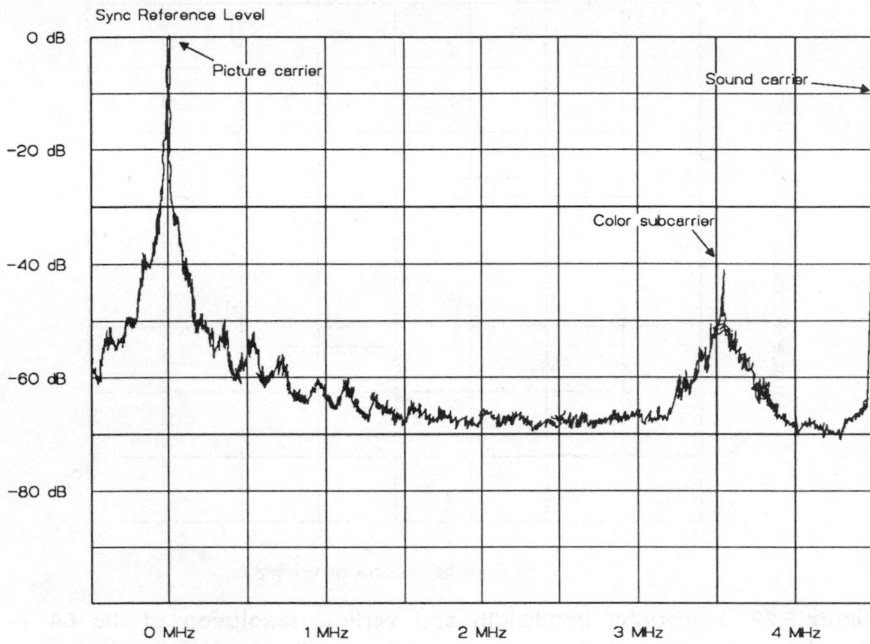

Figure 8.29 Amplitude spectrum of a typical NTSC program video signal averaged over a one-hour period.

The peak near 4.5 MHz in Figure 8.29 is eliminated by transmitting audio as part of the Digital Data signal as well.

The described processing generates signals I and Q of low and evenly distributed average power without any subcarriers. The bandwidth of both I and Q components is 3 MHz.

Transmission encoding includes one more step prior to RF modulation: temporal filtering. The temporal fiter is, in effect, a frame comb filter. The transmitter version is illustrated in Figure 8.30. The signal representing a static image repeats from frame to frame and is reduced in amplitude. This further reduces the SC-HDTV signal as a potential interference source. The receiver comb filter, shown in Figure 8.31, restores the original signal. It also serves an important interference-reducing function. If the SC-HDTV carrier frequency is offset by an odd integer of half-frame rate with respect to the interfering NTSC carrier (approximately 1.75 MHz), this carrier falls in a "trough" of the receiver comb filter and is attenuated substantially.

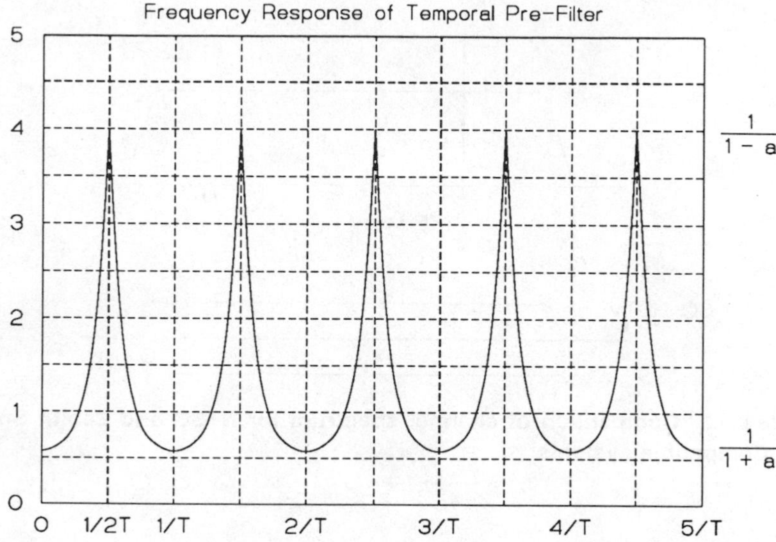

Figure 8.30 Amplitude-frequency response of temporal (frame-comb) filter used in encoding prior to transmission.

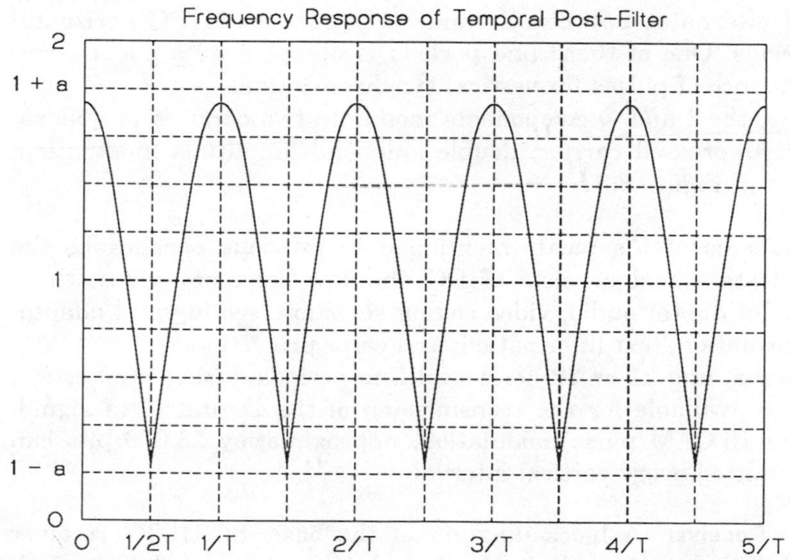

Figure 8.31 Amplitude-frequency response of temporal (frame-comb) filter used in the reception.

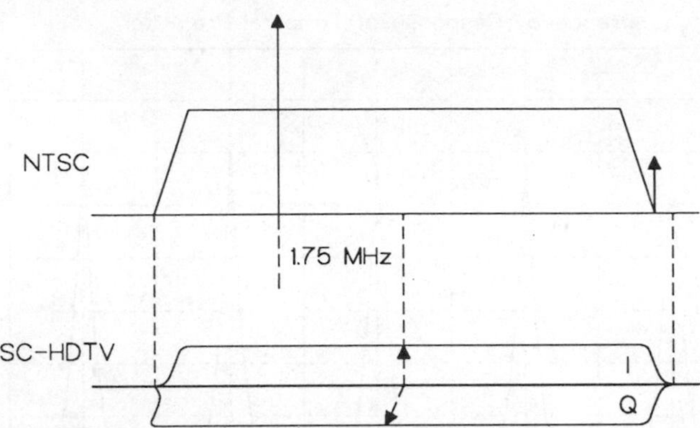

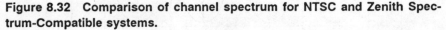

Figure 8.32 Comparison of channel spectrum for NTSC and Zenith Spectrum-Compatible systems.

The I and Q signals include a blank period of approximately 3-μs duration after an active video period of 60.6 μs. The active video period carries the analog coefficients of three SC-HDTV horizontal lines, on the average. Half of the 3-μs blank period is occupied by the synchronizing pulse of a fixed, low amplitude. The vertical blanking interval alternates between 22 and 23 equivalent NTSC horizontal line periods. One of these line periods is occupied by a pseudo-random sequence of pulses for vertical synchronization.

Finally, the I and Q components modulate two carriers in quadrature by suppressed carrier, double sideband amplitude modulation, as shown in Figure 8.32.

Digital Data Signal As partly mentioned in previous paragraphs the Digital Data signal consists of DC and low-frequency video, three channels of digital audio, video coding selection, scaling and adaptation information, text information, and error protection.

The remaining 21 or 22 horizontal line periods during the vertical blank are available for the transmission of the Digital Data signal. By using 16-QAM pulse modulation, approximatley 28,000 bits can be transmitted every vertical interval.

SC-HDTV Receiver A block diagram of the basic SC-HDTV receiver consists of the portion within the dotted lines of Figure 8.33, excluding the block "Switching and Conversion." The block marked "Tuner" covers VHF and UHF, IF and second detector, and yields the 3-MHz I and Q baseband components and the 16-QAM Digital Data signal.

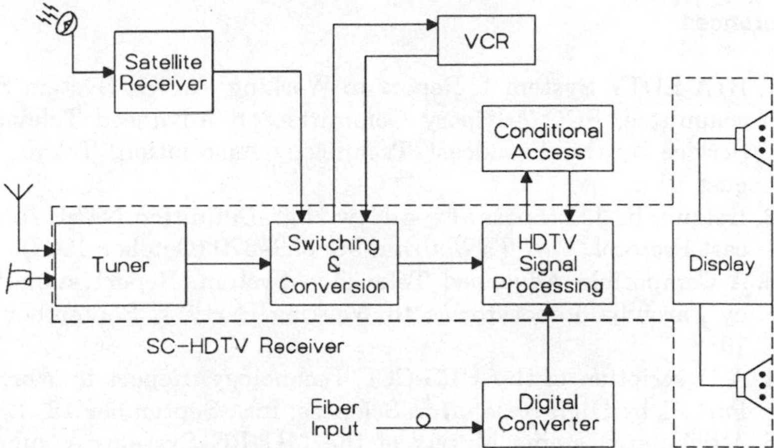

Figure 8.33 Block diagram of receiver with provision for multimedia operation. Inputs are provided for terrestrial broadcasting, direct-broadcast satellite (DBS), and fiber-optic cable.

The "HDTV Signal Processing" block includes A/D conversion, receiver comb filtering, 16-QAM detection, transform/sub-band decoding, and D/A conversion, all combined into VLSI ICs.

As illustrated, the system lends itself to multi-media operation. For cable operation the same basic receiver suffices. Encryption of the Digital Data signal is easily achieved. The Conditional Access box, given the encryption key, allows decryption without loss of signal quality. Optical fiber operation, depending on the signal format, requires a Digital Converter.

Satellite and VCR operation require FM signals. The "conversion" part of the center block in Figure 8.33 is included to this end. The two 3-MHz baseband components, I and Q, delivered by the tuner can be converted into a single 6-MHz component for recording on the VCR. The conversion includes time compression and delay, reminiscent of MAC signal generation. The satellite receiver output can be recorded without conversion. To display the satellite receiver output requires the inverse conversion process. Obviously, the converters are needed at the transmitter side.

There is an advantage for the supplier of encrypted programs in the circuit arrangement of Figure 8.33. The relative position of VCR and Conditional Access Box does not allow recording of the decrypted version of a program. This prevents unauthorized distribution of such programs.

References

1. BTA EDTV System I. Report to Working Part 1, System Subcommittee, FCC Advisory Committee on Advanced Television Service by the Broadcast Technology Association, Tokyo, August 1988.
2. Ireland, R. J. Proposal for a New High-Definition NTSC Broadcast Protocol. *SMPTE Journal* 96: 359–370 (October 1987).
3. A Compatible Advanced Television System. Report submitted by Faroudja Laboratories to Working Party 1, September 15, 1988.
4. A Description of the HRS-CCF Technology. Report to Working Party 1 by High Resolution Sciences, Inc., September 13, 1988.
5. Attributes/Systems Matrix of the MIT-RC System. A submission to Working Party 1 by Massachusetts Institute of Technology, August 29, 1988.
6. Schreiber, W. F. and A. B. Lippman. Single-Channel HDTV Systems, Compatible and Noncompatible, Report ATRP-T-82 Advanced Television Research Program. Media Laboratory, Massachusetts Institute of Technology, Cambridge, MA, March 11, 1988.
7. Schreiber, W. F. OAR: The Open-Architecture Television Receiver, Report ATRP-T-88R, Advanced Television Research Program. Media Laboratory, Massachusetts Institute of Technology, Cambridge, MA, June 12, 1988.
8. Lippmann, A. B., A. N. Netravali, E. H. Adelson, W. R. Neuman, and W. F. Schreiber. Single-Channel Backward-Compatible High-Definition Television Systems, Report ATRP-T-85, Advanced Television Research Program. Media Laboratory, Massachusetts Institute of Technology, Cambridge, MA, March 15, 1988.
9. Development of the MUSE Family Systems. Report submitted to Working Party 1 by NHK (Japan Broadcasting Corporation), Tokyo, November 15, 1988. (MUSE-6 is treated on pages 19–34 of the submission document.)
10. Attributes/Systems Matrix. Submission by Production Services, Inc., to Working Party 1, September 1, 1988.
11. System Description, Advanced Compatible Television. Submission to Working Party 1 by David Sarnoff Research Center, Princeton, NJ, September 1, 1988. (The ACTV-I system is described on pages 4–13 of the submission document.)

12. Isnardi, M. A., C. B. Dietrich, and T. R. Smith. Advanced Compatible Television: A Progress Report. *SMPTE Journal* 98: 484–495 (July 1989).
13. Attributes/Systems Matrix for the MIT-CC System. System Document submitted to Working Party 1 by the Massachusetts Institute of Technology, August 29, 1988.
14. Development of the MUSE Family Systems. Report submitted to WP-1 by Japan Broadcasting Corporation (NHK), November 15, 1988. (The Narrow-MUSE system is described in pages 4–18 of the submission document.)
15. Zenith Spectrum-Compatible HDTV System. Presentation to WP-1 by Consumer Products Group, Zenith Electronics Corporation, September 1, 1988.
16. Bretl, W. The Proposed SC-HDTV Program Production Standard. NAB Convention Proceedings, 44th Annual Engineering Conference, Atlanta, GA, March 29–April 3, 1990.
17. Bretl, W., R. Citta, R. Lee, and P. Fockens. Video Encoding in the Spectral-Compatible HDTV System. IEEE International Conference on Consumer Electronics, Rosemont, IL, June 6–8, 1990.
18. Luplow, W. C. High-Definition Television. HDTV '90 Colloquium, Ottawa, Canada, June 28, 1990.

Chapter

9

Proposed Wide-Channel Advanced Television Systems

9.1 Wide-Channel Systems

The proposed ATV systems listed in Table 9.1 employ channel space wider than 6 MHz. They fall into two broad classes: those employing augmentation channels in conjunction with NSTC-compatible 6-MHz channels, and those intended for the direct-broadcast satellite service. The augmentation channels are divided into two groups: those occupying 3 MHz of channel space, and those employing more than 3 MHz, but not more than 6 MHz. The augmentation channels carry information to increase the vertical and/or horizontal resolution, for the sidepanels of the 16:9 display, and to increase the number of sound channels and/or their quality.

The augmentation channels may be contiguous with the NTSC-compatible channel, i.e., occupying a single channel of 9- or 12-MHz width, or noncontiguous, for example, a VHF-compatible channel with a UHF augmentation channel.

The direct-broadcast satellite channels are intended to frequency-modulate the signal to the transponder, with a video baseband of 9–12 MHz.

9.2 Philips HDS-NA System

The HDS-NA ("High-Definition System for North America") proposal[1–6] by the N. A. Philips Corporation appears in Table 9.1 in three entries. Basic to all these systems is the originating satellite feeder MAC signal, which is described in Section 9.3. The others use augmentation channels of 3 MHz and 4 MHz, respectively. The final version will use the 3-MHz channel.

Table 9.1 Proposed Wide-Channel Advanced and HDTV Systems

Proponent/ name	Scanning (lines/field rate/ fields per frame)	System type	Aspect ratio NTSC	Aspect ratio ATV/HTDV
Wide-Channel Systems with 3-MHz Augmentation Channels				
NHK/ MUSE-9	1125/60/2:1	NTSC	16:9	16:9
Philips/ HDS-NA	525/59.94/1:1 1050/59.94/2:1	NTSC HDTV	4:3	16.9
Wide-Channel Systems with 4- to 6-MHz Augmentation Channels				
NYIT/ Glenn	1125/59.94/2:1	HDTV	5:3	5:3
Osborne	1125/60/2:1	NTSC HDTV	4:3	16:9
Philips/ HDS-NA	525/59.94/1:1 1050/59.94/2:1	NTSC HDTV	4:3 4:3	16:9 16:9
Sarnoff/ ACTV-II	1050/59.94/2:1	NTSC HDTV	4:3	16:9
Satellite Transmission Systems				
NHK/ MUSE-E	1125/60/2:1	NTSC HDTV	16:9	16:9
Philips/ HDS-NA	525/59.94/1:1 1050/59.94/2:1	NTSC HDTV	4:3	16:9
Scientific Atlanta	525/59.94/1:1 and 2:1 1125/59.94/2:1	NTSC HDTV	4:3	16:9

Source: Abstracted from proponents' presentations to the FCC ATAC. *TV Digest* 29(1): 6–7 (January 2, 1989).

The HDS-NA system employs two principal signals. The first, HDMAC, is a satellite-transmitted multiplexed analog component signal from the program origination point to other transmission media, i.e., terrestrial broadcast and cable systems. It can also serve in the direct-broadcast service (Section 9.6.2).

The second signal, named HDNTSC by Philips, is intended for the terrestrial and cable services. It comprises the NTSC-compatible 6-MHz channel and an augmentation channel. The augmentation information is proposed for transmission either in digital or analog form. The interconnections of the HDS-NA system are shown in Figure 9.1, its display elements in Figure 9.2.

9.3 The HDMAC System

The HDMAC signal originates at 525 lines, 59.94 frames per second progressively scanned, with an aspect ratio of 16:9. An alternative is 1050/59.94/2:1/16:9. The video baseband of the origination signal is 16.8 MHz, reduced by the formation of line-difference signals and signal compression to 9.5 MHz before frequency modulation of the satellite up-link signal. This modulation band is wider than those used in other DBS systems (e.g., 8.1 MHz in the MUSE DBS system [Section 2.8]) but can be accommodated in recent designs of transponders.

The luminance resolutions available in the HDMAC satellite signal are shown in Figure 9.3. The 495 horizontal lines per picture height, combined with the vertical 480 lines per picture height, at the 16:9 aspect ratio, offer a total of $16 \times 495 \times 480/9 = 422{,}400$ picture elements per frame.

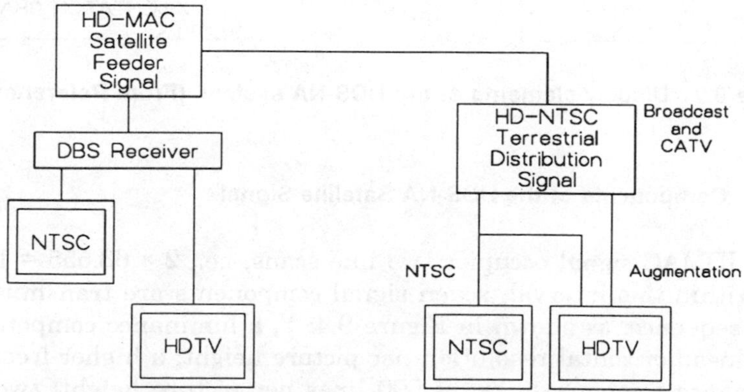

Figure 9.1 System concept of the HDS-NA system. (From Reference 1.)

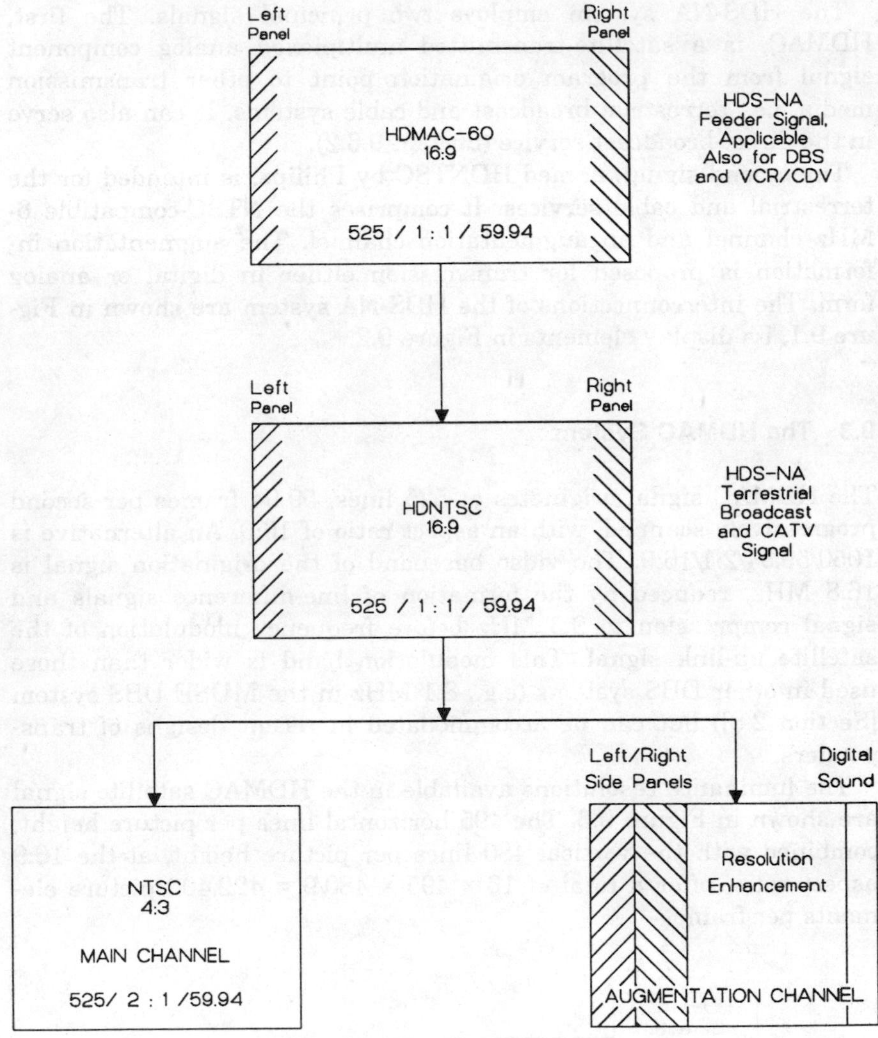

Figure 9.2 Display elements of the HDS-NA system. (From Reference 4.)

9.3.1 Components of the HDS-NA Satellite Signal

The HDMAC signal occupies two line scans, i.e., 2 × 63.555 = 127.11 μs. Within this interval, seven signal components are transmitted in time sequence, as shown in Figure 9.4: *Y*, a luminance component of 330 lines horizontal resolution per picture height; a higher-frequency luminance component, Y_h, of 500 lines per picture height; two line-

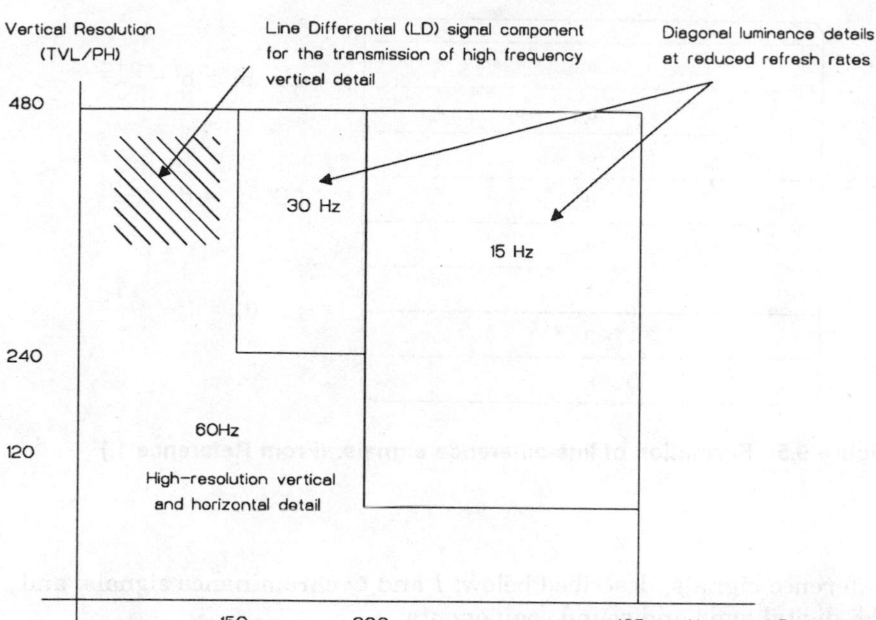

Figure 9.3 Contributions to vertical and horizontal resolutions embodied in the HDMAC satellite signal. (From Reference 1.)

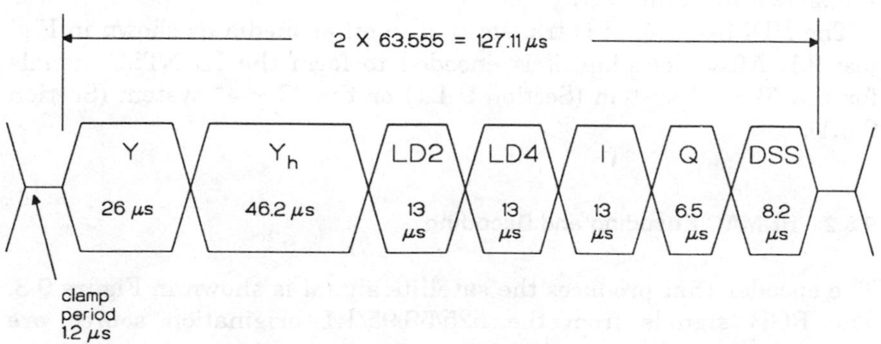

Figure 9.4 Time sequence of the components of the HDMAC signal. (From Reference 6.)

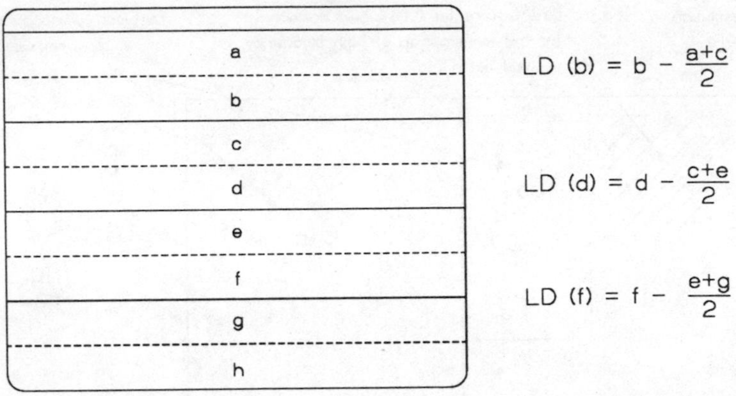

Figure 9.5 Formation of line-difference signals. (From Reference 1.)

difference signals, described below; *I* and *Q* chrominance signals; and the digital sync and sound components.

The line-difference signals are formed from three successive line scans (Figure 9.5), the difference between the center line and an average of the preceding and following lines. The line-difference signals are divided into two groups, containing the left and right sidepanel elements, respectively (Figure 9.6).

An additional function provided in the HDMAC signal is "pan-and-scan" control of the portion of the 16:9 image shown on the 4:3 displays of NTSC receivers. This portion is selected at the program origination point to place the center of interest of the scene within the 4:3 display. The functions performed by the pan-and-scan system are shown in Figure 9.7.

The HDMAC signal is transmitted to other media as shown in Figure 9.1. After decoding, it is encoded to form the HDNTSC signals for the "6 + 3" system (Section 9.4.1) or the "6 + 4" system (Section 9.5.1).

9.3.2 HDMAC Encoding and Decoding

The encoder that produces the satellite signal is shown in Figure 9.8. The RGB signals from the 525/59.95/1:1 origination source are

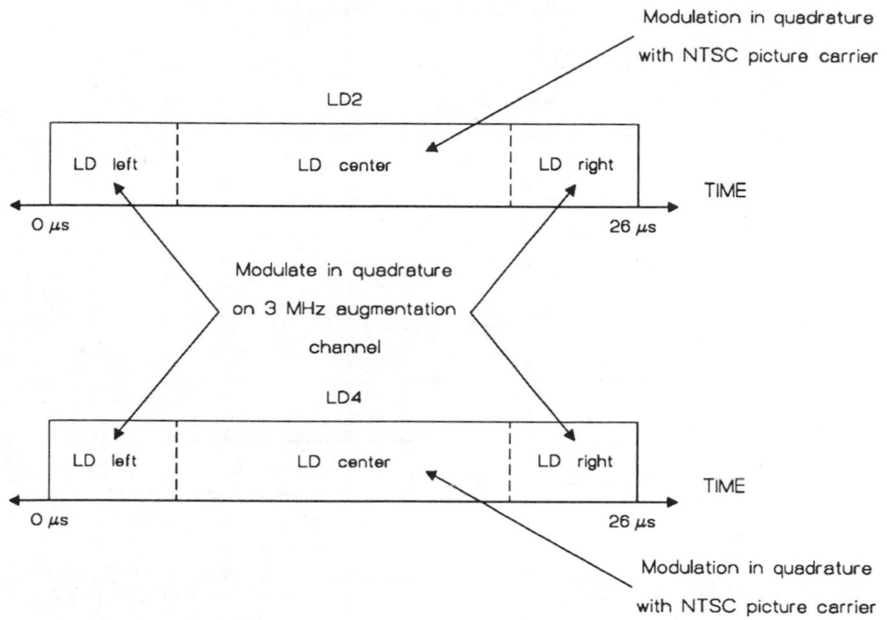

Figure 9.6 Locations and modulations of the line-difference signals. (From Reference 1.)

matrixed to form the *Y, I,* and *Q* signals. The *Y* component is fed to the line-difference generator. After filtering, the *Y* signal is split into high- and low-frequency components. The low-frequency *Y, I,* and *Q* signals are separated into the center and sidepanel segments by the pan-and-scan control. The line-difference, high-frequency luminance, and the center and side-panel segments are combined with the digital audio signals to form the line-time of the augmentation channel (Figure 9.4). The pan-and-scan control is added during the vertical interval. The outputs are the modulating signals for the up-link to the satellite transponder.

The decoder (Figure 9.9) recovers the R, G, and B components and the audio channels. The decoder is contained within the HDS-NA receiver (Figure 9.10) in the DBS version of the system. The decoder output serves as the input for the HDS-NA "6 + 3" and "6 + 4"encoders (Figures 9.11 and 9.20).

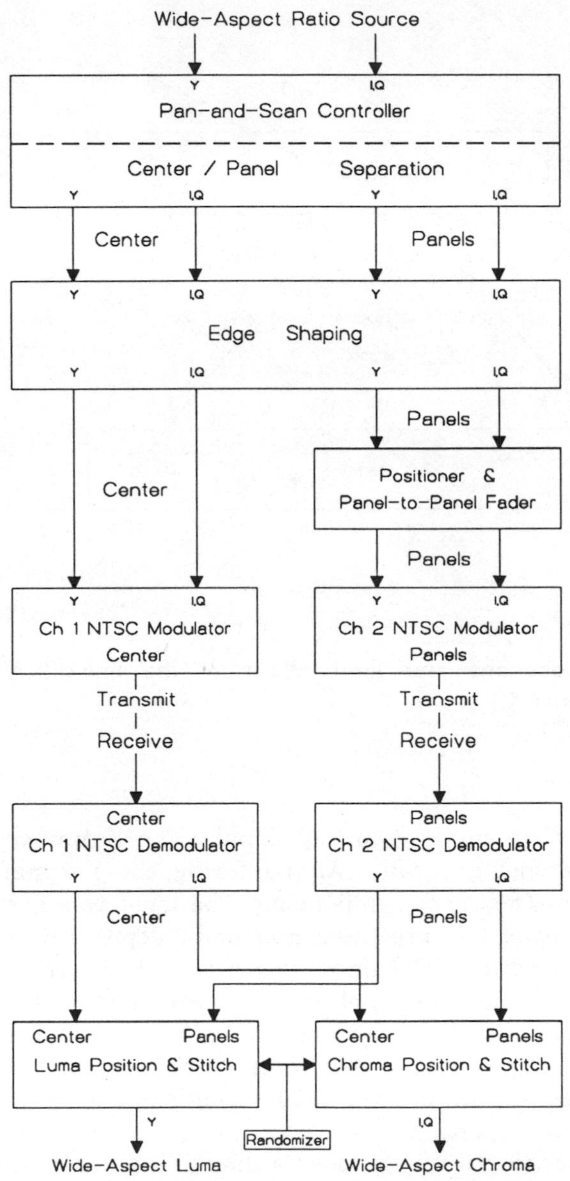

Figure 9.7 "Pan-and-scan" control system for selecting the portion of the 16:9 wide-screen image to be shown on 4:3 displays. (From Reference 5.)

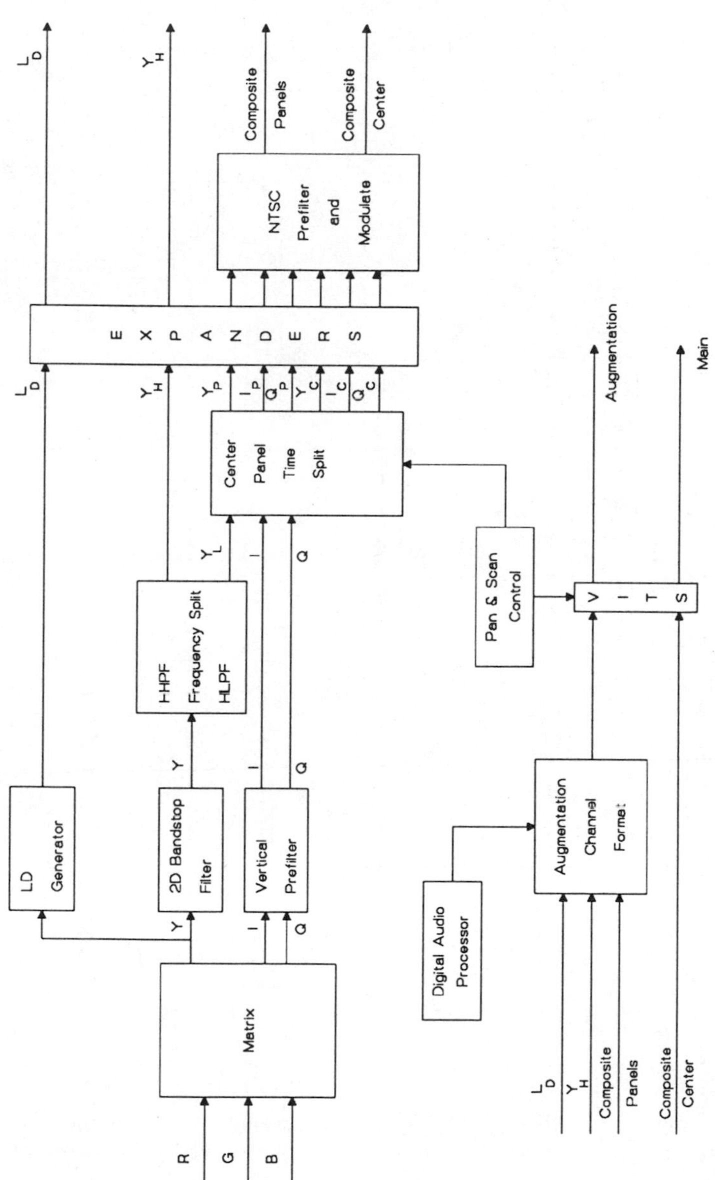

Figure 9.8 HDMAC encoder. (From Reference 1.)

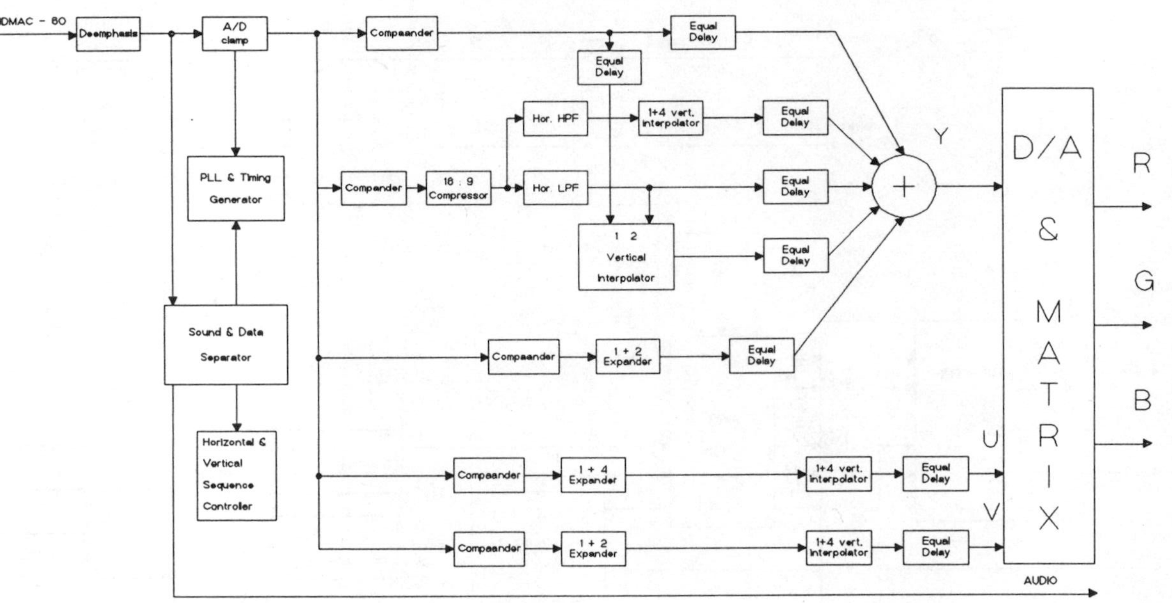

Figure 9.9 HDMAC decoder. (From Reference 1.)

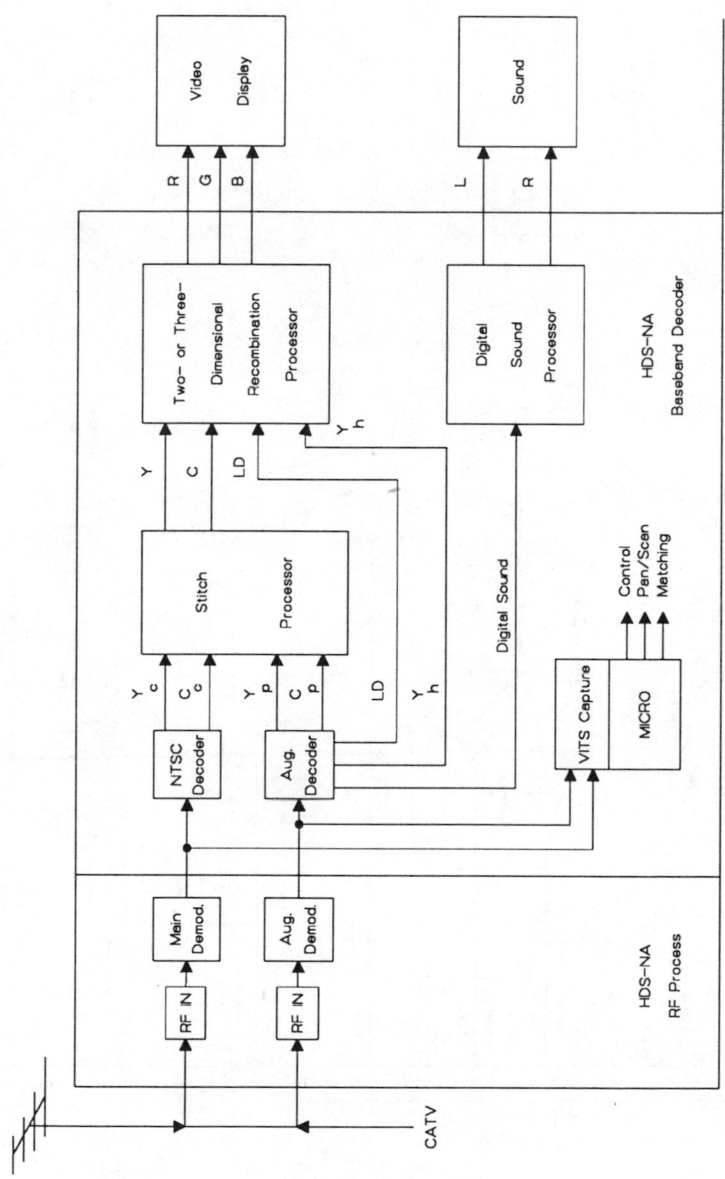

Figure 9.10 Receiver for the HDMAC signal. (From Reference 1.)

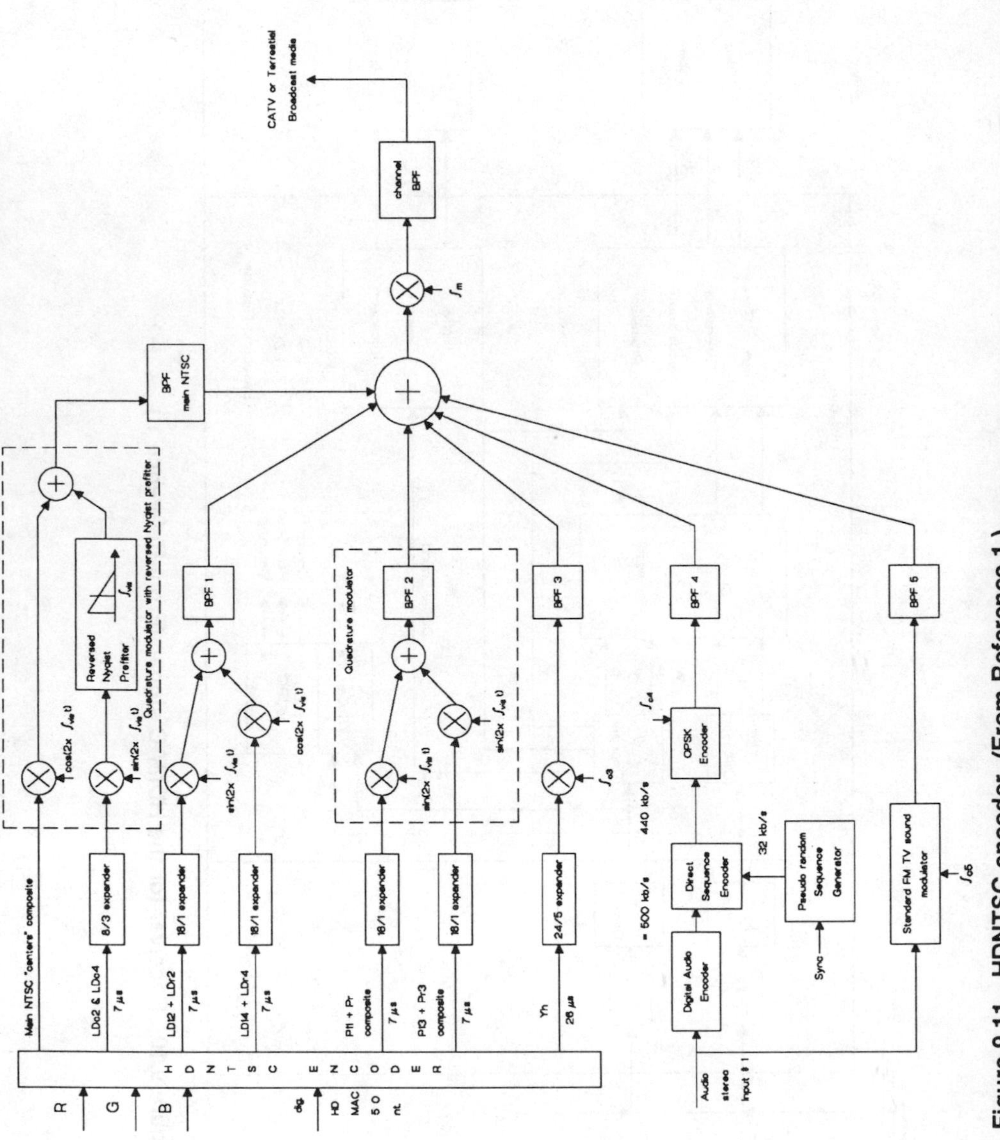

Figure 9.11 HDNTSC encoder. (From Reference 1.)

9.4 Systems with 3-MHz Augmentation Channels

9.4.1 HDS-NA "6 + 3" System

The decoded output of the HDS-NA satellite down-link signal, described above, is the input to the HDNTSC system. Here we describe the "6 + 3" version[1] of the HDNTSC system, which consists of an NTSC-compatible 6-MHz channel and a 3-MHz augmentation channel. The HDMAC output contains the following components, which must be accommodated in the HDNTSC encoder: (1) the line- difference signals for the central and sidepanel areas of the 16:9 image as originated at the source (Figure 9.6); (2) the standard 4.2-MHz luminance signal; (3) the higher-frequency portion, Y_h, of the luminance content as separated from the low-frequency portion in the HDMAC encoder; (4) the sidepanel I and Q chrominance signals; (5) the right and left sidepanel luminance contents of the 16:9 display; and (6) the digital data and audio signals decoded from the HDMAC transmission. In one version of the "6 + 3" system, these components are accommodated in the 6-MHz and 3-MHz channels as shown in Figure 9.12a. The center line-difference signals, LD_c, are modulated in quadrature on the vision carrier of the 6-MHz channel, the remainder of that channel conforming to the NTSC standard. In the 3-MHz channel, three carriers are provided: (1) f_{c3} for digital sound and data; (2) f_{c2}, symmetrically amplitude modulated by the sidepanel luminance and chrominance components; and (3) f_{c1}, a symmetrically modulated carrier carrying the left and right sidepanel line-difference signals. The higher-frequency luminance signal Y_h is also carried as shown.

An alternative method of carrying the augmentation channel information, shown in Figure 9.12b, sends the sidepanel, line-difference, and high-frequency luminance components in time sequence on a single, symmetrically modulated carrier, f_{c2}. The system proposal indicated that both methods were under development when the document was submitted.

The frequency-multiplexed version (Figure 9.12a) is decoded to the line-difference, sidepanel, and higher-frequency luminance signals as shown in Figure 9.13. These are then passed though the HDNTSC decoder to recover the 525/59.94/1:1 RGB components for the display, with two stereo-pair sound channels.

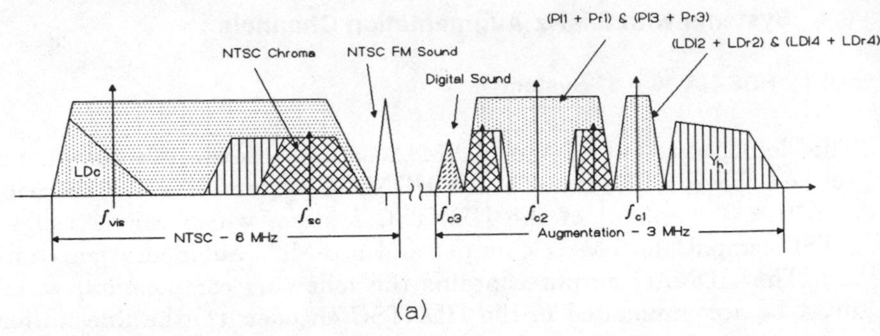

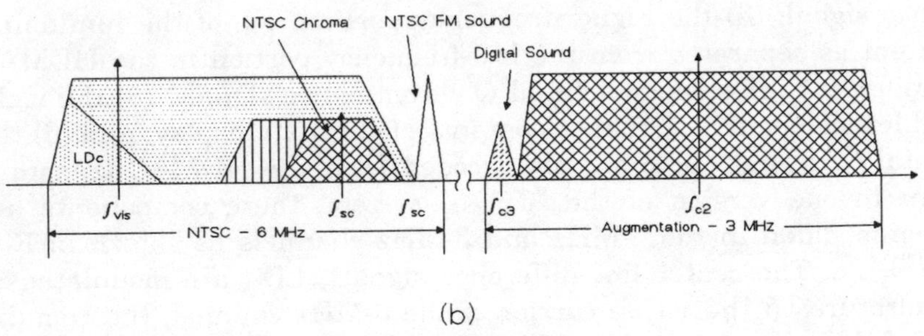

Figure 9.12 Alternative spectra of the "6 + 3" systems: (a) frequency-multiplexed and (b) time-multiplexed. (From Reference 1.)

9.4.2 NHK MUSE-9 System

The NHK MUSE-9 system[7] is one the family of MUSE systems illustrated in Figure 8.10. It uses a 3-MHz augmentation channel in conjunction with the MUSE-6 NTSC-compatible system described in Section 8.2.6. Two channel arrangements are proposed, the first with the 6-MHz and 3-MHz channels contiguous, the second with the two channels separate (Figure 9.14).

As shown in Figure 9.15, the encoder for the augmentation channel receives the moving high-frequency signal from the MUSE-6 encoder and the signals for two additional sound channels, which are digitized, multiplexed, and added to the digital content of the augmentation channel prior to conversion to the analog form for transmission over the 3-MHz channel. Figure 9.16 shows the decoding processes in the MUSE-9 receiver.

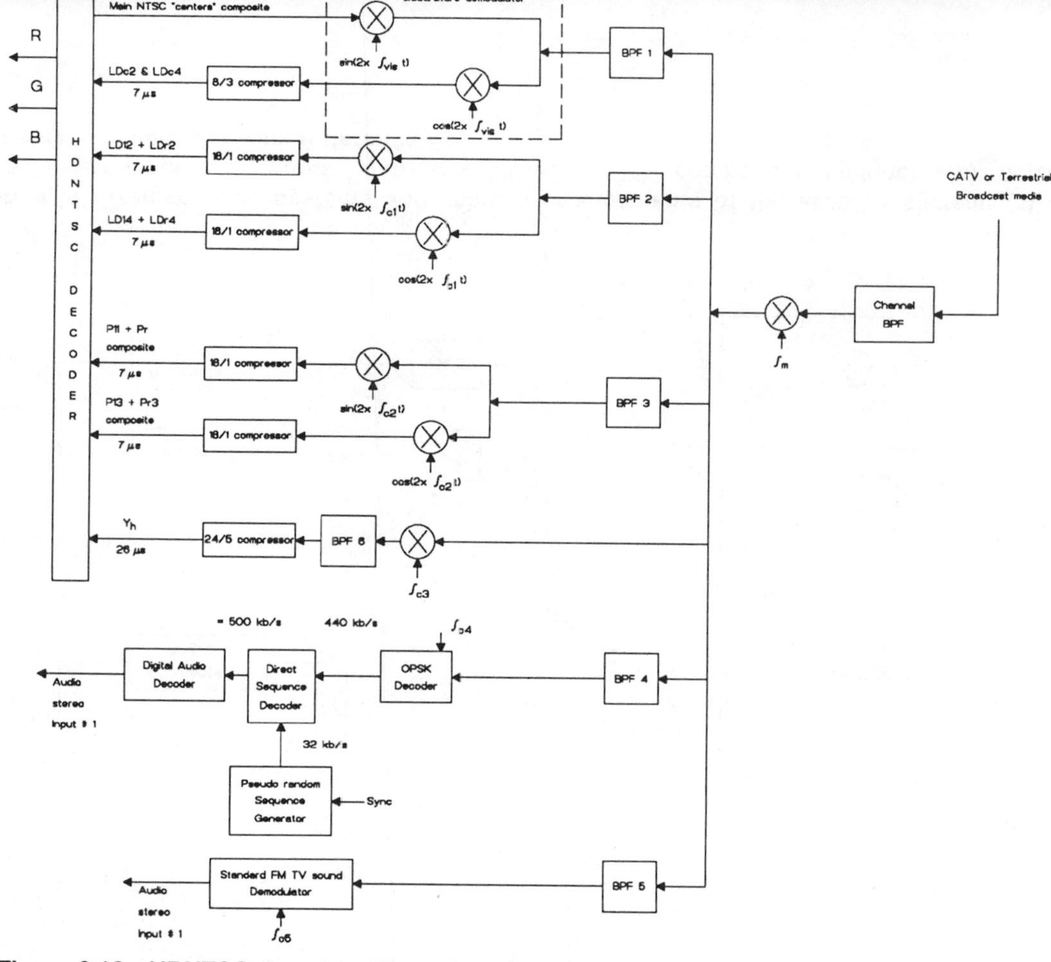

Figure 9.13 HDNTSC decoder. (From Reference 1.)

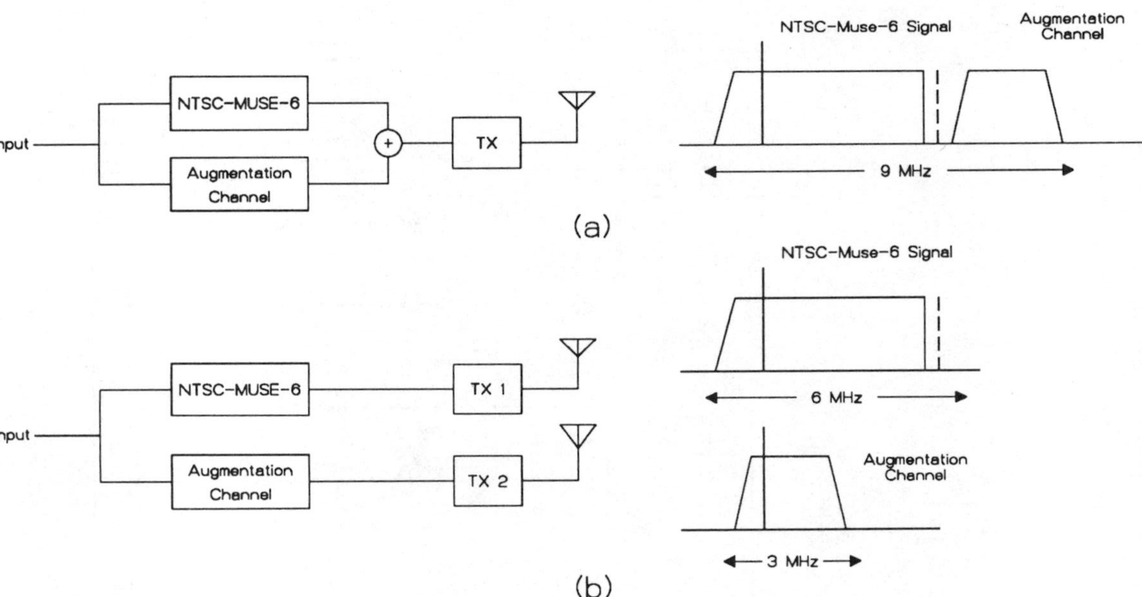

Figure 9.14 Transmission systems and channel arrangements of the MUSE-9 system: (a) augmentation channel contiguous with the MUSE-6 NTSC-compatible channel; (b) separate augmentation channel. (From Reference 7.)

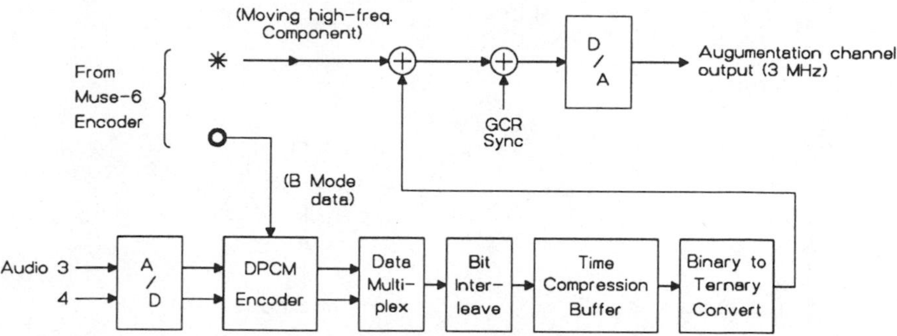

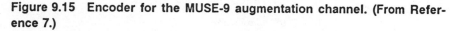

Figure 9.15 Encoder for the MUSE-9 augmentation channel. (From Reference 7.)

Signal Format In common with all the MUSE systems, the MUSE-9 system signal originates in the 1125/60/2:1 format (SMPTE 240M standard). The MUSE-6 system converts to 750 lines, 59.94 Hz. The augmentation channel has the format shown in Figure 9.17. The moving high-frequency components, in the baseband from 4 to 6 MHz, are derived in the MUSE-6 encoder. In the MUSE-6 system this frequency range can be transmitted only for static regions of the image, whereas the video signal component of the MUSE-9

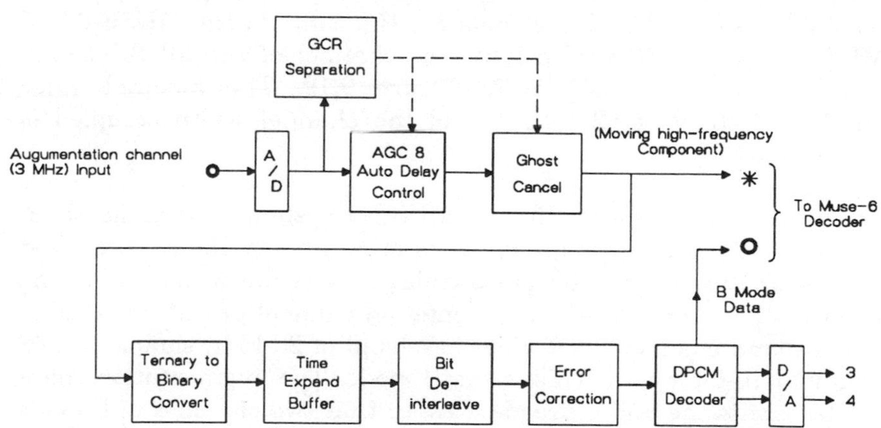

Figure 9.16 Decoder for the MUSE-9 augmentation channel. (From Reference 7.)

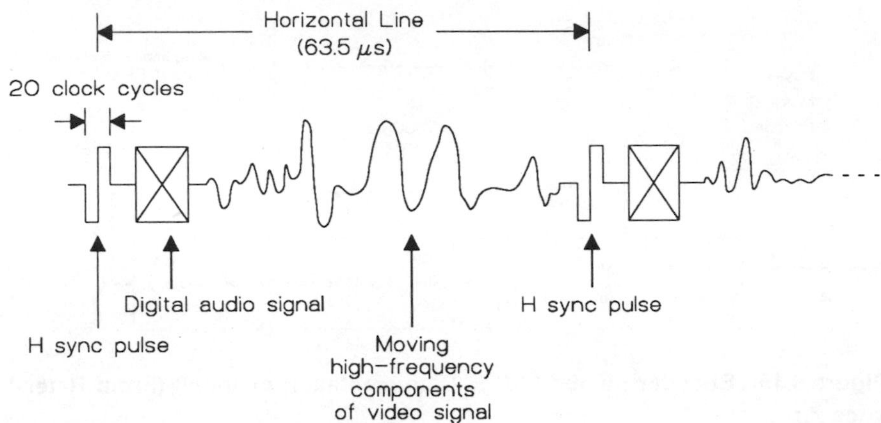

Figure 9.17 MUSE-9 signal format during the 63.5-μs line scan, showing digital sound and moving high-frequency components. (From Reference 7.)

augmentation channel permits transmission of image areas in motion. The vertical and horizontal resolutions available in the display are as shown in Figure 9.18 (compare with Figure 8.11a). The improvement is from 210 to 290 lines per picture width at a vertical resolution of 345 lines per picture height and from 410 to 600 lines per picture width at a vertical resolution of 173 lines per height. The resolutions of the *I* and *Q* signals are the same in the MUSE-6 and MUSE-9 systems (Figure 8.11b). The channel of the MUSE-9 augmentation signal is shown in Figure 9.19. The nominal video baseband extends to 2.1 MHz, and the channel width occupied is 2.81 MHz.

Additional Sound Channels In the MUSE-6 system, two audio channels of baseband 15 kHz are digitized at 32 kb/s in 15-bit words that are compressed to an 8-bit pulse train. This is the A mode of sound transmission. The MUSE-9 augmentation channel provides two additional sound channels at a wider baseband of 20 kHz, sampled at 48 kb/s into 16-bit words. This is the B mode. The augmentation channel transmits the higher frequencies, so that two channels of B-mode quality are available in the MUSE-9 receiver, or four channels of A-mode quality.

Display The MUSE-9 receiver display operates at the same rates as the MUSE-6 receiver (1125/59.94/2:1), with the additional moving-area horizontal resolution shown shaded in Figure 9.18.

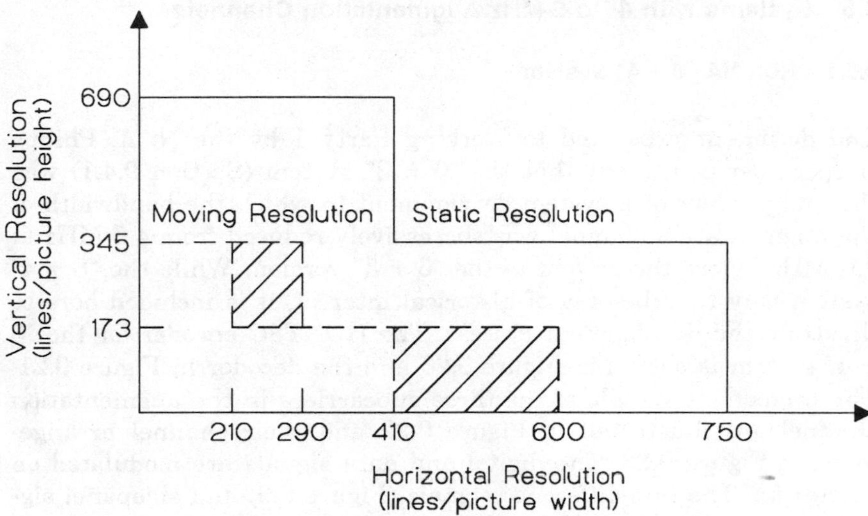

Figure 9.18 Improvement in moving luminance resolution (diagonal lines) provided by MUSE-9 augmentation channel. (From Reference 7.)

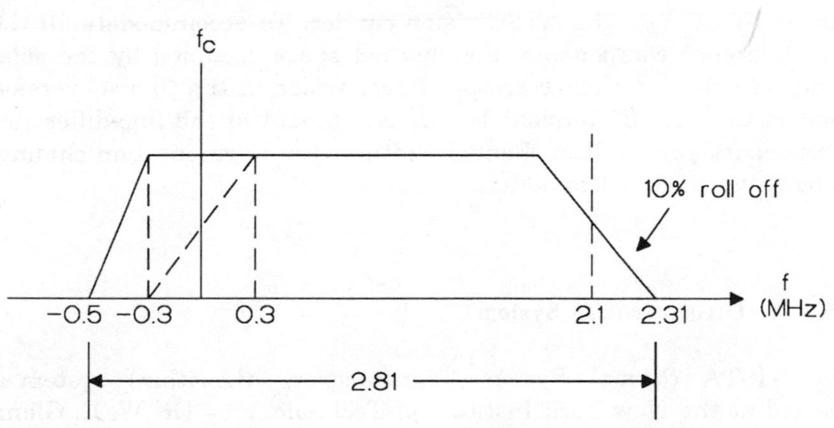

Figure 9.19 Rf spectrum of the MUSE-9 augmentation channel. (From Reference 7.)

9.5 Systems with 4- to 6-MHz Augmentation Channels

9.5.1 HDS-NA "6 + 4" System

The document presented to Working Party 1 by the N. A. Philips Corporation points out that the "6 + 3" system (Section 9.4.1) was the end product of a system development in which the bandwidth of the augmentation channel was successively reduced from 4.6 MHz to 4.0 MHz before the advent of the "6 + 3" version. While the "6 + 4" system may thus be only of historical interest, it is included here to illustrate the development process. The HDNTSC encoder for the "6 + 4" system is shown in Figure 9.20 and the decoder in Figure 9.21. The component signals of the three subcarriers in the augmentation channel are illustrated in Figure 9.22 and their channel arrangement in Figure 9.23. The digital and data signals are modulated on carrier f_{c3}. The line-difference signals (Figure 9.6) and sidepanel signals are quadrature-modulated symmetrically on carrier f_{c1}, occupying 3.78 MHz of the 4-MHz channel, leaving 0.22 MHz for the sound and data signals. This same bandwidth is assigned to the higher-frequency luminance components Y_h as the upper sideband amplitude modulation of carrier f_{c2}.

The spectral arrangement of the 6-MHz NTSC channel and the 4-MHz augmentation channel (Figure 9.23) can be compared with the corresponding diagram for the "6 + 3" system (Figure 9.12a). The "6 + 4" NTSC channel has no added signal content, whereas the "6 + 3" NTSC channel contains the central line-difference signals quadrature-modulated on the NTSC vision carrier. To accommodate all the line-difference components, the channel space occupied by the sidebands of carrier f_{c1} are correspondingly wider in the "6 + 4" version than in the "6 + 3" version. This reassignment of the line-difference components permits the channel width of the augmentation channel to be reduced from 4 to 3 MHz.

9.5.2 NYIT/Glenn VISTA System

The VISTA (Visual System Transmission Algorithm) system[8,9] devised at the New York Institute of Technology by Dr. W. E. Glenn employs a standard NTSC 6-MHz channel and a 6-MHz augmenta-

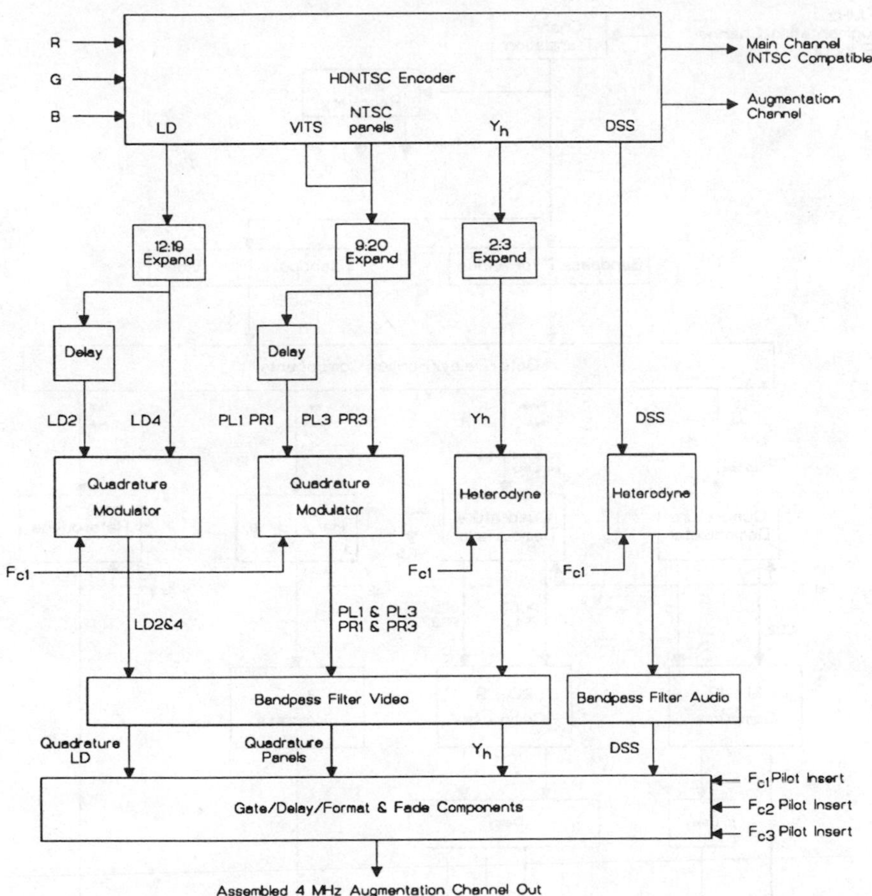

Figure 9.20 Encoder for the HDNTSC "6 + 4" system. (From Reference 1.)

tion channel. A variation employs one of two 3-MHz augmentation channels, in a 6-MHz channel shared by two stations.

Source Processing The source of the VISTA signal is an 1125-line progressively scanned image, at 59.94 fields, at an aspect ratio of 5:3. This is converted to 525/59.94/2:1 for the NTSC transmission. The VISTA HDTV signal is digitized and converted to a lower frame rate.

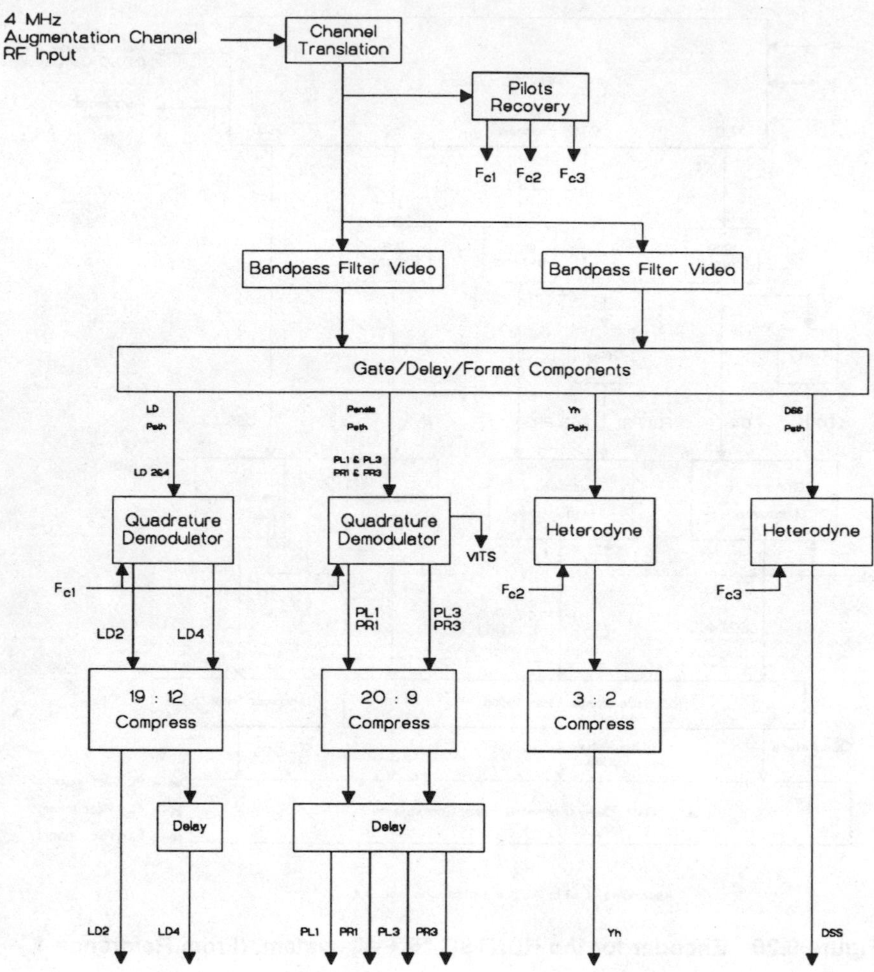

Figure 9.21 Decoder for the HDNTSC "6 + 4" system. (From Reference 1.)

A lower rate of 14.985 frames per second reduces the video baseband from 10.6 MHz to 5.3 MHz. The 5.3-MHz signal modulates the augmentation channel by single-sideband amplitude modulation. This signal distribution is illustrated in Figure 9.24.

Alternative processing reduces the frame rate by four, to 7.4925 per second. At this rate, the video baseband is reduced to 2.65 MHz,

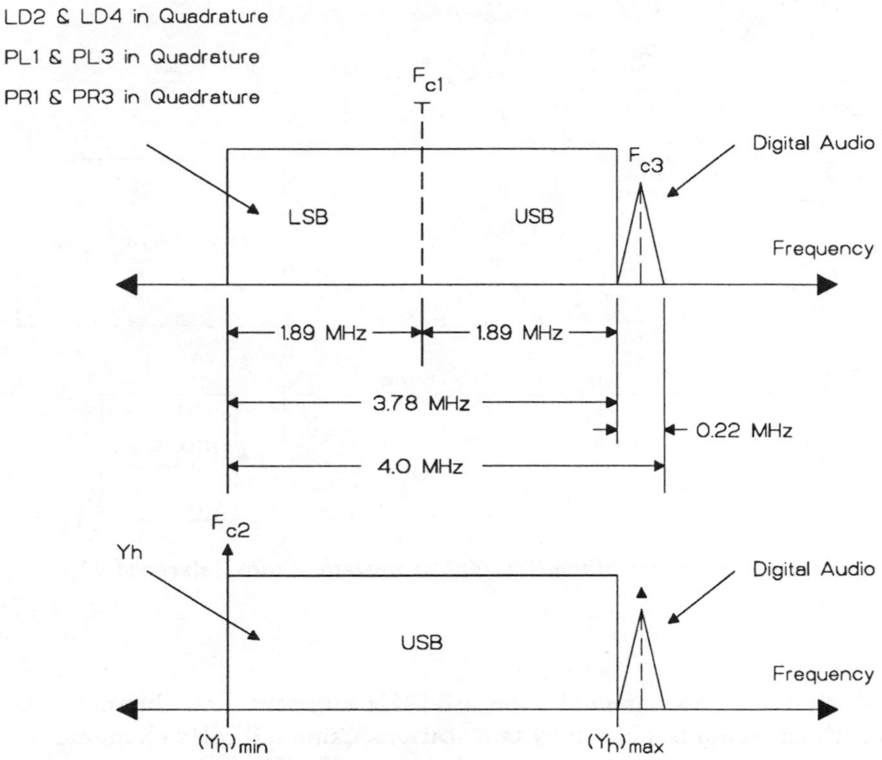

LD2 & LD4 in Quadrature
PL1 & PL3 in Quadrature
PR1 & PR3 in Quadrature
Fc1
Digital Audio
Fc3
LSB
USB
Frequency
1.89 MHz
1.89 MHz
3.78 MHz
0.22 MHz
4.0 MHz
Fc2
Yh
Digital Audio
USB
Frequency
(Yh)min
(Yh)max

Figure 9.22 Signal components included in the 4-MHz augmentation channel. (From Reference 1.)

(Pl1 + Pr1) & (Pl3 + Pr3)
NTSC Chroma
NTSC FM Sound
Digital Sound
LD2 & LD4
Yh
fvis
fsc
fc3
fc2
fc1
NTSC - 6 MHz
Augmentation - 4 MHz

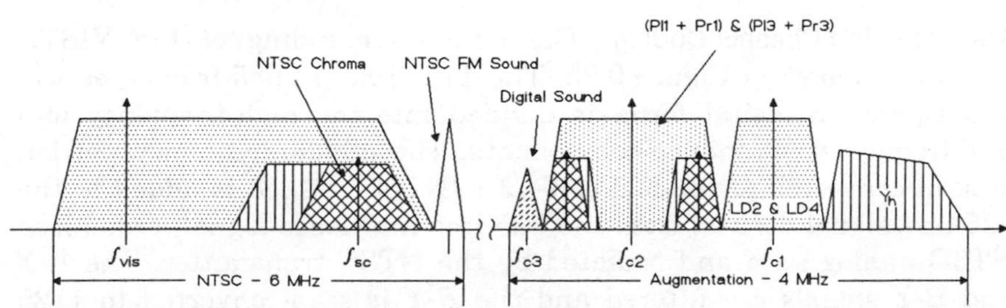

Figure 9.23 Channel contents of the "6 + 4" system, frequency-multiplexed version. (From Reference 1.)

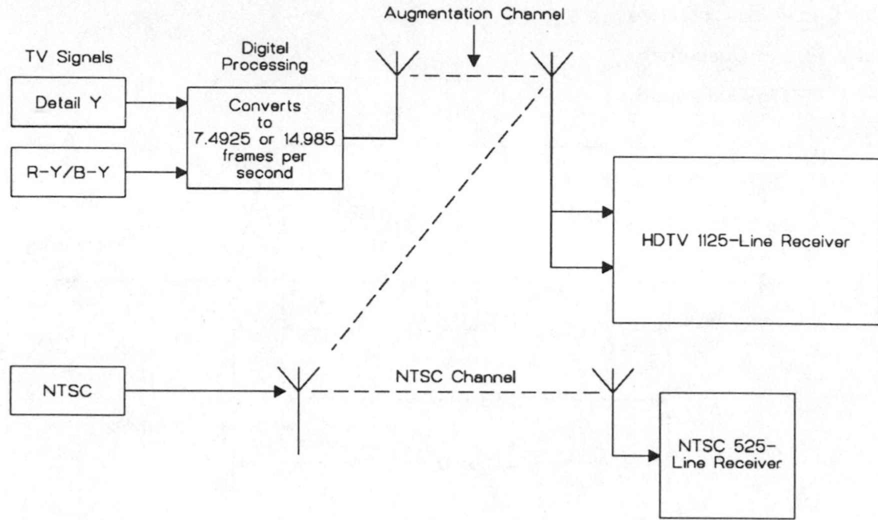

Figure 9.24 Elements of the NYIT/VISTA system. (From Reference 9.)

which can be accommodated on a 3-MHz augmentation channel, two of which would be shared by two stations using a 6-MHz channel.

A special camera has been built for the VISTA system, combining a standard NSTC camera tube producing an RGB output at the 525/59.95/2:1 standard, and a single high-definition tube producing a progressivly scanned luminance signal of 1125 lines, at the lower frame rate of 14.985 per second.

Augmentation Channel Coding The detailed encoding of the VISTA system is shown in Figure 9.25. The 1125-line, 14.985-frame-per-second signal, in digital form, is divided into the high-frequency and mid-frequency luminance components. The middle-frequency portion is scan-converted to the 525/59.94/2:1 (NTSC), which is added to the 525-line signal derived from the 525-line interlaced signal, coded into NTSC analog form and radiated by the NTSC transmitter. The *R-Y* and *B-Y* signals are filtered and the *R-Y* is scan-converted to 1125 lines, progressively scanned for the chrominance component of the augmentation signal.

The decoding at the VISTA HDTV receiver is shown in Figure 9.26. A combed frame-store recovers the low-frequency luminance and chrominance signal, to which are added the high-frequency detail signals from the augmentation channel. The output provides the

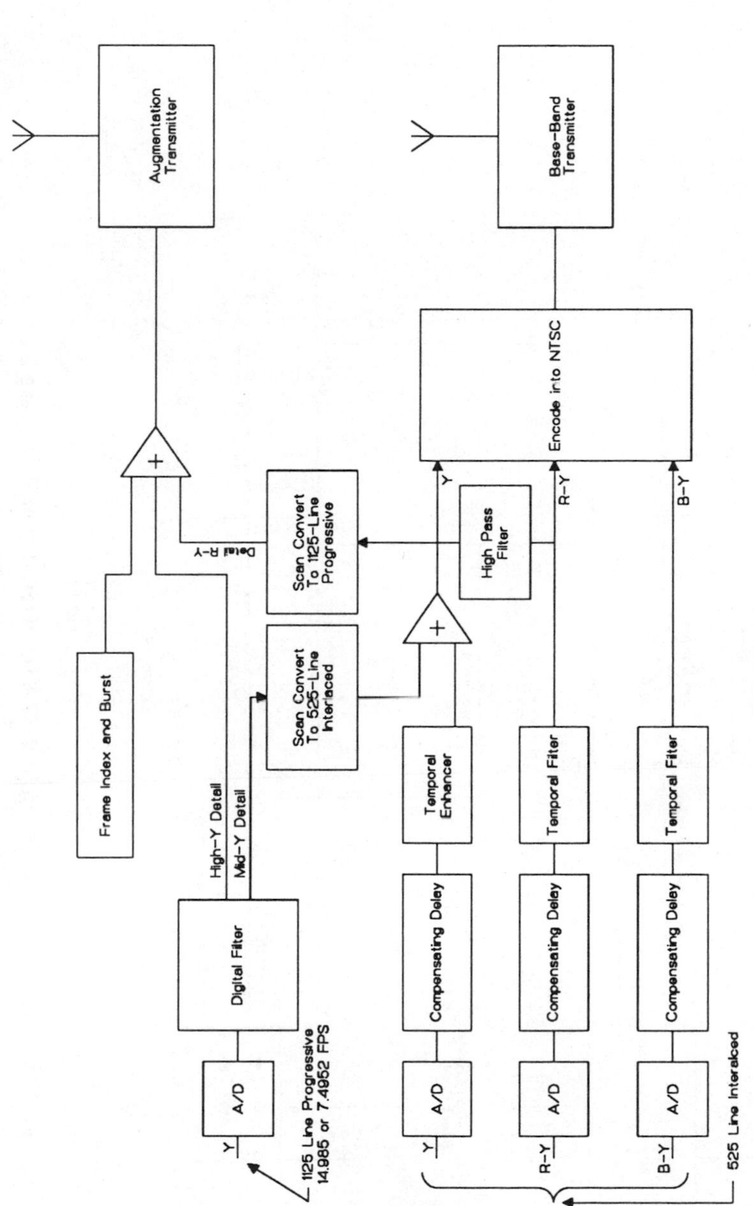

Figure 9.25 Details of processing in VISTA transmitter. (From Reference 8.)

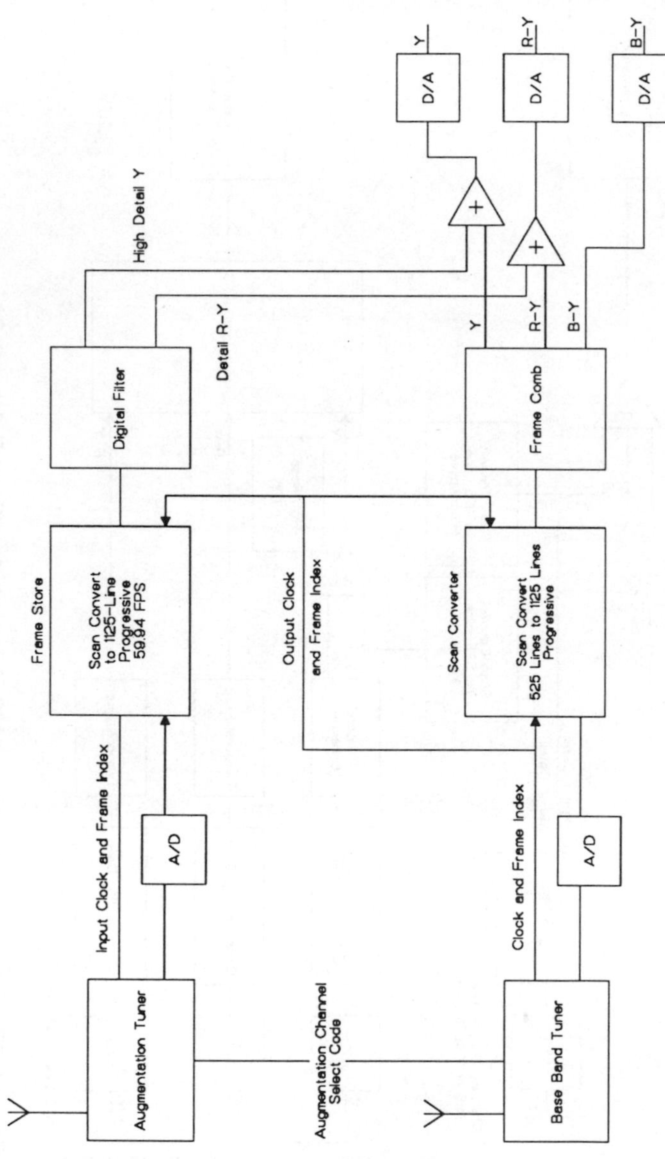

Figure 9.26 Details of processing in VISTA receiver. (From Reference 8.)

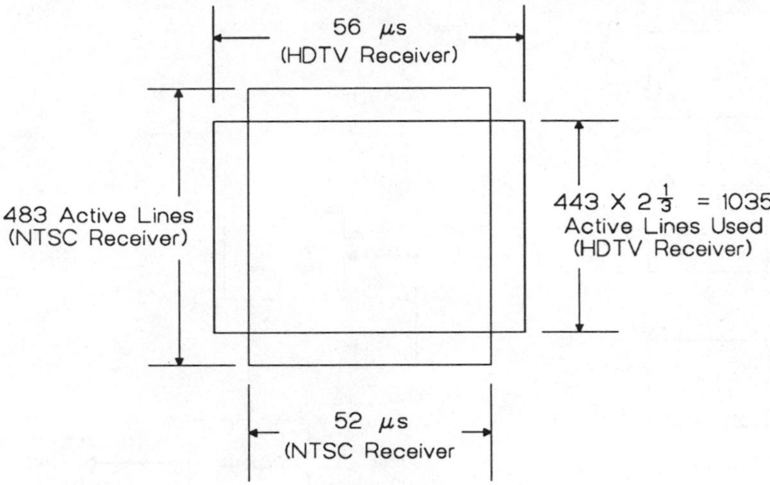

Figure 9.27 Line times and active vertical lines employed in NTSC and VISTA-HDTV receivers. (From Reference 8.)

HDTV luminance and chrominance signals at 1125/59.94/2:1, which are matrixed to form the RGB signals for the HDTV display.

Allocation of Scan Elements to NTSC and HDTV Displays The lines available for vertical resolution and the line times occupied by the NSTC and HDTV displays are shown in Figure 9.27. It is evident that a larger portion of the line time, 56 μs, is used in the HDTV display. The augmentation channel accommodates additional luminance content for 4 μs during the line retrace time, following the color burst.

9.5.3 Osborne System

The Osborne Compression System (OCS) for Advanced Television Systems[10] was submitted to Working Party 1 by Osborne Associates. Three embodiments are disclosed, for a single 6-MHz channel, for a 6-MHz channel augmented by a 3-MHz channel, and for two 6-MHz channels. The latter is the system listed in Table 9.1, described here.

In the OCS system, an "artificial HDTV" signal is subtracted from the HDTV input, and the difference between them is the "error signal" transmitted on the 6-MHz augmentation channel. The NTSC-compatible channel is employed by NTSC receivers, while both 6-MHz channels are decoded for the HDTV receiver. These functions

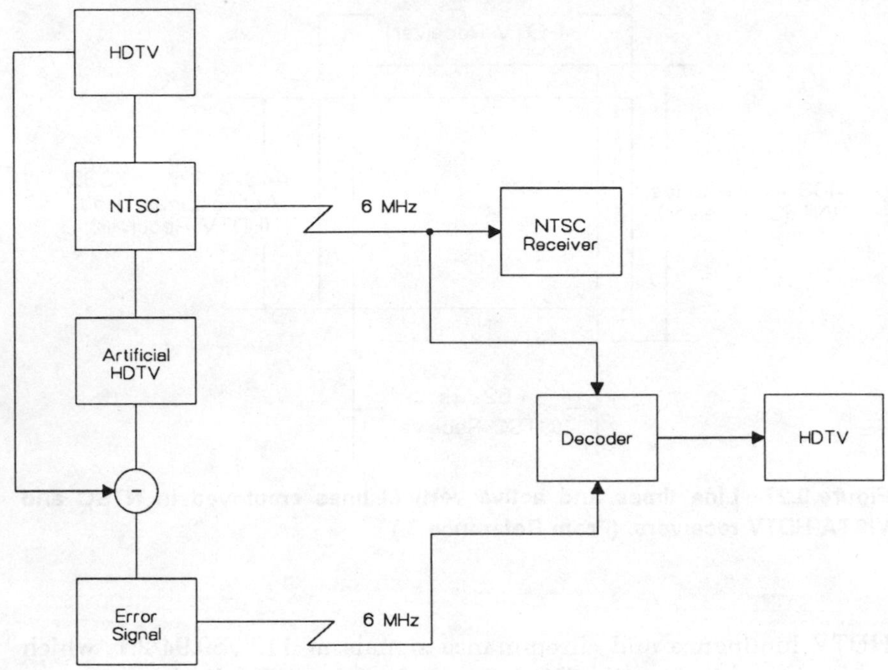

Figure 9.28 Elements of the OSC system. (From Reference 10.)

of the OCS system are shown in Figure 9.28. The HDTV signal is subsampled to form the corresponding NTSC image and this version is transmitted on the NTSC channel.

The NTSC signal serves to form the artificial HDTV signal, as shown in Figure 9.29. Each picture element of the NTSC image serves as a picture element of the artificial HDTV image, but spaced out to form every fourth element. This sparse arrangement of picture elements is averaged and smoothed to provide the missing elements by a prediction algorithm. After subtraction from the initial HDTV signal (e.g., 1125/59.94/2:1/16:9), the error signal has low amplitudes and is digitally encoded for transmission on the augmentation channel.

At the OCS receiver the inverse processing recovers the error signal, and the NTSC signal from the compatible channel is used to form the artificial HDTV image by the same processes used at the transmitter. The sum of the error signal and the artificial image recovers the initial HDTV input of the transmitter. This serves, after

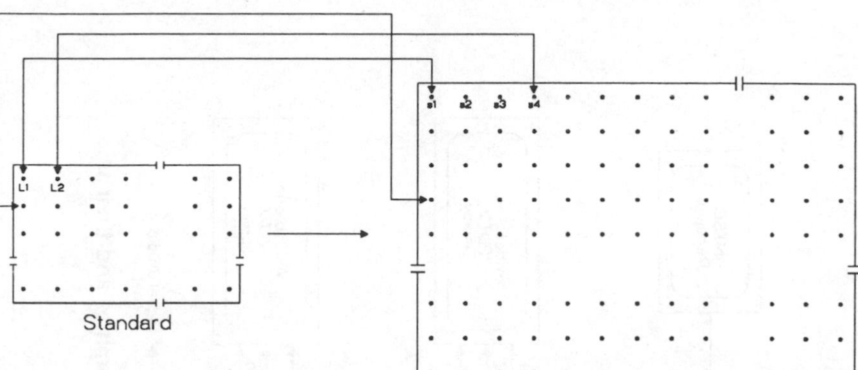

Figure 9.29 Formation of the artificial HDTV image from the NTSC image in the OCS system. (From Reference 10.)

conversion to analog form, to drive the display of the receiver. No detail is presented on the formation of the artificial HDTV images at the transmitter and receiver. Evidently the processing must be precisely identical in transmitter and receiver to ensure that the fine detail of the HDTV image is preserved.

9.5.4 Sarnoff Research Center ACTV-II System

The ACTV-II system is an evolutionary extension of the ACTV-I system[11,12] (Section 8.9). Whereas the ACTV-I receiver (Figure 8.18) has a wide-screen EDTV display of 525-lines progressively scanned, the ACTV-II receiver offers an HDTV display that reproduces the 1050/59.94/2:1 output of the camera. The aspect ratio given in the proposal is 5:3, but it is stated that this is to be changed to 16:9.

The ACTV-II system is intended for terrestrial broadcasting and cable systems. It would be introduced after the ACTV-I system is in place, when large high-definition displays of adequate brightness and contrast are available, and when spectrum space is allocated for the augmentation channel.

Figure 9.30 shows the elements of the ACTV-I and ACTV-II systems. The 6-MHz augmentation channel is formed, in the transmitter, by decoding the ACTV-I signal to form the Y', I', and Q' signals, which are subtracted from the Y, I, and Q signals from the wide-screen 1050/59.94/2:1 source. The luminance difference signals

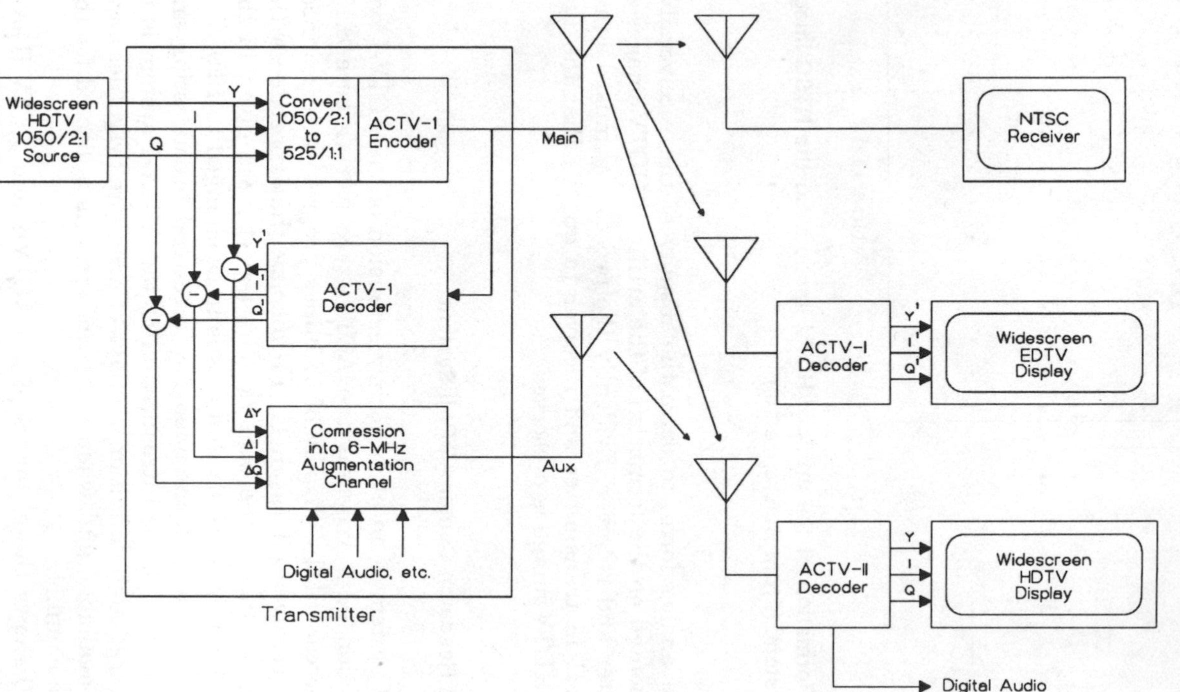

Figure 9.30 Formation of the ACTV-I and ACTV-II signals in the ACTV-II transmitter and their use in NTSC, EDTV, and HDTV receivers. (From Reference 11.)

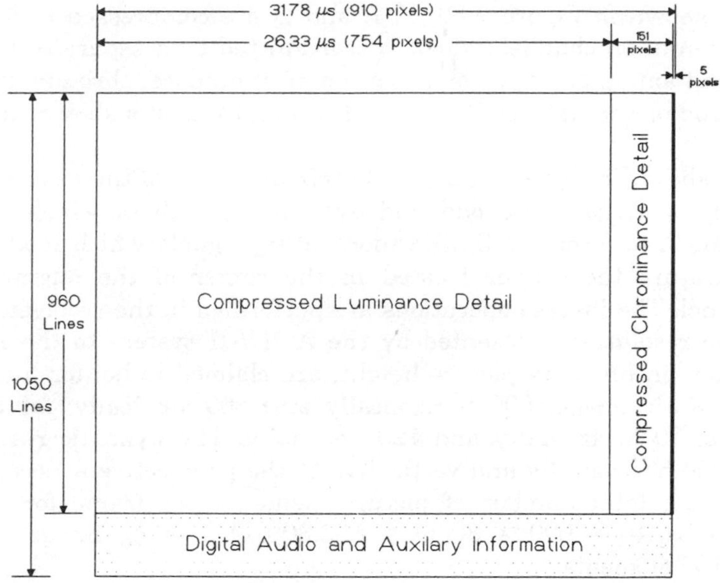

Figure 9.31 Typical frame allocation of luminance, chrominance, and audio/data components of the ACTV-II augmentation signal. (From Reference 11.)

occupy a total bandwidth of 18 MHz. A low-frequency segment of this band, 0 to 6 MHz, is transmitted during moving areas of the scene. The high-frequency segments from 6 to 18 MHz are compressed to form a 0- to 6-MHz signal that is transmitted in both static and moving areas. In the static areas, the low- and high-frequency segments are transmitted on alternate frames.

The chrominance difference signals are filtered to a maximum of 2.4 MHz. After time compression by a factor of five increases the bandwidth to 12 MHz, each chrominance component is divided into a low-frequency segment (0–6 MHz) and a high-frequency segment (6–12 MHz). In static portions of the scene, the low- and high-frequency segments are transmitted in alternate frames. In moving areas, only the low-frequency components are sent. Typically, these chrominance transmissions occupy a time interval of 151 picture elements, as shown in Figure 9.31. The selection between static and dynamic portions of the scene is made in the ACTV-I encoder.

As shown in Figure 9.30, Y, I, and Q are compressed to form the augmentation channel, which is transmitted by a separate transmitter and antenna. A typical allocation of luminance, chrominance, digital audio, and data in the frame lines and times is shown in Figure 9.31.

As shown in Figure 9.32, the 6-MHz baseband of the augmentation signal is divided into odd and even lines, each of which is time-expanded to form two 3-MHz modulating signals which modulate in quadrature the carrier located in the center of the augmentation channel. The inverse operations are performed in the receiver.

The resolutions presented by the ACTV-II system to the receiver display, in lines per picture height, are claimed to be: for luminance of a static image, 600 horizontally and 800 vertically; for chrominance, 80 horizontally and 120 vertically. The dynamic resolutions are 250 horizontally and vertically. At the prospective aspect ratio of 16:9, the total number of picture elements per frame for a static image is 16 × 600 × 800/9 = 853,300. A moving image displays 111,100 elements.

No specifications for the sound and data components are proposed for the ACTV-I and ACTV-II systems. These are to be adopted after field trials of other systems.

9.6 Satellite Direct-Broadcast Systems

9.6.1 NHK MUSE-E System

The MUSE system (identified as "MUSE-E" in Table 9.1) is the basic direct-broadcast satellite system[13] of the 1125/60/2:1 NHK system. Its principle of operation, decoding, and encoding methods are treated in Sections 2.8 and 7.7. As of 1989 it was the only HDTV system offered to the public in Japan.

The MUSE system is not directly compatible with the conventional systems, but can be transferred to NTSC, PAL, or SECAM by a converter, which is used in Japan to convert MUSE to the NTSC system in use there, at the 59.94-MHz field rate.

The method of improving the constant-luminance performance of the MUSE system is illustrated in Figure 9.33. An inverse gamma correction is inserted prior to matrixing. At the output of the matrix unit, the chrominance signals are low-pass filtered and passed through a symmetrical gamma correction, and these signals are combined in the MUSE encoder. The luminance component is then passed through the 2.2-exponent gamma correction, the reverse of

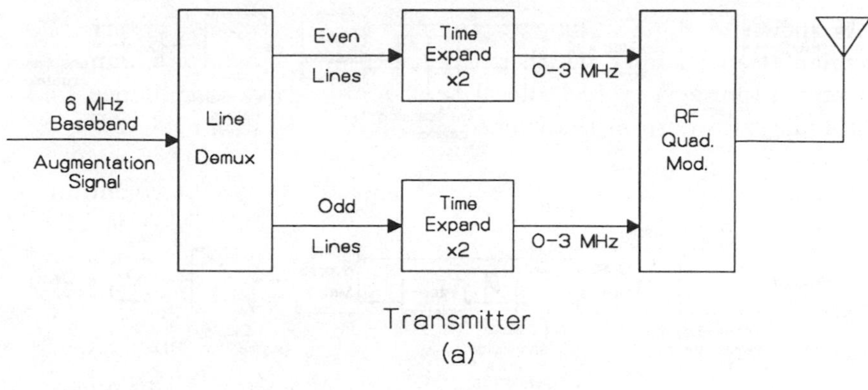

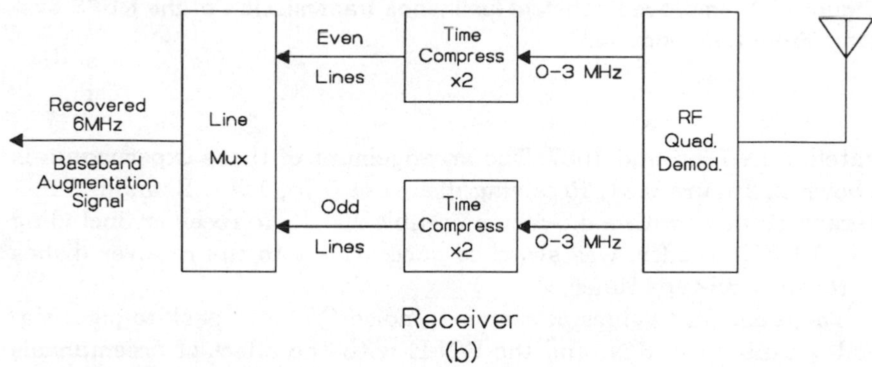

Figure 9.32 Modulation and demodulation of the ACTV-II augmentation signal. (From Reference 11.)

the inverse correction already applied. The digital form of the signal is then converted to analog form and transmitted as frequency modulation of the up-link of the satellite system. At the receiver (Figure 9.33b) the inverse operations are performed to recover the display RGB signals. The transmission of luminance between the matrix unit and the coders is linear in both transmitter and receiver, producing sharper rendition.

Since the principal destination of the MUSE signal is a satellite dish antenna on the premises of the domestic viewer, a matter of importance in the design of the MUSE system was a determination of the diameter of the dish. A diameter of 0.75 meter (about 30 inches) was an objective. To test the performance of this and larger dishes, a number of trials were performed with the BS-2B Japanese

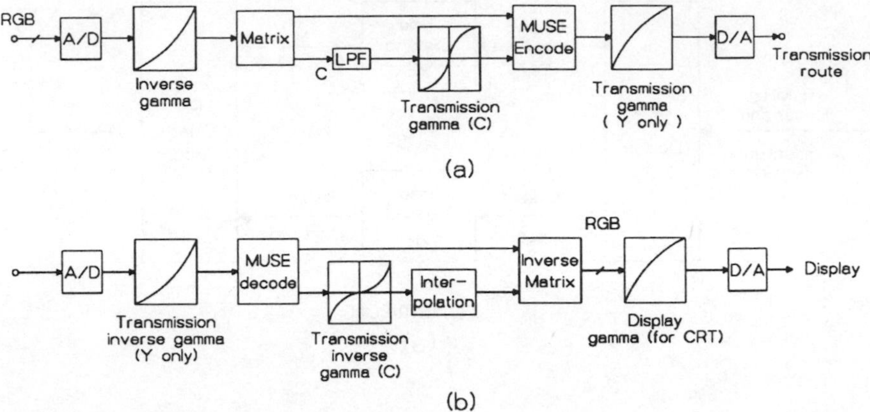

Figure 9.33 Improved constant-luminance transmission of the MUSE system. (From Reference 13.)

satellite in 1986 and 1987. The arrangement of these experiments is shown in Figure 9.34. Receiving dishes of 0.75, 1.0, 1.2, and 1.6 meters were used with a 4.5-meter up-link dish. The receiver, including the MUSE decoder, was switched successively to the receiver dishes through a BS-type tuner.

The measured values of carrier-to-noise C/N and peak-to-peak signal to rms noise S/N, and the S/N-E with the effect of preemphasis and deemphasis of 9.5 dB, were:

Antenna diameter (meters)	*C/N* (dB)	*S/N* (dB)	*S/N-E* (dB)
0.75	17.8	29.96	39.46
1.0	20.3	32.17	41.67
1.2	22.0	33.22	42.72
1.6	23.2	34.61	44.11

The 39.46-dB signal-to-noise ratio is deemed adequate for the MUSE DBS service. Other designers of DBS systems have stated that such systems must provide adequate signal-to-noise ratio when the dish radius is substantially smaller than 0.75 meter.

While designed primarily for the DBS service, the MUSE system can be converted, prior to transmission, for terrestrial broadcast or cable systems.

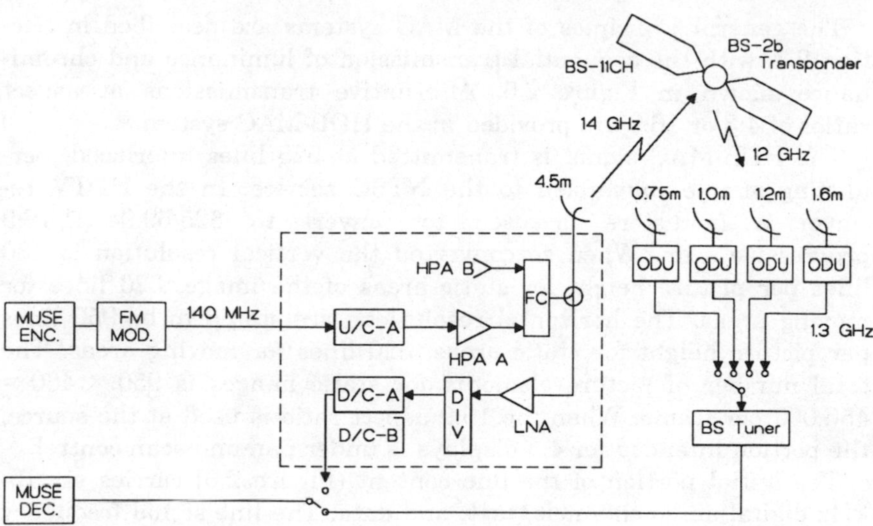

Figure 9.34 Arrangement of tests to determine the carrier-to-noise and signal-to-noise performance of MUSE receiving dishes of 0.75–1.6 meter diameter. (From Reference 13.)

9.6.2 Philips HDS-NA System

The third entry of the Philips HDS-NA system in Table 9.1 is listed under satellite DBS systems. This is the HDMAC system described in Section 9.3 and illustrated in Figures 9.3 to 9.10. In addition to serving as a direct-broadcast medium, HDMAC may also be used as a distribution medium for other satellite systems in the so-called "fixed-satellite service" (FSS).

9.6.3 Scientific Atlanta System

The Scientific Atlanta HDB-MAC ("High-Definition B-MAC") system[14,16] is a noncompatible service intended for direct-broadcast and fixed satellite use. It is based on the B-MAC[16] service now standardized in Australia and widely used for business applications of television. It is closely related to the D2-MAC system[15] used for DBS service in Europe.

The general principles of the MAC systems are described in Section 2.7, with the sequential transmission of luminance and chrominance shown in Figure 2.6. Alternative transmissions at aspect ratios of 4:3 or 16:9 are provided in the HDB-MAC system.

The HDB-MAC signal is transmitted at 525 lines interlaced, permitting simple conversion to the NTSC service. In the HDTV receiver a field-store is used to convert to 525/59.94/1:1/16:9 progressive scan. When so converted the vertical resolution is 480 lines per picture height for static areas of the image, 320 lines for moving areas. The horizontal resolutions are stated to be 950 lines per picture height for static areas, 320 lines for moving areas. The total number of picture elements for static images is 950 × 480 = 456,000 per frame. When the 16:9 aspect ratio is used at the source, the portion intended for 4:3 displays is under pan-and-scan control.

The initial portion of the line content (Figure 2.6) carries six 15-kHz digital-audio channels, text, and data. The line signal frequency modulates the up-link signal on the 24-MHz or 27-MHz transponder channel. This signal permits display at 525 lines sequentially scanned at 16:9 aspect ratio, with a luminance baseband of 18 MHz. Although not designed for cable service, it is pointed out that the up-link signal may be accommodated on two contiguous cable channels.

The source baseband is reduced to 10.7 MHz by the spectrum folding shown in Figure 9.35. The source baseband is recovered at the HDTV receiver by a field-store line doubler that provides 18 MHz of luminance bandwidth and 5 MHz for chrominance. The HDB-MAC signal space is as shown in Figure 9.36.

9.7 European DBS HDTV System

9.7.1 HDMAC System

The nations of the European Economic Community have jointly established a research and development organization known as EUREKA. One of the projects of EUREKA, EU-95, is to establish standards for a European HDTV system[15–23] compatible with the conventional D-MAC and D2-MAC[15] packet DBS systems in use there. The organizations principally involved in the project are Philips

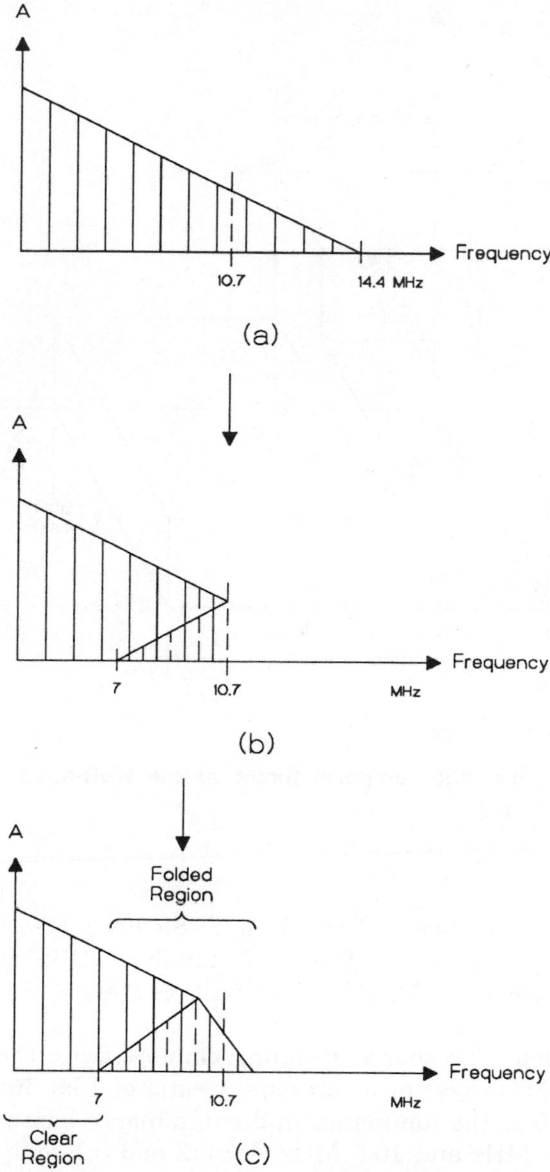

Figure 9.35 Spectrum folding in the Scientific Atlanta HDB-MAC system: (a) initial video spectrum; (b) folded spectrum; (c) transmitted spectrum. (From Reference 14.)

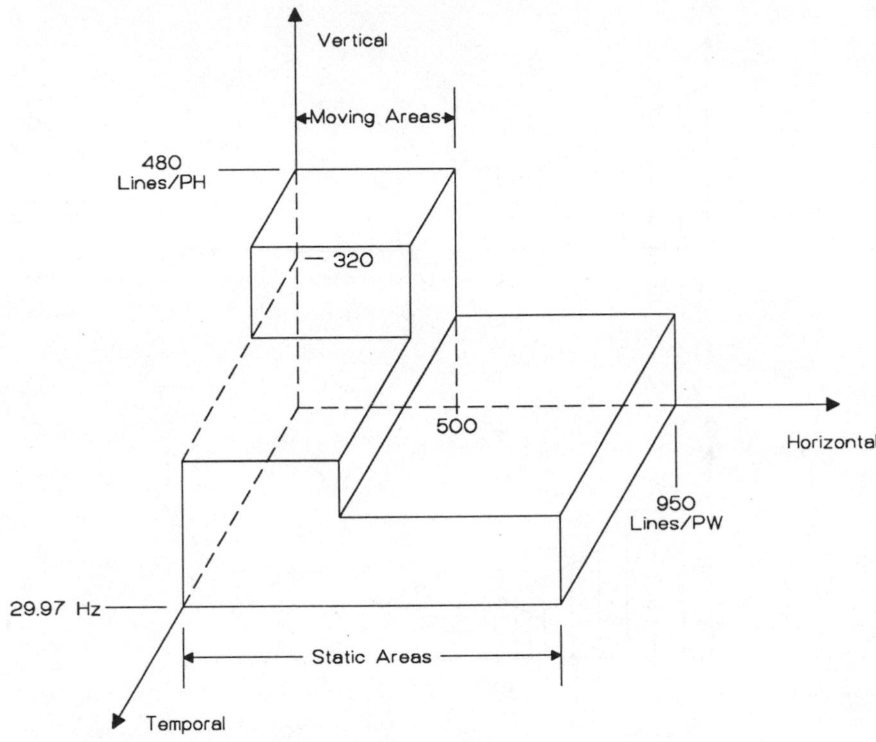

Figure 9.36 Spatial and temporal limits of the HDB-MAC signal space. (From Reference 14.)

(Netherlands), Thomson (France), and Siemens (West Germany). Since the project is one of the MAC family of DBS systems, it is generally known as HDMAC (high-definition MAC).

Source Scanning The source scanning employs twice the line rate of the 625-line MAC system at an aspect ratio of 16:9, interlaced, i.e., 1250/50/2:1/16:9. The luminance and chrominance basebands are, respectively, 21 MHz and 10.5 MHz. By 3:2 and 3:1 compressions, respectively, a common sampling frequency of 20.5 MHz is achieved and this imposes a maximum modulation bandwidth of 10.125 MHz. This is the up-link frequency modulation of the satellite signal, using the 27-MHz bandwidth available in Europe. Additional information (sound/data/DATV) is carried by multiplex in the vertical and horizontal blanking intervals.

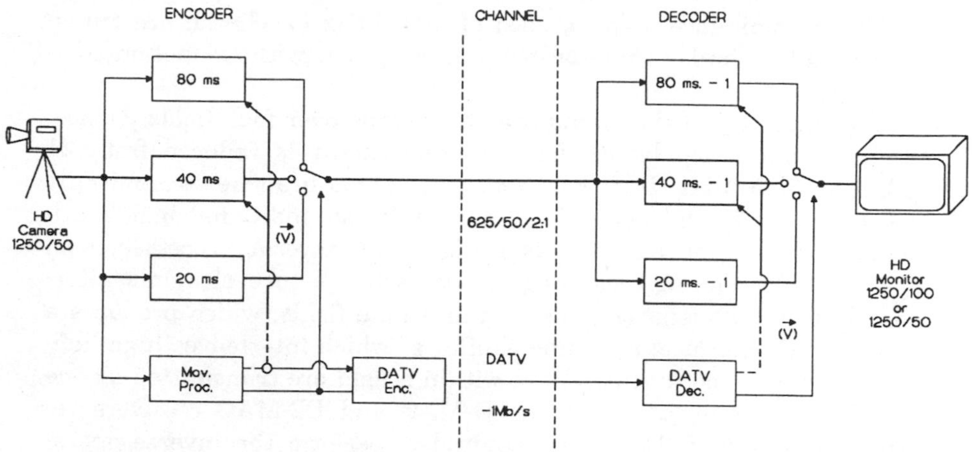

Figure 9.37 Elements and transmission paths of the Eureka-95 HDMAC system. (From Reference 17.)

Digital Assistance Signal An essential element of the HDMAC system is the DATV (digitally assisted television) signal[22] transmitted during the vertical field interval of 1/50 second = 2000 μs. This time permits a maximum bit rate of 1 Mb/s, but after allowance for error correction, a maximum bit rate of 960 kb/s is available. A principal function of the DATV signal bit stream is to keep the branch-switching, described below, in precise synchronism at the transmitter and receiver.

Bandwidth Reduction To transmit the 21-MHz luminance baseband compatibly to the D-MAC and D2-MAC receivers, bandwidth reduction by a factor of approximately four is required. This is accomplished by the coding and decoding processes shown in Figure 9.37. The encoder has "branches" for three degrees of motion: an 80-ms (4 fields) branch for stationary and slowly moving areas of the scene; a 40-ms (2 fields) branch for moving areas; and a 20-ms (1 field) branch for rapid motion and sudden scene changes. These branches are switched to the transmission channel by the motion processor. The switching signals are also transmitted to the receiver via the DATV channel, where the branch in use at a particular time, after decoding, is connected to the receiver for processing and display at the 1250/50/2:1/16:9 rates of the camera or telecine equipment at the transmitter.

The chrominance signals, each of 10.5 MHz baseband, are transmitted after similar three-branch encoding, but without motion compensation.

The encoding in the 80-ms branch extends over four fields. Hence, the luminance bandwidth for stationary areas is reduced from 21 MHz to 5.25 MHz. But the 40-ms and 20-ms branches extend only over two fields and one field, respectively, so additional bandwidth reduction is required. This is achieved by several processes, e.g., "quincunx" scanning (scanning of successive picture elements alternately from two adjacent lines) on alternate fields, which produces a "synthetic" interlace; and "line shuffling" which interleaves high-definition samples so that two lines within a field are transmitted as one MAC/packet line, to which the D-MAC and D2-MAC receivers respond compatibly. The clues required to perform the inverse operations at the receiver are transmitted over the DATV channel.

Times and Lines Occupied by the HDMAC System Figure 9.38 shows the elements of the HDMAC signal structure[18] in terms of the times occupied during the 64-μs line-scanning time and the image line structure. At the left are shown the sound and data signals multiplexed during the horizontal blanking interval. The DATV data signals occupy the vertical (field) blanking times. Two sets of the HDMAC vision signal are shown, transmitted on the successive fields. Times for teletext data are also provided during the vertical blanking intervals. The luminance and chrominance signals are transmitted in time sequence, compatibly with the D-MAC and D2-MAC packet systems (Figure 2.6).

Timetable The HDMAC system was under continued development at the time of publication of this book. It was anticipated that agreement on the HDMAC standards might be obtained among the EEC nations by 1992, and that the HDMAC service might be available by 1995.

9.7.2 Presentation of ATV and HDTV Equipment at NAB Convention, May 1989

In the 1988 National Association of Broadcasters Convention, only the NHK MUSE system was demonstrated in hardware form, others being shown only in computer simulations. By the 1989 convention[24] many other systems had been reduced to hardware form and were demonstrated. Included were: the MUSE variations MUSE-6, MUSE-9,

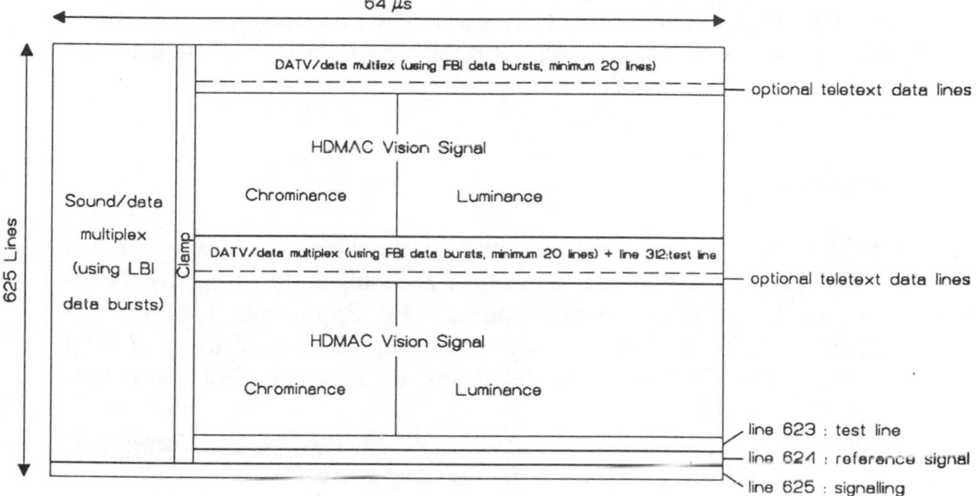

Figure 9.38 Luminance, chrominance, DATV, and sound/data components of the HDMAC image lines and scan times. (From Reference 18.)

and Narrow-MUSE; the Philips HDS-NA system; the Sarnoff Labs ACTV systems; the Zenith simulcast system; and the Scientific Atlanta HDB-MAC system. The 1125/60/2:1 system (SMPTE 240M Standard) was demonstrated with a complete line of equipment for program production, including switching and special effects.

Also shown were systems intended for use in motion picture theaters. An example was the Mitsubishi projection system, with 1000-line definiton on a 200-inch wide screen. Another was Eidophor's projection system on a 12-by-20-foot screen, with a contrast ratio greater than 100:1.

At this meeting the president of the Philips Laboratories stated his opinion that, of the 19 systems proposed to Working Party 1 (shown in Tables 8.1 and 9.1), only six were viable then and would be subject to test by the Advanced Television Test Center in 1990.

Evidence of the systems to be tested by the ATTC was given when the Chairman of the FCC Advanced Television Advisory Committee called a meeting of proponents in September 1989 to make specific commitments for the starting date of testing and for the posting of performance bonds. The organizations invited to attend were: Faroudja Laboratories (Section 8.2.3), Massachusetts Institute of Technology (Sections 8.2.5 and 8.3.1), Japan Broadcasting Corporation (NHK) (Sections 8.2.6, 8.3.2, 9.4.2, and 9.6.1), New York Institute of Technology (Section 9.5.2), Philips Laboratories (Sections 9.2,

9.3, 9.4.1, 9.5.1, and 9.6.2), Production Services, Inc. (Section 8.2.7), Sarnoff Research Center (Sections 8.2.8 and 9.5.4), and Zenith Electronics Corporation (Section 8.3.3).

References

1. HDS-NA: High Definition System for North America—An Advanced TV Emission System Proposal. Submitted by North American Philips Corporation to WP-l, September 1, 1988.
2. Toth, A. G. and M. Tinsberg. Hierarchical Evolution of High Definition Television. *IEEE Trans. on Broadcasting* 33(4): 124–129 (December 1987).
3. Toth, A. G., M. Tinsberg, and C. W. Rhodes. NTSC Compatible HDTV Emission System. *IEEE Trans. on Consumer Electronics*. 34(1): 40–47 (February 1988).
4. Toth, A. G. High Definition Television—A North American Perspective, Conference Record, International Broadcast Conference—Brighton, 1988. Institution of Electrical Engineers, London.
5. Cavallerano, A. P. Systems and Technological Details of Terrestrial/Cable NTSC Compatible HDTV, Conference Record of International Conference on Consumer Electronics—1989. Institute of Electrical and Electronics Engineers, New York, June 1989.
6. Basile, C. An HDTV MAC Format for FM Environments. International Conference on Consumer Electronics, Institute of Electrical and Electronics Engineers, New York, June 1989.
7. Development of the MUSE Family Systems. Report submitted to WP-1 by the Japan Broadcasting System (NHK), November 15, 1988. (The MUSE-9 System is described in pages 35–45 of the submitted document.)
8. VISTA System—System Description Submitted to Working Party 1 by New York Institute of Technology, August 28, 1988. Includes brochure presenting VISTA (VIsual System Transmission Algorithm) system diagrams.
9. Glenn, W. E. and K. G. Glenn. HDTV Compatible Transmission System. *SMPTE Journal* 96(3): 242–252 (March 1987).
10. The Osborne Compression System for Advanced Television Systems. Proposal submitted by Osborne Associates, Inc., to Working Party 1, August 26, 1988. Attached to the proposal is a copy of United States Patent 4,665,436, dated May 12, 1987, and issued to inventors James A. Osborne and Cindy Seiffert.

11. System Description, Advanced Compatible Television. Submission to Working Party 1 by David Sarnoff Research Center, September 1, 1988. (The ACTV-II system is described on pages 13–18 of the submitted document.)
12. Isnardi, M. A., C. B. Dietrich, and T. R. Smith. Advanced Compatible Television: A Progress Report. *J. SMPTE* 98: 484–495 (July 1989).
13. Ninomiya, Y., Y. Ohutsuka, Y. Izumi, S. Gohshi, and Y. Iwadate. An HDTV Broadcasting System Utilizing a Bandwidth Compression Technique—MUSE. *IEEE Trans. on Broadcasting* 33(4): 130–160 (December 1987).
14. HDB-MAC System. Proposal submitted to Working Party 1 by Scientific Atlanta, August 30, 1988.
15. Sabatier, J., D. Pommier, and M. Mathiue. The D2-MAC-Packet System for All Transmission Channels. *J. SMPTE* 94: 1173–1179 (November 1984).
16. Raven, J. G. High Definition MAC: The Compatible Route to HDTV. *IEEE Trans. on Consumer Electronics* 34: 61–63 (February 1988).
17. HDMAC Bandwidth Reduction Coding Principles, Draft Report AZ-11, International Radio Consultative Committee (CCIR), Geneva, January 1989.
18. Conclusions of the Extraordinary Meeting of Study Group 11 on High Definition Television, Document 11/410-E. International Radio Consultative Committee (CCIR), Geneva, June 1989.
19. Lucas, K. B-MAC: A Transmission Standard for Pay DBS. *SMPTE Journal* 94: 1166–1172 (November 1984).
20. Arragon, J. P., J. Chatel, J. Raven, and R. Story. Instrumentation for a Compatible HDMAC Coding System Using DATV. Conference Record, International Broadcasting Conference, Brighton—1988, Institution of Electrical Engineers, London.
21. Vreeswijk, F. W. P., F. Fonsalas, T. I. P. Trew, C. Carey-Smith, and M. Haghiri. HDMAC Coding for High Definition Television Signals. International Radio Consultative Committee (CCIR), Geneva, January 1989.
22. Story, R. Motion Compensated DATV Bandwidth Compression for HDTV. International Radio Consultative Committee (CCIR), Geneva, January 1989.
23. Story, R. HDTV Motion-Adaptive Bandwidth Reduction Using DATV. BBC Research Department Report, RD 1986/5.
24. Rapid Maturation of HDTV Seen at NAB. *TV Digest* 29(19): 4–6 (May 8, 1989).

Chapter

10

Picture-Signal Generation and Processing

10.1 HDTV System Requirements*

As noted in Section 1.3, the greater number of picture elements (pixels) in a high-definition television display, compared to the current 525- and 625-line interlaced systems, dictates the use of a wider video bandwidth, in order to transmit the increase in horizontal resolution, and a greater number of scanning lines for a comparable increase in vertical detail. Table 10.1 tabulates the number of pixels for various television systems.

If progressive rather than 2:1 interlaced scanning is used in camera-signal generation in order to avoid the introduction of artifacts with image motion, the bandwidth requirement is doubled. Furthermore, the number of pixels in each scanning line is increased by the use of a wider aspect ratio than the present 3:4 conventional television screen presentation.[1]

These more stringent video-system parameters necessitate significant improvements over the components used in 525- and 625-line

*Portions of Sections 10.1 and 10.2 have been adapted from Reference 4 (courtesy Sony Advanced Systems).

Table 10.1 Data on Picture Elements in Various Television Systems

	CCIR Standard			
Quantity	M U.S. NTSC	B/G CCIR PAL	L France SECAM	1125/60 U. S. SMPTE 240M
Information elements per frame	280,000	400,000	480,000	2,000,000
Picture elements per raster	215,000	300,000	360,000	1,600,000
Picture elements per line	446	520	620	1,552
Picture-element length[a]	30	26	21	11
Picture-element width[a]	21	17	17	10

[a]Based on raster height = 10,000 units.
Source: Benson, K. B. (Ed.). *Television Engineering Handbook.* McGraw-Hill, NewYork, 1986, p. 4.8, modified for SMPTE 240M.

systems for signal sources, signal processing, transmission and reception, and picture displays. The major requirements are:

1. Image transducers with photosensitive elements having a higher resolving power and a significantly higher signal output to maintain an adequate signal-to-noise ratio with the higher wideband preamplifier noise level.
2. More precise scanning registration among camera color channels, compatible with the increase in horizontal and vertical resolution.
3. More precise positioning of camera and receiver vertical scanning to preserve the higher vertical resolution resulting from the increase in scanning lines.
4. Video-signal processing and system components with greater bandwidth, and with no degradation of analog-signal amplitude and phase-frequency characteristics.
5. Picture-viewing displays with a smaller scanning-beam spot size for reproduction of the greater number of lines and horizontal picture elements.
6. Wide-screen displays to accommodate the wider aspect ratio.

Figure 10.1 Sony HDC-300 second generation high-definition color camera. (Courtesy Sony Advanced Systems, Reference 4.)

Although these requirements can be met with the basic system configurations found in high-quality cameras and video-processing equipment designed for 525- and 625-line signal standards, new equipment designs of the camera and major components are required. Fully equipped Sony and BTS cameras using photoconductive tubes are shown in Figures 10.1 and 10.2, respectively. This chapter discusses the characteristics requiring more rigorous specifications and how these are met in hardware proposed for HDTV systems.

10.1.1 Amplitude Transfer Characteristic

NTSC Specifications The FCC Rules and Regulations,[2] April 1, 1972, for 525-line NTSC Color Transmissions specified: "gamma-corrected red, green, and blue voltages suitable for a color picture tube . . . having a transfer gradient (gamma exponent) of 2.2 associated with each primary color." No tolerance was set on the value of gamma. In practice, picture tube gammas are considerably higher, as much as

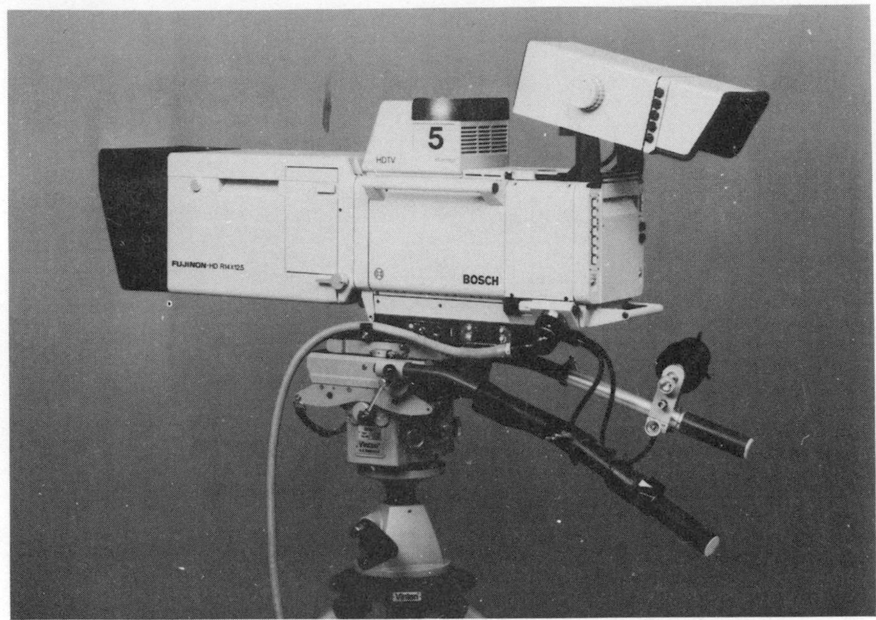

Figure 10.2 BTS KCH-1000 multistandard high-definition color camera. (Courtesy BTS Bosch Television Systems Inc.)

3.0. The gamma correction in cameras has taken note of this fact and, in addition, generally has deviated from a uniform power law to correct for ambient light present in most viewing environments and to provide greater detail in low-light areas of a scene.

SMPTE 240M HDTV Production System Standard In recognition of these important considerations, the SMPTE 1125/60 HDTV Production System Standard[3] provides in mathematical terms (see Figure 10.3) a more detailed specification of the transfer characteristic for a reference camera normalized to reference white, as follows:

$$V = 1.1115 \times L^{(0.45)} - 0.1115$$

$$V = 4.0 \times LV = 4.0 \times L$$

where

L = light input to camera or light output from reproducer

V = video-signal output from camera or video-signal input to reproducer

SMPTE 240M-1988

SMPTE Standard for television—signal parameters—1125/60 high-definition production system

Page 1 of 5 pages

1. Scope

This standard defines the basic characteristics of the video signals associated with origination equipment operating in the 1125/60 high-definition television production system. As this standard deals with basic system characteristics, all parameters are untoleranced.

2. Scanning Parameters

The video signals represent a scanned raster with the following characteristics:

Total scan lines per frame	1125
Active lines per frame	1035
Scanning format	Interlaced 2:1
Aspect ratio	16:9
Field repetition rate	60.00 Hz
Line repetition rate (derived)	33750 Hz

3. System Colorimetry

The system is intended to create a metameric reproduction (visual color match) of the original scene under conditions of equal color temperature and luminance between the original scene and its reproduction. To this end, the combination of a camera's optical spectral analysis and linear signal matrixing shall match the CIE color-matching functions (1931) of the reference primaries. Further, the combination of a reproducer's linear matrixing and reproducing primaries shall be equivalent to the reference primaries. (See Appendix A1.)

Colorimetric analysis and signal amplitude transfer function are defined in the following sections:

3.1 Chromaticity of Reference Primaries:

G: $x = 0.310$ $y = 0.595$
B: $x = 0.155$ $y = 0.070$
R: $x = 0.630$ $y = 0.340$

where

x and y are CIE 1931 chromaticity coordinates.

3.2 Reference White. The system reference white is an illuminant which causes equal primary signals to be produced by the reference camera, and which is produced by the reference reproducer when driven by equal primary signals. For this system, the reference white is specified in terms of its 1931 CIE chromaticity coordinates, which have been chosen to match those of CIE illuminant D_{65}:

$x = 0.3127$ $y = 0.3291$

3.3 Opto-Electronic Transfer Characteristic of Reference Camera:

$$V_c = 1.1115 \times L_c^{(0.45)} - 0.1115 \quad \text{for } L_c \geq 0.0228$$
$$V_c = 4.0 \times L_c \quad \text{for } L_c < 0.0228$$

where

V_c is the video signal output of the reference camera normalized to the system reference white, and L_c is the light input to the reference camera normalized to the system reference white.

Approved March 14, 1988

Figure 10.3 SMPTE 240M Standard for Television-Signal Parameters 1125/60 High-definition Production System. (Courtesy Society of Motion Picture and Television Engineers, from Reference 3.) ***(Continued)***

3.4 Electro-Optical Transfer Characteristic of Reference Producer:

$L_r = [(V_r + 0.1115)/1.1115]^{(1/0.45)}$ for $V_r \geqq 0.0913$

$L_r = V_r/4.0$ for $V_r < 0.0913$

where

V_r is the video signal level driving the reference reproducer normalized to the system reference white, and L_r is the light output from the reference reproducer normalized to the system reference white.

4. Reference Clock

Signal durations and timings are specified both in microseconds and in reference clock periods:

Reference clock periods in total line: 2200
Reference clock frequency (derived): 74.25 MHz

5. Video Signal Definitions

The image is represented by three parallel, time-coincident video signals. Each incorporates a synchronizing waveform. The signals shall be either of the following sets:

Color Set	Color Difference Set
E_G' — green	E_Y' — luminance
E_B' — blue	E_{PB}' — blue color difference
E_R' — red	E_{PR}' — red color difference

where

$[E_G'\ E_B'\ E_R']$ are the signals appropriate to directly drive the primaries of the reference reproducer (being non-linearly related to light levels as specified in Secs. 3.3 and 3.4), and $[E_Y'\ E_{PB}'\ E_{PR}']$ can be derived from $[E_G'\ E_B'\ E_R']$ through a linear matrix.

Specifically,

$E_Y' = (0.701 \times E_G') + (0.087 \times E_B') + (0.212 \times E_R')$

E_{PB}' is amplitude-scaled $(E_B'\text{-}E_Y')$, according to

$$E_{PB}' = \frac{(E_B'\text{-}E_Y')}{1.826}$$

and E_{PR}' is amplitude-scaled (R-Y), according to

$$E_{PR}' = \frac{(E_R'\text{-}E_Y')}{1.576}$$

where the scaling factors are derived from the signal levels given in Sec. 6.3, and the transformation equations which follow.

The derived transformation between the two sets is

$$\begin{bmatrix} E_G \\ E_B \\ E_R \end{bmatrix} = \begin{bmatrix} 1.000 & -0.227 & -0.477 \\ 1.000 & 1.826 & 0.000 \\ 1.000 & 0.000 & 1.576 \end{bmatrix} \begin{bmatrix} E_Y \\ E_{PB} \\ E_{PR} \end{bmatrix}$$

$$\begin{bmatrix} E_Y \\ E_{PB} \\ E_{PR} \end{bmatrix} = \begin{bmatrix} 0.701 & 0.087 & 0.212 \\ -0.384 & 0.500 & -0.116 \\ -0.445 & -0.055 & 0.500 \end{bmatrix} \begin{bmatrix} E_G \\ E_B \\ E_R \end{bmatrix}$$

6. Video and Synchronizing Signal Waveforms

The combined video and synchronizing signal shall be as shown in Fig. 1. For illustrative purposes, a video signal of the form E_Y', E_G', E_B', or E_R' is shown. The details of the synchronizing signal are identical for the E_{PB}' and E_{PR}' color-difference signals.

6.1 Timing

6.1.1 The timing of events within a horizontal line of video is illustrated in Fig. 1(a) and summarized in Table 1. All event times are specified, in terms of the reference clock period, at the midpoint of the indicated transition.

Table 1
Timing of Events of a Video Line

	Reference Clock Periods
Rising edge of sync (timing reference)	0
Trailing edge of sync	44
Start of active video	192
End of active video	2112
Leading edge of sync	2156

6.1.2 The durations of the various portions of the video and sync waveforms are illustrated in Fig. 1 (b), (c), and (d), and summarized in Table 2.

Figure 10.3 SMPTE 240M Standard for Television-Signal Parameters 1125/60 High-definition Production System. ***(Continued)***

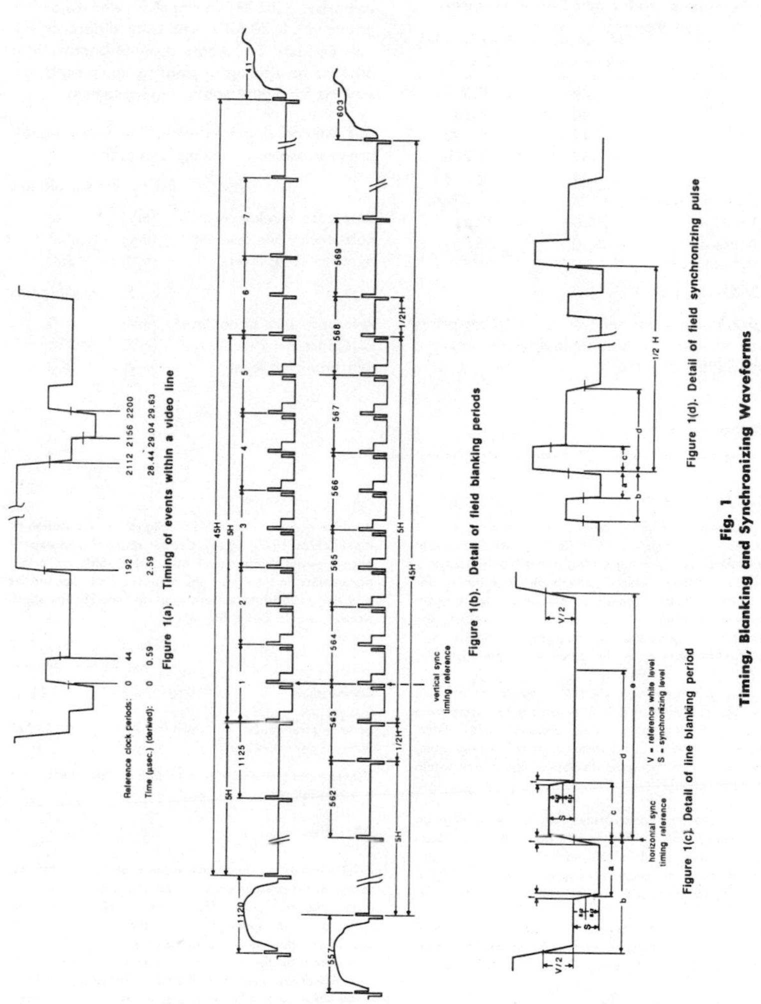

SMPTE 240M-1988

Figure 10.3 SMPTE 240M Standard for Television-Signal Parameters 1125/60 High-definition Production System. ***(Continued)***

Table 2
Duration of Video and Sync Waveforms

	Reference Clock Periods	Time (Derived) (μsec)
a	44	0.593
b	88	1.185
c	44	0.593
d	132	1.778
e	192	2.586
f (Sync rise time)	4	0.054
Total line	2200	29.63
Active line	1920	25.86

6.2 Bandwidth

6.2.1 The color set [E_G' E_B' E_R'] comprises three equal-bandwidth signals whose nominal bandwidth is 30 MHz.

6.2.2 The color difference set [E_Y' E_{PB}' E_{PR}'] comprises a luminance signal E_Y' whose nominal bandwidth is 30 MHz, and color difference signals E_{PB}' and E_{PR}' whose nominal bandwidth is 30 MHz for analog originating equipment, and 15 MHz for digital originating equipment.

6.3 Analog Representation. The video signals are represented in analog form as follows:

		E_Y', E_G', E_B', E_R' Signals
Reference black level	(mV)	0
Reference white level	(mV)	700
Synchronizing level	(mV)	±300

		E_{PB}', E_{PR}' Signals
Reference zero signal level	(mV)	0
Reference peak levels	(mV)	±350
Synchronizing level	(mV)	±300

Appendix

(This Appendix is not part of the SMPTE Standard, but is included for information only.)

A1. System Colorimetry

The parameter values in Sec. 3 are based on current practice and technical constraints. It is recognized that the availability of a wider color gamut is highly desirable in an originating system. Furthermore, it is useful, for purposes of picture processing, to have available video signals proportional to light levels. In particular, the encoding of linear signals, commonly identified as a "constant-luminance" system, is believed to be desirable.

In order to achieve a practical implementation of these desirable characteristics, it is necessary to incorporate non-linear processing (according to the equations of Sec. 3) at various points in the system, within the originating production plant, along the distribution chain, and within the home receiver.

It has not yet been demonstrated that this processing can be achieved with an appropriate balance of precision and cost. Further study of this matter is required in order to demonstrate the feasibility of incorporating such signal processing into commercial equipment where appropriate.

With respect to color gamut, it is felt that the system should embrace a gamut at least as large as that represented by the following primaries:

G: $x = 0.210$ $y = 0.710$
B: $x = 0.150$ $y = 0.060$
R: $x = 0.670$ $y = 0.330$

A2. Digital Representation

It will be necessary to define the digital representation of the 1125/60 HDTV signal. Current practice and experience have not clarified the most suitable value for certain parameters, in particular, the number of bits per sample and the coding law. An appropriate form for the specifications would be the following:

		E_Y', E_G', E_B', E_R' Signals	E_{PB}', E_{PR}' Signals
Quantization	(bits)	[ ]	[ ]
Coding law		[ ]	[ ]
Sampling frequency	(MHz)	74.25	37.125
Samples per active line		1920	960

The sampling structure is orthogonal; line, field, and frame repetitive.

A3. Tolerances

A3.1 Parameter Aim Points. Tolerances are not affixed to any of the parameters specified in this document since it specifies aim points for system design rather than detailed system specifications. Furthermore, it was not possible to determine comprehensive tolerances for all parameters on the basis of current information. It was concluded that tolerancing should be left to future documents which give detailed specifications for specific components or equipment operating in the 1125/60 HDTV system.

Figure 10.3 SMPTE 240M Standard for Television-Signal Parameters 1125/60 High-definition Production System. ***(Continued)***

A3.2 Line Repetition Rate. Initially, a tolerance was specified for this parameter, to serve as a fundamental tolerance on all system timing. This tolerance was moved to the Appendix to leave the standard untoleranced. The stated value, with tolerance, is:

Line repetition rate (derived) 33750.00 Hz ± 10 ppm

A3.3 Synchronizing Signal. The synchronizing signal specified in this standard is based on time durations, which are toleranced. For the purposes of this standard, counts of reference clock periods were chosen as the primary time specification, and time durations in microseconds are given as derived, quoting no tolerances. For information, the original tolerances on these parameters are:

a		0.593 ± 0.040 μsec
b		1.185 ± 0.040 μsec
c		0.593 ± 0.040 μsec
d		1.778 ± 0.040 μsec
e		2.586 ± 0.040 μsec
f	(sync rise time)	0.054 ± 0.020 μsec
S	sync pulse amplitude	300 ± 6 mV
	amplitude difference between positive and negative-going sync pulses	<6 mV

A4. Relationships Between Basic and Derived Parameters

Certain parameters have been determined as basic and fundamental system parameters. The values of all other system parameters can be derived from those chosen as basic. The purpose of this Appendix is to describe and define the derivations.

Line Repetition Rate (L):

$L = S \times F/2$

where

F = field repetition rate (Sec. 2),
and S = total scan lines per frame (Sec. 2).

Reference Clock Frequency (C):

$C = L \times R$

where

L = line repetition rate (Sec. 2, derived above), and R = reference clock periods in total line (Sec. 4).

Transformation Matrices Between Component Sets (Sec. 5):

The transformation matrices can be calculated from the chromaticity coordinates of the reproducer primaries and the chromaticity of reference white (i.e., the color reproduced when the reference reproducer is driven by equal primary signals), according to well-known methods.

Stated briefly, the equation for Y can be found as follows:

$$Y = (G \times J_g \times y_g) + (B \times J_b \times y_b) + (R \times J_r \times y_r)$$

where

J_g, J_b, and J_r are derived as follows:

$$\begin{bmatrix} J_r \\ J_g \\ J_b \end{bmatrix} = \begin{bmatrix} x_r & x_g & x_b \\ y_r & y_g & y_b \\ z_r & z_g & z_b \end{bmatrix}^{-1} \begin{bmatrix} x_w/y_w \\ 1 \\ z_w/y_w \end{bmatrix}$$

and

x_r, y_r, z_r are the chromaticity coordinates of the red primary,

x_g, y_g, z_g are the chromaticity coordinates of the green primary,

x_b, y_b, z_b are the chromaticity coordinates of the blue primary,

x_w, y_w, z_w are the chromaticity coordinates of reference white.

Figure 10.3 SMPTE 240M Standard for Signal Parameters 1125/60 High-definition Production System.

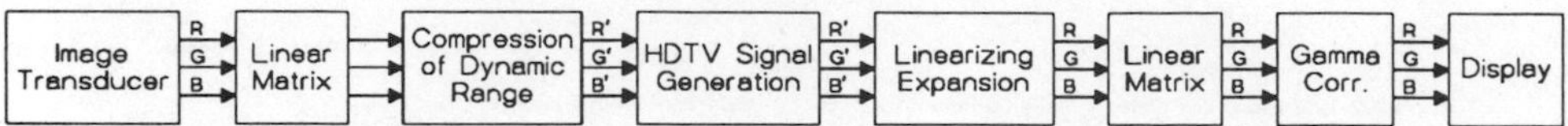

Figure 10.4 Block diagram for proposed camera and display linear-matrix systems for colorimetric correction in future HDTV displays. (Courtesy Sony Advanced Systems, Reference 4.)

This characteristic is intended to implement a reasonably close approximation to the precorrection required to conform to SMPTE Step-One C colorimetry by the gamma curve of a conventional NTSC camera and monitors with no video-signal processing. The precise definition of the curve, however, anticipates future requirements for restructuring a linear set of component signals anywhere in the overall HDTV system, such as digital processing or signal processing in future display systems.

In the implementation of Step-Two, the transfer curve has an additional role of compression and expansion control for improved signal-to-noise ratio in transmission and in recording on magnetic tape or optical disk. This does not avoid the need for precise complementary *linearizing* of the signal at the display system for the camera gamma correction. This, in turn, allows appropriate linear matrixing to be incorporated to provide accurate color reproduction in the display system whether it be, for example, phosphor, liquid-crystal, or laser (see Figure 10.4).

Extension of Dynamic Range Modification of gamma-correction circuits curve to enhance detail in low lights is described in the first paragraph of this section. A similar problem exists for highlights beyond the limited 100-IRE units range from reference black to reference white level. The SMPTE has recognized the importance of capturing the information beyond normal peak white, particularly to provide a large-screen HDTV image which compares subjectively with the wide dynamic range of competing motion-picture film. Consequently, the SMPTE Working Group on High-Definition Production has set up an ad hoc group to produce an addendum to 240M in the form of a Recommended Practice.

Knee Circuits The effective dynamic range in the Sony and BTS cameras is extended by means of highlight-compression circuits.[4] Saticon and Plumbicon pickup tubes have an essentially linear transfer characteristic up to the level at which the beam reserve is exhausted. This point may be as much as 400% above the normal

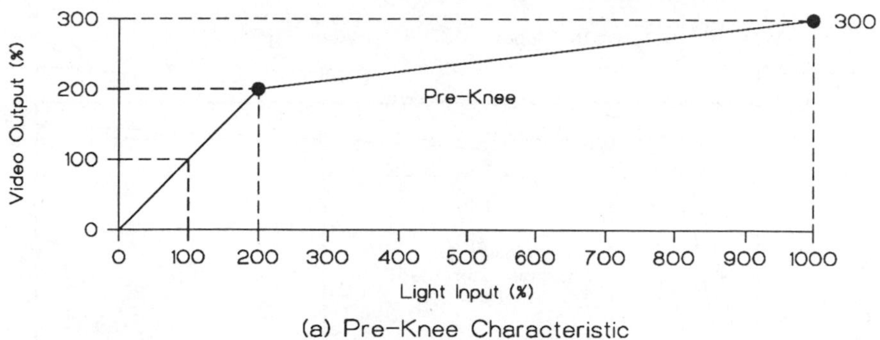

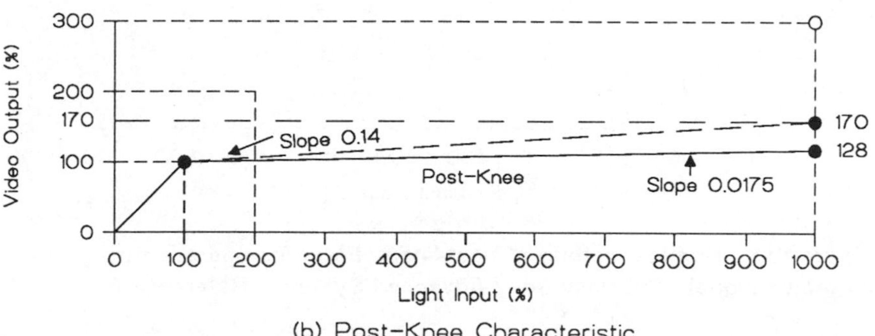

Figure 10.5 Transfer characteristic for Sony HDC-300 camera: (a) pre-knee and (b) post-knee. (Courtesy Sony Advanced Systems, Reference 4.)

operating level for peak-white video. Rather than sharply clipping all signals above reference white, gain-reduction electronic circuits, operating on levels several times the reference-white level of 100 IRE units, are used to compress the higher-level signals to within the 100-unit range.

In the Sony camera, the first compression system, called the *knee,* immediately follows the first stage of video preamplification, prior to any processing of the video signal (see Figure 10.5a).

The primary purpose of this circuit is to reduce the dynamic-range requirements of all subsequent video-signal circuits. The pre-knee characteristic introduces a reduction in dynamic range of about 3:1 (see Figure 10.5a).

The second knee system, or *post-knee,* follows the normal video-processing operations of black-and-white level setting, shading, and gamma correction (Figure 10.5b). The purpose is to further compress

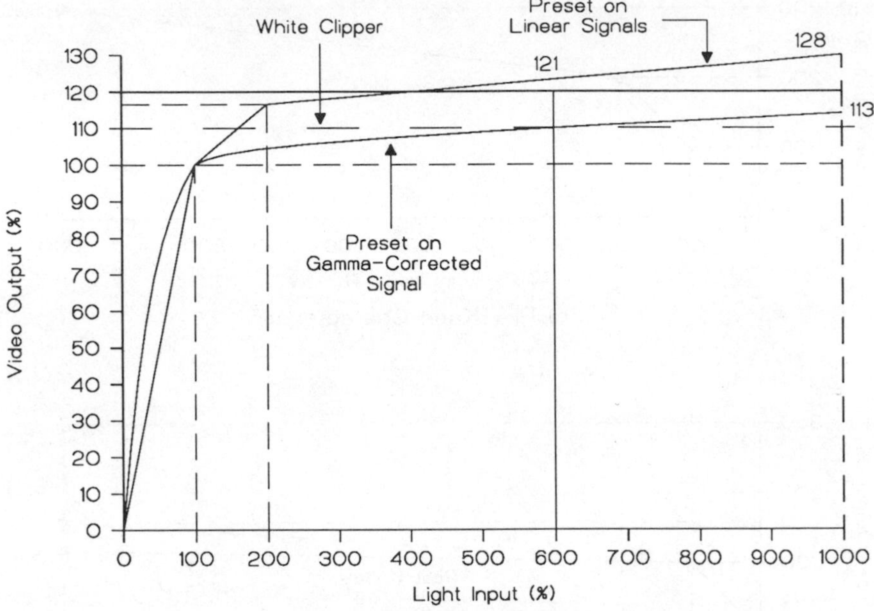

Figure 10.6 Nominal setting of post-knee circuit for operation on gamma-corrected signal. (Courtesy Sony Advanced Systems, Reference 4.)

all video information extending above nominal peak white to within the maximum range of 100 IRE units set by the white clippers. The clippers normally are set at 105–110 IRE units to prevent the overload of subsequent transmission circuits and any attendant distortion or streaking.

The knee system can be preset and switched to an automatic mode or adjusted manually by the video operator. (The characteristic in preset operation is shown in Figure 10.11.)

Figure 10.6 shows the transfer characteristic of the post-knee system, in a nominal setting, with a gamma-corrected signal. It can be seen that, with the white clipper set at 110 units, approximately a 600% dynamic range of scene illumination can be handled by the camera signal-processing system. This range can be lowered, or extended to as much as 1000%, by adjustment of the level, or knee point, at which the post-knee circuit becomes active. Operational experience to date has indicated that a knee-point range of 45–120% is ample for virtually all studio and field conditions. The effective slope of the combined pre-knee and post-knee systems is 0.017 above the 20% level.

Table 10.2 Primary-Color Coordinates for NTSC and SMPTE Specifications

Color	FCC/NTSC		SMPTE 240M		SMPTE C	
	x	y	x	y	x	y
Red	.67	.33	.630	.340	.62	.34
Green	.21	.71	.310	.595	.31	.59
Blue	.14	.08	.155	.070	.17	.07

The sophisticated techniques described in this section are intended to facilitate the handling of wideband video signals with an extended dynamic range, resulting from extreme conditions of staging and lighting. They are influenced by the limitations of the television system and the ongoing development of HDTV videotape recording and reproduction equipment, and in picture reproduction equipment. Thus, while much of the information concerns fundamental technologies, future changes may be seen in practical implementation.

10.1.2 Camera Colorimetry

Standardization System colorimetry is specified in terms of the picture-display reproduction of primary colors. The FCC Rules and Regulations for 525-line broadcast service call for gamma-corrected red, green, and blue voltages suitable for a color picture tube having primary colors in accordance with the "x and y chromaticities in the CIE system of specification" listed in Table 10.2. The SMPTE 240M Standard for 1125/60 HDTV has recommended a greater gamut of color response than that of the present FCC Rules and Regulations for picture-tube reproduction primaries or the SMPTE Type-C studio Standard. The x, y coordinates tabulated in Table 10.2 are shown in Figure 10.7.

The SMPTE Ad Hoc Group on Colorimetry encountered considerable difficulty in realizing a practical approach to implementing this wider gamut for HDTV. The primary problems were:

1. Manufacturing feasibilty of appropriate display phosphors.
2. Resolving the inevitable compromises in camera operation necessary to achieve adequate sensitivity for normal studio and field environments.

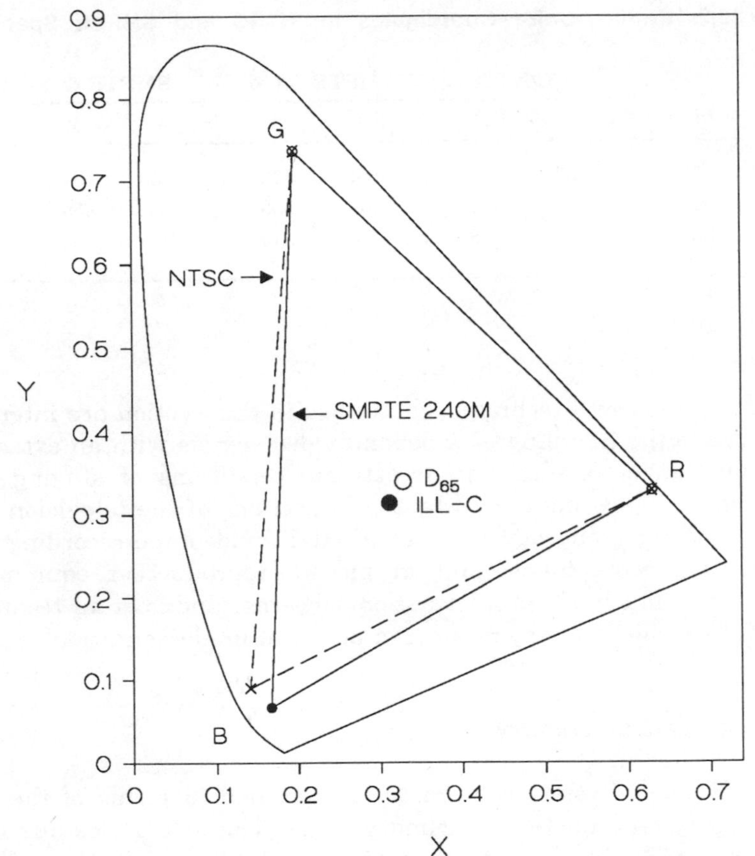

Figure 10.7 SMPTE 240M wide-gamut color characteristics compared to NTSC primaries. (Courtesy Sony Advanced Systems, Reference 4.)

Two-Step Approach The dilemma was resolved by agreement on a two-step procedure which would allow equipment manufacturers adequate lead time to develop both camera and display technologies.

In Step-One, SMPTE called for implementation as soon as possible by all display manufacturers to a set of HDTV studio-monitor phosphors in accordance with the recently adopted, albeit of a narrower gamut, SMPTE Type-C coordinates. These are shown in Figure 10.8.

This important interim step would allow a reasonable time interval for manufacturers to develop appropriate nonlinear processing and linear matrixing to correct their HDTV display systems and in

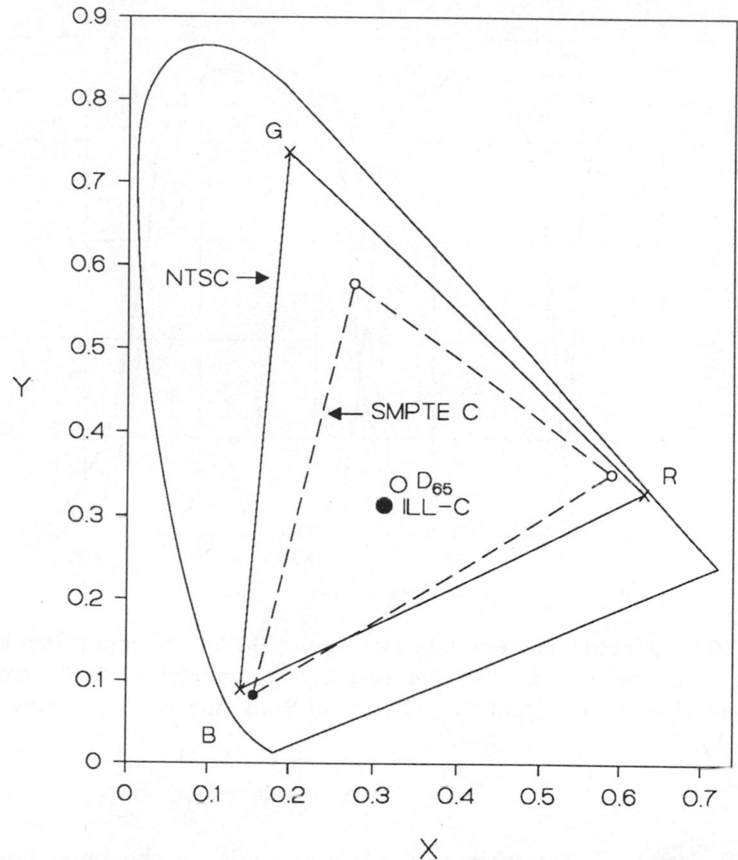

Figure 10.8 SMPTE Type-C color primaries compared to NTSC (FCC). (Courtesy Sony Advanced Systems, Reference 4.)

the future to undertake further development of phosphors having the wider gamut of color emission. During this interim period, manufacturers would be encouraged to conform to the SMPTE Type-C colorimetry recommendations for the basic "taking" characteristics.

An overlapping second step would involve HDTV cameras capable of encompassing the wider color gamut recommended by SMPTE 240M and future HDTV displays evolving toward this Recommended Practice with advances in manufacturing technology. Concurrent with phosphor development, processing and matrixing techniques would be developed to allow more faithful reproduction of the

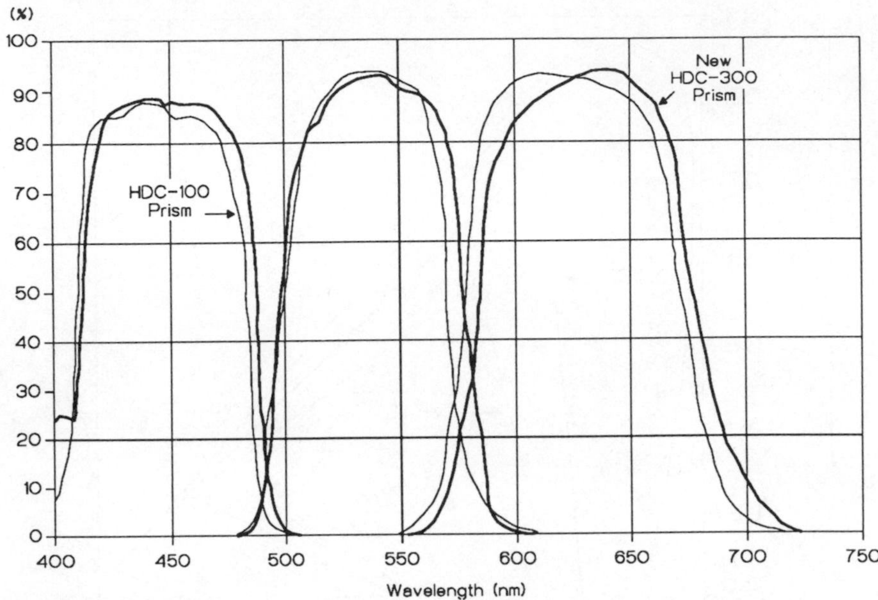

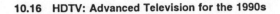

Figure 10.9 Spectral characteristics of Sony HDC-300 camera prism beam splitter. The characteristics of the earlier experimental HDC-100 camera prism are shown for comparison. (Courtesy Sony Advanced Systems, Reference 4.)

SMPTE 240M colors within the inherent limits of the phosphors in use.

Camera manufacturers and television program producers are faced with a second dilemma, in that cameras designed for initial use with Step-One phosphors quite likely will still be in regular use when both Step-One and Step-Two displays are in the marketplace. For this reason, it was decided to design cameras to have switchable capability to meet the two colorimetry criteria.

This required, in the case of Sony, the design and manufacture of a unique optical block and a switchable linear matrix to comply with the specifications of either Step-One SMPTE Type C or Step-Two SMPTE 240M. Spectral responses of the new prism and the earlier design intended specifically for 240M requirements are shown in Figure 10.9. Figure 10.10 shows the ideal wide-gamut SMPTE-240M spectral characteristics compared with those of SMPTE Type C and NTSC. All three curves are based upon D6500 for white.

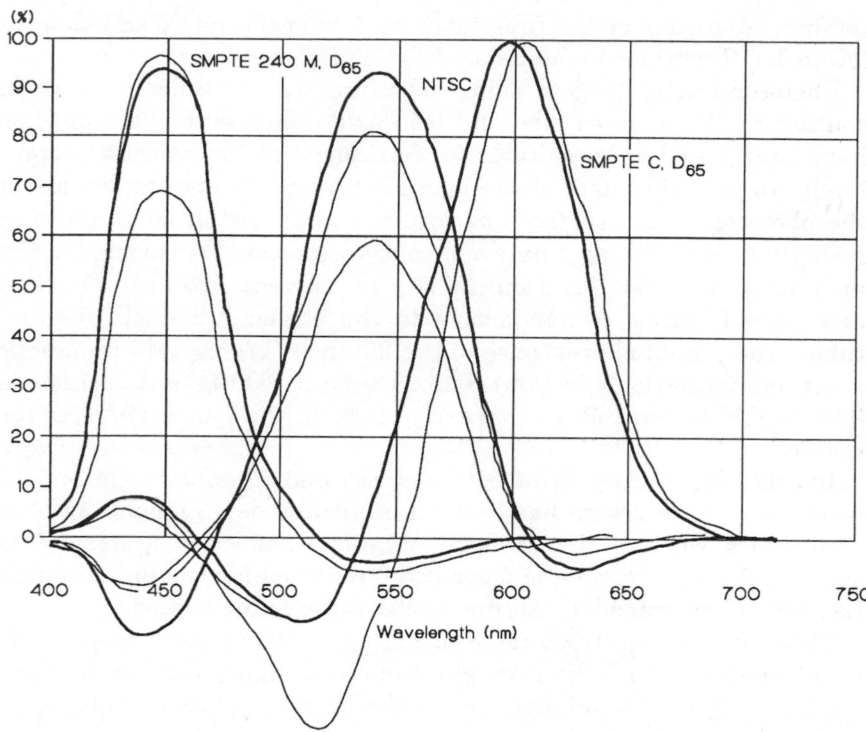

Figure 10.10 SMPTE 240M wide-gamut color primaries compared with SMPTE Type C standard. (Courtesy Sony Advanced Systems, Reference 4.)

The responses chosen for the prism permit the use of minimum-value matrix coefficients, thus minimizing the inherent reduction in signal-to-noise ratio.

10.2 Image Transducers

10.2.1 Photoconductive Tubes

Performance Characteristics Since their introduction in the early 1960s, Saticon and Plumbicon tubes and their associated components have undergone continuing development to overcome the shortcomings of the initial designs. Improvements in photocathode materials and gun design have permitted a reduction in size from the 1.25-inch

(30-mm) diameter of the first tubes to 1 inch (25 mm), and then to 2/3 inch (17 mm) for high-quality 525/625-line service.

Photoconductive targets using selenium with antimony trisulfide coating on the scanned side and lead-oxide photosensitive materials have been found to be suitable for Saticons and Plumbicons, respectively. Improved control of the manufacturing process and doping of the photoconductive surface has resulted in a substantial increase in resolution, as well as improved color response. For example, the modulation-transfer function (MTF) of current 2/3-inch (17-mm) diode-gun Saticons is comparable to the earlier 1.25-inch (30-mm) tubes. The amplitude response of the 2/3-inch (17-mm) MS (magnetic focus, electrostatic deflection) Saticon tube at 800 television lines is 33% in the center and an average of 20% in the four corners of the raster.[5]

In addition, packaging of both Saticon and Plumbicon tubes in a practical camera design has been simplified by developments in electron optics wherein the external magnetic deflection system,[6,7] in some designs the magnetic focus coil,[4] required for the earlier tubes has been supplanted by internal electrostatic configurations.

Thus, use of electrostatic electron-optics technology has led to more compact and lighter weight camera designs, with mechanical alignment limited to adjustments of the beam-splitting and objective-lens optical assemblies.

Diode-Gun Design One of the most significant improvements has been the diode gun. The basic design has been improved since its introduction and further improved in subsequent application to HDTV tubes.[5] The conventional triode gun, shown in Figure 10.11, produces a spread and crossover of scanning-beam electrons after they leave the control grid (G_1). The major shortcommings are: (a) a variation in beam landing and a loss in resolution across the scanned area, and (b) an increase in beam resistance. The high beam resistance and the long RC time-constant of the beam and the photocathode target capacitance primarily determines the field-to-field image-charge retention or lag. A reduction in either target capacitance or beam resistance reduces lag.

The viable options to decrease the capacitance are to (a) decrease the scanned area, (b) increase the thickness, or (c) decrease the dielectric constant of the photocathode material.

Decreasing the scanned area obviously is not applicable to HDTV service because of the reduction in resolution. Increasing the thickness and decreasing the dielectric constant of the photoconductive

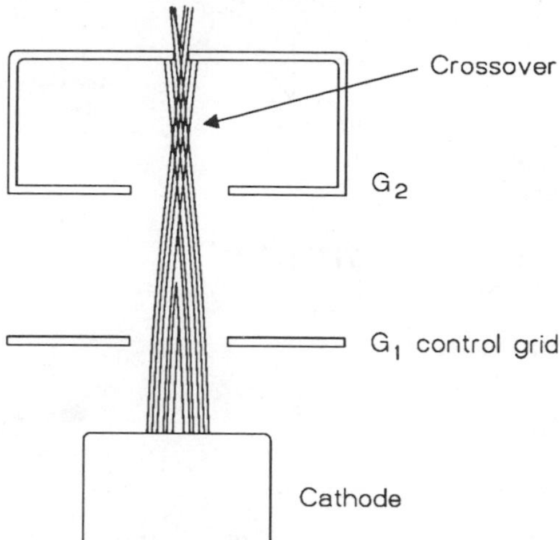

Figure 10.11 Cross-section of a conventional triode gun showing the high electron density crossover region that can contribute to high beam resistance. (From Benson, K. B. [Ed.]. *Television Engineering Handbook.* McGraw-Hill, New York, 1986. Courtesy McGraw-Hill.)

layer depends upon the materials available with adequate sensitivity and color response. An alternate, more attractive solution is to use a diode gun.

The diode gun used in Saticons and Plumbicons actually is a triode gun operated in a positive control-grid (G_1) mode. As shown in Figure 10.12, the gun structure permits the control grid (G_1) to be operated at a positive potential relative to the electron-emitting cathode. The scanning beam is formed at a small aperture in the third electrode (G_2). The control grid provides coarse limiting of the beam, with the small-diameter beam defined by G_2. This prevents a space charge from causing a highly concentrated beam of electrons as in Figure 10.11. The smaller and uniform-diameter beam produced by a diode gun reduces the beam resistance and increases the resolution. In other words, a greater number of pixels can be resolved, a basic requirement for HDTV (see Table 10.1).

A variation on the basic diode-gun configuration (a triode gun operated in a positive G_1 mode) is shown in Figure 10.13. The scanning beam is formed in a small beam-defining aperture in a third electrode.

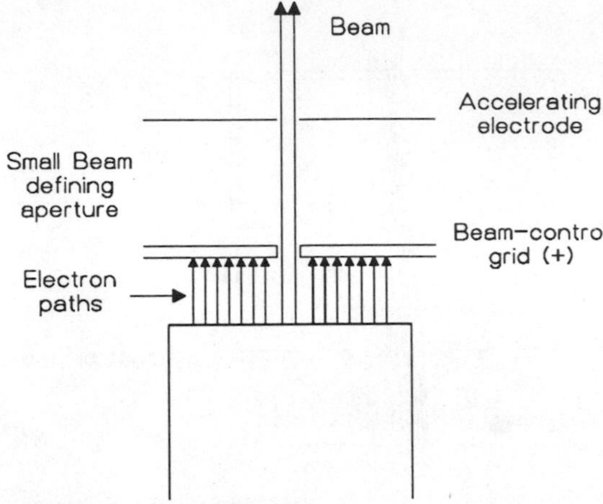

Figure 10.12 Diode-gun configuration. (From Benson, K. B. [Ed.]. *Television Engineering Handbook*. McGraw-Hill, New York, 1986. Courtesy McGraw-Hill.)

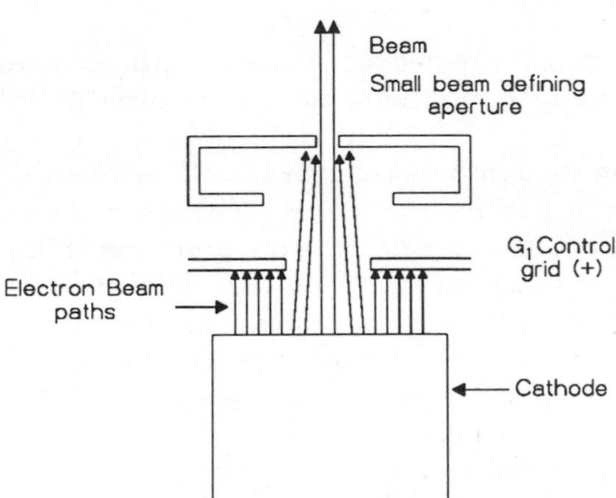

Figure 10.13 Diode gun with positive grid added before final beam-defining aperture. (From Benson, K. B. [Ed.]. *Television Engineering Handbook*. McGraw-Hill, New York, 1986. Courtesy McGraw-Hill.)

Choice of Pickup Tube Nevertheless, with appropriate aperture correction and improvements in the target and gun design, the resolution of 2/3-inch tubes is marginal, at best, for HDTV, except in applications such as electronic news gathering (ENG) where a trade-off of noise level or resolution for a reduction in size and weight is warranted. This dictates the use of 1-inch tubes, where the higher modulation-transfer function (MTF) of the larger photocathode, combined with improvements in tube design, provides the increased resolution to meet HDTV system requirements. Furthermore, the aperture-correction requirement is reduced, thus providing a higher signal-to-noise ratio in the order of 47 dB over a bandwidth of 20 MHz,[4] the minimum required for HDTV.

Significant improvements in the Plumbicon, in addition to the incorporation of the diode-gun design, have resulted in (a) a reduction in lag, (b) a color response covering the full visible spectrum with negligible degradation of the red region, and (c) increased photocathode sensitivity.

Photoconductive Target HDTV advanced-television systems specify a wider aspect ratio than the 1.33 (3:4) of conventional television and "flat" motion pictures. The first NHK systems, and the first Zenith proposal, employed an aspect ratio of 1.67 (5:3), while the SMPTE 240M 1125/60 proposal for an ANSI Standard specifies an aspect ratio of 16:9 (1.78). The decreased height of the image frame on the camera-tube photocathode, for a fixed value of illumination of the faceplate, lowers the camera sensitivity by almost 10% (see Figure 10.14).

Fortunately, a new enhanced Saticon photoconductive layer developed by Hitachi in conjunction with NHK has provided an average sensitivity improvement of 2 times in red and 1.5 times in green, with no change in blue. This translates into an effective improvement in luminance sensitivity of 1.6 times. This figure is calculated as follows:

$$R: 2 \times 0.3 = 0.6;\ G: 1.5 \times 0.59 = 0.89;\ B: 1 \times 0.11 = 0.11;$$

$$Y: 0.6 + 0.89 + 0.11 = 1.6.$$

Thus, the improvement in red and green Saticon output signals has not only made up for the reduction in sensitivity resulting from the wider image format, but in addition has produced an overall

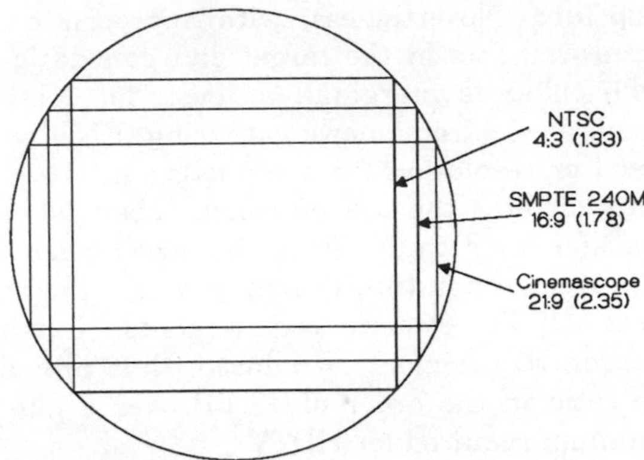

Figure 10.14 Utilization of photocathode for NTSC 4:3 (1.33) and SMPTE 240M 16:9 (1.78) aspect ratios. Cinemascope motion-picture film 21:9 is shown also for comparison.

increase in luminance sensitivity of 1.6 times. Taking into account the 10% reduction resulting from the wider aspect ratio, the effective overall increase in luminance sensitivity is 1.44 [(1.6 × (100% − 10%)].

Low-Light Sensitivity The optical system is another component in camera sensitivity. Under scene conditions of low light level, or when a smaller lens aperture (larger *f*-stop) must be used for greater depth of field, the efficiency of the prism block providing color separation is the limiting factor. The first NHK/Sony HDTV camera used a block with an aperture rating of *f*/1.6. In the Sony HDC-300 this was replaced with a prism having a faster rating of *f*/1.2, thus permitting the use of faster objective lenses for a gain in overall sensitivity. On the other hand, the BTS KHC-1000, in order to bridge the transition from 525/625-line applications to HDTV, and to permit the use of either 25-mm or 30-mm Plumbicons and Saticons, uses a prism block with a somewhat slower rating of *f*/1.5.

Noise Level The loading capacitance of the coupling circuit between the pickup tube and the preamplifier input is another critical factor in the effective sensitivity of a camera. This consists of the output capacity of the pickup tube photocathode, the coupling lead, and the preamplifier input. A lower output capacitance of 3.5 pF was achieved in the new 25-mm Saticon tube. This, combined with the figure of 5 pF for a new field-effect transistor developed by Sony and improvements in pickup tube output-coupling layout and preamplifier circuitry, reduces stray capacitances to a minimum. The high-frequency loading is lower, and the high-frequency peaking necessary for flat amplitude-frequency response is decreased. The lower peaking increases the signal-to-noise ratio by reducing the $1/f$ noise-spectrum level.

The net result of these improvements in input circuitry allows the nominal signal current for peak white in the green channel to be reduced from 400 nA in earlier cameras to 300 nA with no degradation in signal-to-noise performance. This results in a 1.33 increase in luminance sensitivity. Combined with the previous figure of 1.6, the total sensitivity increase (for equal S/N) is over two times, the equivalent of one f-stop in objective-lens speed.

Lag The rate of decay of the video signal when illumination is changed abruptly or cut off is called *lag*. It is affected by several factors, namely, incomplete discharge of the photocathode at the time of scan, target-storage capacitance, scanning-beam impedance, signal level, value of beam current selected for operation, and, if used, the bias-light level.

A photoconductor is a capacitor that stores the charge carriers generated by light at each point on the projected image. These charges are integrated during the interval between successive scans of each point of the image. Immediately after the light is applied, a small charge is developed which increases, in level with time, the maximum level, which is not reached until the end of the 1/30-second frame-rate period (1/25 second for 625-line/25-frame systems).

When the light is removed or cut off, the signal level does not drop to zero immediately because a charge is stored on each element of the charge image on the photoconductor. The image is in proportion to the integrated illumination of each point since the last scan. Ideally, after removal of the light, the charge should decay linearly during the period of a frame. Depending upon target-storage capacitance, signal level, beam current, and bias light, if used, the duration of this linear decay in charge level will vary. The application of bias light to the target can have a substantial effect on lag; a bias-light signal of 20–30 nA can reduce the lag signal measured in the third field after light cutoff by a factor of as much as 2:1.

In addition to the linear decay of charges, after each successive scan, some long-term nonlinearly decaying level of charge, called a *trapping effect*, will remain. The length of time for retention of this residual charge will vary with different photocathode materials.

The charge from all of these effects remaining at the next scan and subsequent scans after removal of the light is termed *photoconductive lag*. A measure of lag is the rate of decay of the video signal appearing as tailing on moving objects, or as an image retention after the cutoff of the light of an image. If the exposure to light is short, the lag will be low and will be determined primarily by the time-constant (RC) of the beam resistance multiplied by photoconductor capacitance. Furthermore, if the exposure to light is longer, the lag will be greater because of a combination of photoconductive lag and the significant addition of *trapping effects*.

Resolution Improvements for HDTV Service A reduction in the thickness of the photoconductor layer, because of less lateral light dispersion in the layer, results in an increase in resolution, or modulation depth. The conventional Saticon target thickness is in the order of 4–6 μm. The newer tubes intended for HDTV use have a reduced layer thickness of 2 μm. The greater capacitance of a thinner layer, unfortunately, increases the lag. Consequently, the increase in modulation depth provided by a thinner target layer, without appropriate precautions in tube design, is achieved at the expense of an increase in lag. In the development of pickup tubes to meet the higher resolution demands of HDTV, these undesirable lag effects from the higher target capacitance have been reduced by the use of a diode gun. The increased capacitance is offset by the lower beam resistance of the diode gun, resulting in a shorter [(beam resistance) × (target capacitance)] time constant and less lag.

Electrostatic Beam Deflection Two types of electron-optical systems have been developed for high-resolution systems. In both of these the conventional external magnetic deflection coils have been replaced with an electrostatic system built into the pickup tube. This type of deflection has several advantges over magnetic systems:

- Uniform resolution over the scanned area.
- Minimum of geometric distortion because of uniform field.
- Low shading.
- Shorter flyback because of no resonant energy storage in the defection coils.
- No eddy-current losses and resultant line-start nonlinearity.
- High registration stability because of no heating of deflection electrodes, compared to heating of defection coils.

The principal trade-offs for these advantages are as follows:

- Deflection amplifiers are required to generate voltages of up to 300 peak-to-peak.
- The defection voltages must be biased and dc restored to the Grid 3 voltage of about 400–600 volts. Centering is accomplished by shifting the bias level. This requires capacitive coupling of the defection voltages and clamping on a reference level during retrace.
- A signal-to-noise ratio of at least 100 dB in the deflection amplifiers is necessary in order to avoid random variations in horizontal scanning-line spacing and jagged reproduction of vertical lines.

Electrostatic Focus In the NHK/Hitachi 1-inch Saticon, the focusing field is provided by an internal electrostatic lens. The major advantages are:

- A reduction in size of the tube assembly.
- Elimination of power dissipation from a focus coil and the accompanying temperature rise in the camera head, thus reducing the cooling requirements and permitting a very compact packaging of the camera head.

Electromagnetic Focus Alternatively, several other manufacturers have elected to use magnetic focusing with electrostatic deflection. The primary advantage is to avoid limiting the pickup tubes to one type of Saticon. Magnetic-focus, static-deflection (MS) tubes are available in both Saticons and Plumbicons from several manufacturers, namely, Saticon II and III, Plumbicon, and with both triode and diode guns, although it is hard to justify not using a tube with a diode gun.

10.2.2 Charge-Coupled-Device Image-Storage Technology

Initial Developments In the first solid-state CCD camera system, demonstrated publicly in 1967, a video signal was generated by sampling the charge voltages of an array of diode photoelectric picture elements (pixels) directly in an x and y (horizontal and vertical) scanning pattern. In the early 1970s a major improvement was achieved with the development of the CCD or, in operation, a *charge-transfer* device. The photosensitive action of a simple photodiode was combined in one component with the charge-transfer function and metal-oxide capacitor storage capability of the CCD. In the sequence of operation, the photo-generated charges are transferred to a *metal-oxide semiconductor* (MOS) capacitor for subsequent readout as signals corresponding to pixels.

Thus, rather than directly sampling the instantaneous charge on each photosensitive picture element, the charges are stored for readout either as a series of picture-scanning lines in the *interline transfer* system, or as image fields in the *frame-transfer* system.

Interline-Transfer Structure The early CCD imagers were *interline-transfer* devices in which vertical columns of photosensitive picture elements are alternated with vertical columns of sampling gates (see Figure 10.15a). The gates in turn feed registers to store the individual pixel charges. The vertical storage registers then are sampled one line at a time in a horizontal and vertical scanning pattern to provide an output video signal.

Frame-Transfer Structure This structure differs from the interline system in that at the end of each field exposure, the entire charge pattern is transferred through vertical CCD shift-register columns into the storage register, as shown in Figure 10.15b. This action frees the image register to begin sampling the next field. Concurrently, during this time interval, the charge packets in the storage register are transferred one row at a time into the output register, through which they are transferred sequentially to the output stage.

When all rows of the storage registers have been read out, it is ready to receive the next parallel transfer from the image register, and the entire cycle is repeated for the second field to provide a complete frame of video.

Developments in recent years have increased the level of CCD performance to where it is competitive with photoconductive cameras and acceptable for an increasing variety of high-quality 525/625-line broadcast television applications.[8,9] The primary degradation limiting use of CCDs in broadcast-type applications has been that of vertical smear on excessive highlights. The problem is common to both frame-transfer and interline designs, albeit the principles of operation differ.[10]

In the frame-transfer design, during the high-speed frame shift, some charge-packets pass through the highlight area and acquire an additional spurious charge. This contamination during the vertical movement produces a vertical smear on the highlight. This unwanted charge can be avoided by blocking the high-level exposure

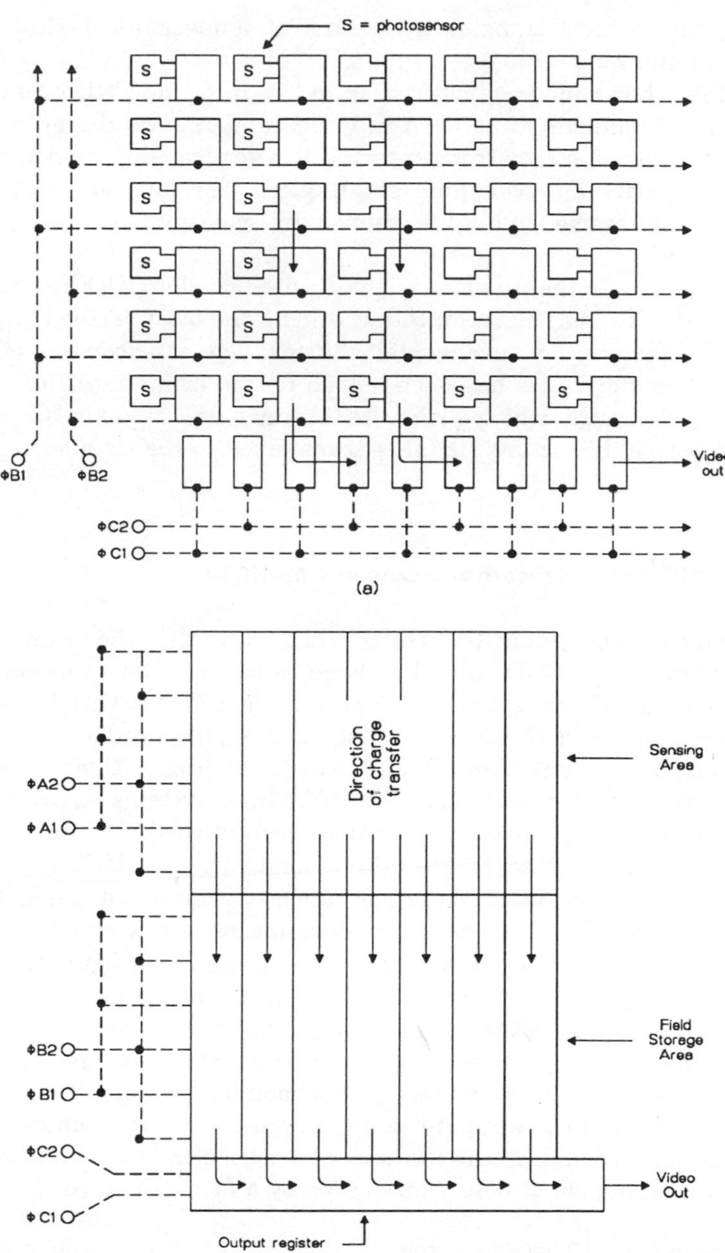

Figure 10-15 CDD imager architectures: (a) interline-transfer structure; (b) frame-transfer structure. (From Benson, K. B. and J. Whitaker. *Television and Audio Engineering Handbook*. McGraw-Hill, New York, 1990. Courtesy McGraw-Hill.)

during the vertical transfer by means of a mechanical shutter or electrical blanking.

On the other hand, vertical smear in the interline CCD on normal exposure of highlights has been reduced to about –80 dB by a drain to divert stray electrons from entering the vertical-shift register, and by low-impurity *P*-type doping techniques. This has not eliminated smear on excessive highlights caused by exposures several *f*-stops over normal.

One solution to the problem in the frame-transfer CCD has been to blank the CCD from light excitation during the brief vertical-transfer period by means of a mechanical shutter. The shortcoming of both types of transfer CCDs has been solved by the frame interline transfer (FIT) structure, which drains these unwanted signals out of the system before the picture signal is transferred to the storage section.

10.2.3 CCD Image Sensors and Cameras for HDTV

Resolution Capability of CCDs Until very recently, the number of pixels provided by CCD chips has been inadequate for the resolution requirements of the systems proposed for HDTV. The maximum number of pixels in 11-mm chips has been slightly under 400 (Reference 3) for both frame-transfer and interline configurations.[8]

This is adequate for 525- and 625-line systems transmitting 210,000 and 300,000 pixels per frame, respectively. Currently available CCDs are limited in resolution capability for HDTV applications. Thus, while photoconductive cameras remain the mainstay of 525/625-line local and network broadcasting service, virtually all new cameras use solid-state sensors in a three-channel configuration.

In order to meet the requirement of up to as many as 2,000,000 pixels for an HDTV system with CCDs, the straightforward solution would appear merely to increase the number of sensor elements (see Table 10.1). Unfortunately, this poses complex manufacturing problems. According to a state-of-the-art axiom in the semiconductor industry, merely doubling the number of sensor elements would reduce the production yield of defect-free chips by a factor of 10 to 1.

Image Sensor with 2,000,000 Pixels Nevertheless, the ongoing efforts to increase the number of pixels that can be provided in a manufacturable CCD sensor appears to be near fruition. The Toshiba Corporation of Japan has announced the development of a solid-state sensor providing 2,000,000 pixels in a chip 16.2 × 10.5 mm

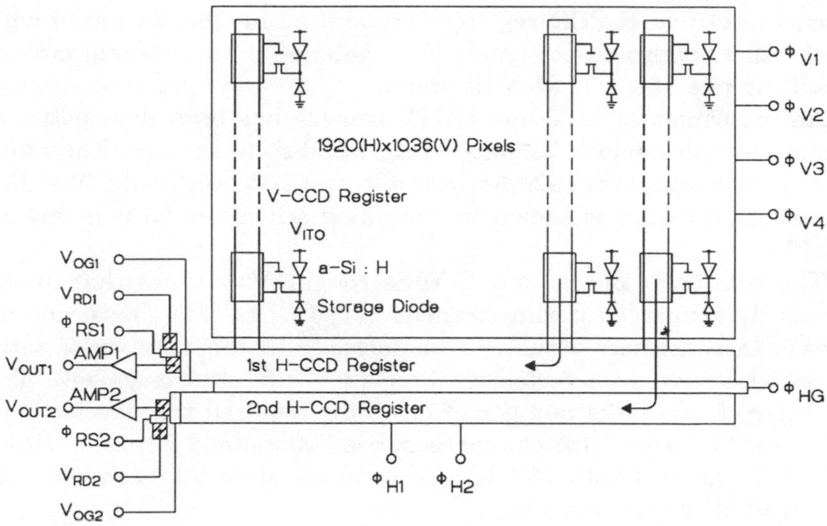

Figure 10.16 Simplified schematic diagram of 2,000,000-pixel image sensor. (Courtesy Toshiba Corp., Reference 11.)

(0.638 × 0.414 inch).[11] The imaging area for a 9:5 (1.8) aspect ratio is 14.0 × 7.8. This is compatible with a 25-mm (1-inch) format for a typical camera optical system composed of an objective lens and three-channel beam splitter.

The structure for a photoconductive layered solid-state imaging device (PSID) is similar to that of an interline-transfer (IT) CCD sensor (see Figure 10.16). However, the photoconversion layer is overlaid with amorphous silicon, which achieves a 100% pixel aperture ratio. The sensitivity and saturation signal charge can be increased because the photoconversion layer and the charge-transfer layer are structured separately.

Initially, the signal charges in the amorphous silicon layer are gathered in the storage diodes, where they then are transferred to the vertical CCD registers during a vertical-blanking period. The horizontal-CCD (H-CCD) register is constructed with dual H-CCD registers, resulting in a halving of the driving frequency. The charges from the V-CCD registers are alternatively transferred to individual H-CCD registers. Each H-CCD register is driven at 37.125 MHz, which is one-half of the the signal readout rate. Output signals from the double H-CCD registers are shifted by 180° in phase with each other. Both output signals are combined in an external circuit, resulting in a single 74.25-MHz signal.

An experimental 1125-line HDTV camera has been demonstrated which uses three such 2,000,000-pixel solid-state sensors. The number of active pixels is 1920 horizontally and 1036 vertically. The 2M-pixel image sensor is shown in simplified schematic form in Figure 10.16.

The sensitivity of the new CCD is roughly five times that of the presently available interline-transfer (IT) CCDs. This significant increase in sensitivity results in an increase in the number of saturated electrons by a factor of three (2×10^5 electrons/pixel). The transfer of this large number of electrons required the development of a new high-speed switching-resonance 700 method for the horizontal CCD registers with 75% less power dissipation than conventional voltage-buffer drive circuits.

The new CCD image sensor and the accompanying circuit technology has provided a horizontal resolution of 1000 television lines. The spectral-response characteristics are shown in Figure 10.17. The sensitivity of the experimental camera provides a signal-to-noise ratio of

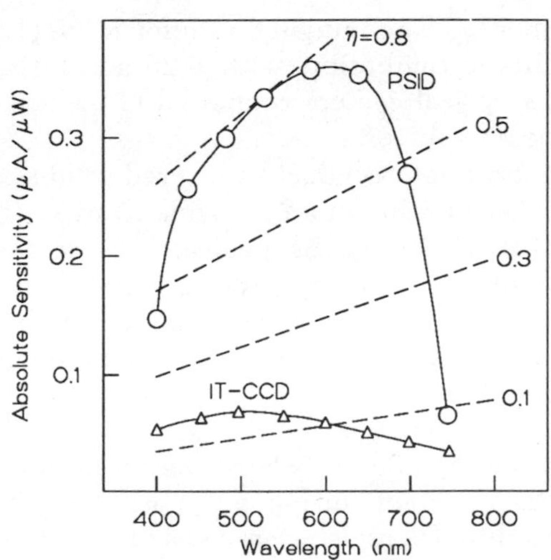

Figure 10.17 Spectral-response characteristics of CCD HDTV camera. (Courtesy Toshiba Corp., Reference 11.)

52 dB with a light level of 2000 lux and an *f*/6.6 objective lens. Thus, the performance characteristics of the CCD camera are suitable for normal studio and field television production requirements.

10.3 Signal Processing

10.3.1 Digital Image Enhancement

The wider bandwidth and greater number of lines used for HDTV systems, compared to 525/625-line systems, result in a bulkier hardware package if conventional delay-line techniques and discrete components are used. In order to realize a practical physical size, the use of digital systems and LSI of components is dictated. This approach has been employed by Sony in their second-generation HDC-300 camera, used by NHK in Japan, and for HDTV demonstrations and program production in the United States on the proposed SMPTE 1125/60 standards.[10]

System Configuration Figure 10.18 is a block diagram of the digital enhancement system. Rather than deriving separate horizontal and vertical detail correction from only the green signal, and thus neglecting the need to enhance the resolution of the blue and red components, a choice is available to the video operator to meet the creative requirements of the program and lighting directors. Accordingly, the horizontal and vertical detail signal can be derived from any of the following signals or combination of signals:

R + G + B
G + R
G + B
R, G, or B

Horizontal Correction The second choice available to the production personnel is a variety of aperture-correction characteristics corresponding to the visual characteristics of different types of picture origination. For example, HDTV can be used to produce all sorts of programs, each containing a broad range of scene content and requiring a variety of image *looks,* some the choice of the director, and some to be compatible with 35-mm motion-picture film. This is accomplished by the use of a dual horizontal-correction curve with a choice available of the correction mix.

Vertical Correction A two-line digital delay system is used in a more or less classic configuration to provide three time-coincident representations of three adjacent television lines (see Figure 10.18). These are used in one linear combination to formulate a combed vertical detail signal in the conventional manner and in a second to generate a combed wideband signal from which the horizontal signal is structured.

The separate horizontal and vertical detail signals are summed linearly, with a remote gain control on the vertical component provided to permit artistically subjective adjustments of the H-to-V detail ratio. The combined horizontal and vertical detail signal, after additional processing and delay, is added back to each of the R, G, and B signals.

Two separate horizontal-correction curves are digitally synthesized according to the two Z-transforms shown in Figure 10.19. The two Z-transforms produce cosine-shaped correction curves, one with a peak at 18.5625 MHz, and the second at 37.125 MHz. Each of these two signals is limited separately and then is fed to a digital mixer.

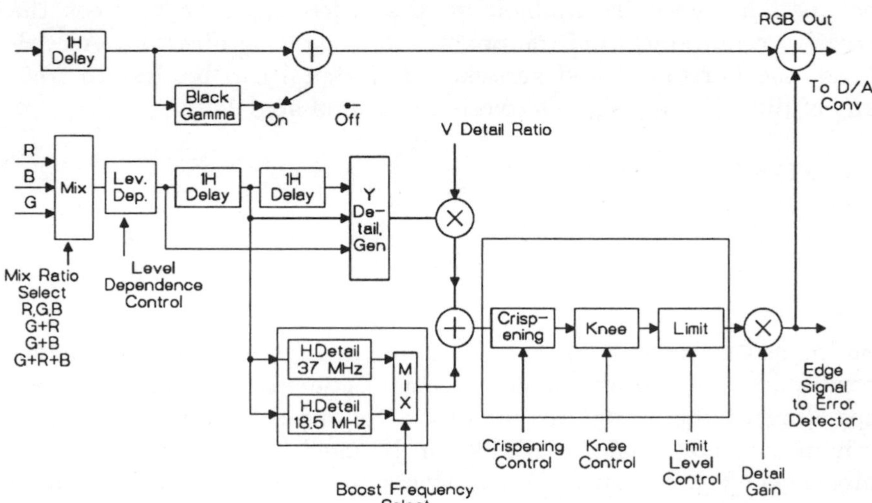

Figure 10.18 Block diagram of HDC-300 digital enhancement system. (Courtesy Sony Advanced Systems, Reference 4.)

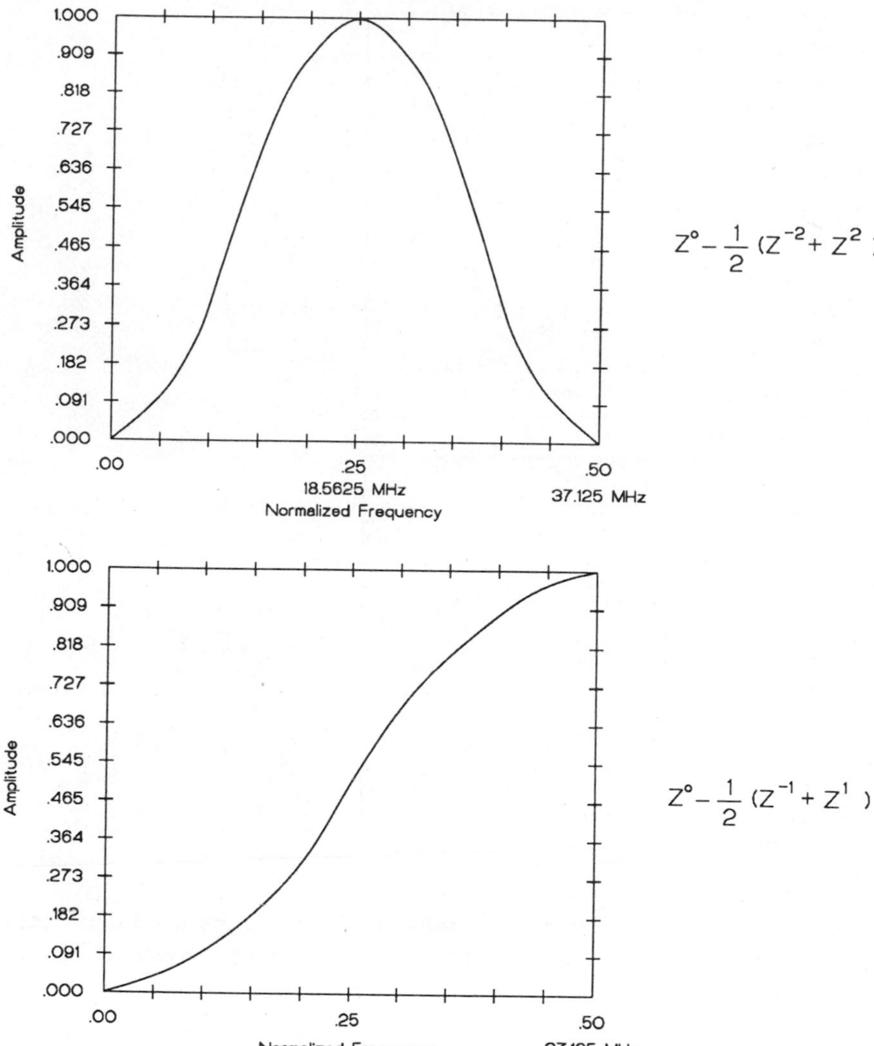

Figure 10.19 Z-transforms of two horizontal aperture-corrector curves. (Courtesy Sony Advanced Systems, Reference 4.)

The coefficients of the separate Z-transforms can be selected remotely before addition. Figure 10.20, showing two typical summations that can be selected, indicates the flexibility in control over the horizontal-detail correction signal.

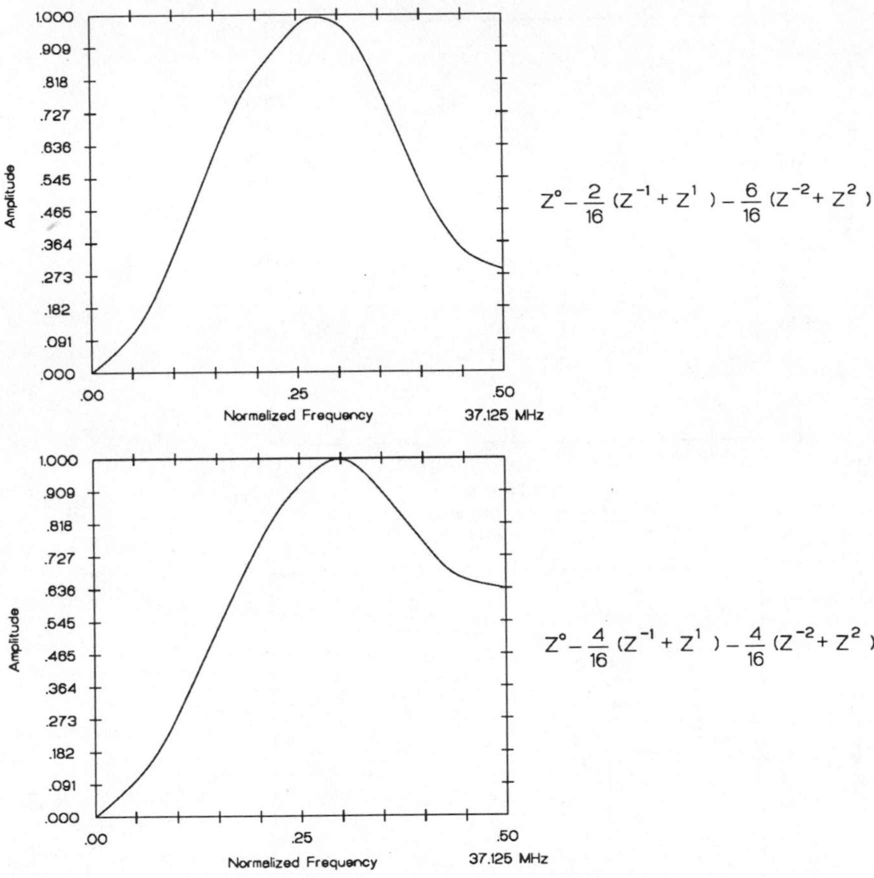

Figure 10.20 Digital coefficient control of Z-transforms producing composite correction signal. (Courtesy Sony Advanced Systems, Reference 4.)

The combination digital detail signal is passed through a nonlinear processing circuit where three separate adjustments can be made by digital remote control from the setup panel in accordance with the transfer characteristic shown in Figure 10.21. This processed detail signal is fed through a remote gain-control system before summation with the main R, B, and G signals.

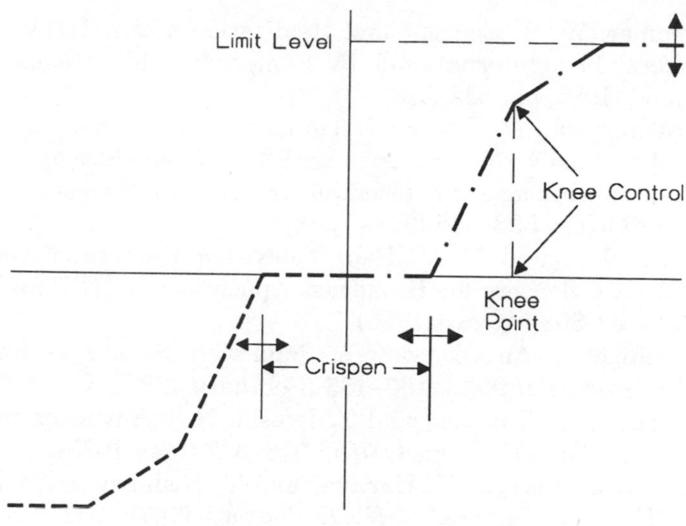

Figure 10.21 Transfer characteristic of nonlinear processor for enhancement system. (Courtesy Sony Advanced Systems, Reference 4.)

References

1. Stain, R. A. The Shape of Things to Come. *SMPTE Journal* 560–567 (July 1988).
2. Excerpts from FCC Rules and Regulations, Subpart E—(Television Stations), Part 73—(Radio Broadcast Services). Federal Communications Commission, Washington, DC.
3. SMPTE Standard for Television-Signal Parameters—1125/60 High-definition Production System. *SMPTE Journal* 98(9): 723–727 (Sept. 1989); *SMPTE Journal* 99(5): 401, 402 (May 1990).
4. Thorpe, L. J. The HDC-300—A Second Generation HDTV Camera. *SMPTE Journal* 99(5): 364–375 (May 1990).
5. Kurashige, M., N. Egami, S. Okazaki, F. Okano, and J. Kumada. 2/3-Inch MS (Magnetic-focus Electrostatic-deflection) Type Saticon for Hand-held Color Cameras. IEEE Transactions on Consumer Electronics, February 1987, pp. 39–46.

6. Klemmer, W. H. Concept and Realization of an HDTV Studio Camera. 14th International TV Symposium, Montreaux, Switzerland, 1985, pp. 382–396.
7. Kurashige, M., N. Egami, K. Tanioka, and K. Shidara. Supersensitive HDTV Camera Tube with the Newly-developed High-amplitude-response Photoconductor (HARP) Target. *SMPTE Journal* 97(7): 538–545 (July 1988).
8. Ad Franken and N. V. Rao. Television Camera Tubes and Solid-state Sensors for Broadcast Aplications. *SMPTE Journal* 95(8): 799–804 (August 1986).
9. Sadashige, K. An Overview of Solid-state Sensor Technology. *SMPTE Journal* 96(2): 180–185 (February 1987).
10. Thorpe, L., E. Tamura, and T. Iwaski. New Advances in CCD Imaging. *SMPTE Journal* 97(5): 378–387 (May 1988).
11. Ide, Y., M. Sasuga, N. Harada, and T. Nishizawa. An HDTV 3-CCD Color Camera. *SMPTE Journal* 99(7): 532–537 (July 1990)

Chapter

11

Picture-Signal Production and Distribution

11.1 Studio Production

11.1.1 Initial HDTV Program Originations

Production Standards Agreement on an HDTV broadcast standard has not been reached at the time of this publication. Nevertheless, primarily as a result of the diligent efforts of the Society of Motion Picture and Television Engineers (SMPTE) to encourage industry to reach agreement on production standards, starting in 1986 and culminating in late 1988 with the proposed SMPTE standard 240M-1125/60,[1] limited HDTV production has been under way in the United States since 1987 to meet the increasing demands by the motion-picture industry for programs and features to be shot and edited on videotape and released on 35-mm motion-picture film.[2]

Although the 1125/60 standard may not be the ultimate answer for all theatrical and television markets, these early efforts using the proposed SMPTE 240M production standard have provided the operating experience necessary to permit component and system require-

Contributions to this chapter are by C. Robert Paulson, AVP Communications, (Reference 10).

development hardware has been replaced with equipment built to production designs. Thus, manufacturers are better prepared to provide studio and field equipment suitable for HDTV program origination and postproduction editing when broadcast standards are approved and HDTV for the home viewer is a reality.

Program Distribution Concurrently, in order to satisfy the more than 1600 existing television broadcasting stations,[3] cable television systems, direct-broadcast satellite (DBS) service, and the videocassette and videodisc market, the output of HDTV production facilities operating on the proposed SMPTE 240M-1125/60 standards must be down-converted to FCC 525/59.94 standards.[4]

In addition, edited HDTV videotape is transferred to 35-mm film for theatrical and 625-line television release[5,6] as well as an alternate format for 525-line broadcast.

11.2 Studio Conversion to HDTV

11.2.1 Component Replacement

Although the video system concepts and equipment configurations applicable to 525-line production have not changed, the transition from this service to HDTV has required production companies to replace virtually all components such as cameras, recorders, monitors, and postproduction editing facilities. Fortunately, there is no shortage of hardware for HDTV production operations. As early as April 1989 at the NAB Convention in Las Vegas, more than 35 equipment manufacturers worldwide displayed manufacturing designs of camera and display equipment (see Chapters 10 and 12), most of which was for the SMPTE-proposed 1125/60 system and intended for television production, postproduction, business and industrial uses, and tape-to-film transfer.[7] Concurrently, a variety of video distribution and terminal equipment has been available which is capable of providing the greater bandwidth required for HDTV signal transmission.

11.2.2 System Requirements

To accommodate wideband HDTV signals, an upgrading of existing transmission and distribution facilities is necessary. This involves additional equalization and amplification in distribution systems to compensate for the high-frequency losses in coaxial-cable interconnection among components. In some complex installations, the equalization and accompanying amplifier gains required for 75-ohm coaxial-cable distribution can be excessive, for example, over 15 dB for 1000 feet of RG/59. Thus, for cable runs of over 100 feet, the use of coaxial cable is prohibitive.

As a consequence, there is a rapid move to convert to fiber-optic cabling. Not only are there many technical considerations favoring the use of fiber cables, but, in addition, fiber cabling is extremely cost-effective. There is no need for equalization, and the low cost of fiber-optic cable at 1/10 that of coaxial cable more than offsets the additional cost of fiber terminal equipment to convert to or from electrical and optical signals. (Fiber-optic technology is discussed in Section 11.3.)

11.2.3 Coaxial-Cable Distribution

525-Line System Equalization The maximum bandwidth required for 525-line NTSC signal transmission is 4.2 MHz. To amplify or process a signal with a complex waveform occupying this bandwidth, in addition to a uniform amplitude-frequency response characteristic, it is essential that the phase relationship among all periodic Fourier components of the waveform be the same at the input and output. This characteristic is termed *envelope*, or *group*, delay. A nonuniform delay will introduce transients or ringing on pulses with fast rise or fall times. Variations in envelope delay usually accompany a sharp cutoff of bandpass. In order to avoid a build-up of envelope delay in cascaded circuits and filter networks, baseband systems usually are designed with a cutoff bandpass of 8 MHz.

The roll-off of response at 8 MHz in coaxial-cable interconnection and distribution circuits is significant. For example, at a cutoff frequency of 8 MHz, 1000 feet of conventional 75-ohm coaxial cable, such as RG/59, exhibits a loss of about 10 dB[8] (see Table 11.1). Therefore, although the cable runs in most studio installations are under 1000 feet, equalization of the loss is necessary.

Correction of a high-frequency roll-off of this magnitude is well within the amplification capabilities of current amplifier designs, requiring a voltage gain of roughly three, which can be corrected at the receiving end of the cable by a distribution amplifier (DA) providing selectable, fixed values of equalization for the high-frequency loss in coaxial cable.

Table 11.1 Attenuation of 75-Ohm Coaxial Cable

Frequency, MHz	4	5	6	8	10	30	45	60
Loss, dB/100 ft	0.6	0.7	0.8	0.9	1.0	1.6	2.0	2.4

Source: Courtesy Dynair Electronics, Inc.

HDTV System Equalization On the other hand, the nominal 30 MHz bandwidth of HDTV noncomposite camera signals and composite encoded signals places more stringent requirements on the amplitude-frequency response characteristics of coaxial-cable distribution systems, as shown in Table 11.1. Transmission of an HDTV signal over 1000 feet of coaxial cable introduces a loss of 16 dB at the signal bandwidth *limit* of 30 MHz. Adding to this a headroom figure comparable to the 6- to 8-MHz figure noted above for 525-line systems, the bandwidth requirement is 45–60 MHz. This translates into 24 dB of equalization, or an amplifier voltage gain of 16, for flat response over 1000 feet of coaxial cable.

Such a low level of high-frequency signal components would be subject to radio-frequency interference (RFI). This indicates that equalization be performed at the sending end of cables, an impractical solution because of the high power capability required in the distribution amplifier.

Consequently, it is clearly evident that for HDTV signal distribution, the following system requirements must be met:

1. Distribution amplifiers at the terminus of coaxial transmission lines provide adjustable high-frequency equalization up to more than 20 dB.
2. The bandwidth of DAs and terminal equipment be at least 40 MHz, preferably 60 MHz.
3. Extreme care be taken in the design and construction of video installations to avoid stray pickup of spurious signals and crosstalk in video-distribution system cabling.
4. Over distances greater than the maximum of a few hundred feet encountered among racks and consoles in equipment or studio-control rooms, fiber-optical cable, rather than coaxial cable, be used.

11.2.4 System and Component Requirements

Amplitude-Frequency Response Video components and subsystems are specified within close tolerances, typically within ±0.25 dB or less, over the nominal bandwidth of the transmitted signal. The luminance-signal bandwidth of the NTSC 525-line system is 4.2 MHz, whereas 30 MHz is proposed for SMPTE 240M 1125/60 and other HDTV systems.

Prior to bandwidth limiting for transmission, in order to reduce to a minimum any transients and ringing which would be caused by a build-up of phase distortion, or envelope delay, below the sharp cutoff of the bandpass, a gradual roll-off of response above the system cutoff must be provided in all components.

Envelope Delay In order to avoid envelope delay over the nominal signal bandpass, without resorting to complex envelope-delay correction circuits such as those used in visual-signal transmitters, the usual practice is to design studio video systems with a significant high-frequency headroom. In 4.2-MHz 525-line systems of cascaded amplifiers and signal-processing equipment, a component bandwidth of at least 6–8 MHz is necessary. Consequently, the active elements must be designed to meet this requirement and compensation must be provided for the high-frequency roll-off of the passive interconnecting coaxial-cable runs.

11.2.5 Active Component Specifications

Distribution Amplifiers Until the late 1980s, most video equipment was designed to meet the 525/625-line requirements of a maximum distribution system bandwidth of 10 MHz. Presently, a few manufacturers have expanded the specifications of their lines to meet the anticipated broadband requirements of HDTV and a current market of program producers shooting and posting on de facto high-definition standards. Other equipment designed for RGB computer systems and for wideband military applications has been found to be well suited for use in HDTV systems.

The salient specifications for one manufacturer's 7-output, unity-gain, video-distribution amplifier listed in Table 11.2 typifies the requirements. Note that the response is flat within 0.25 dB to the

Table 11.2 Distribution Amplifier Specifications

Frequency response	±0.25 to 30 MHz (±0.03 to 6 MHz)
Group delay	<2 ns at 30 MHz (<0.5 ns at 6 MHz)
Output isolation	>35 dB to 30 MHz (>47 dB to 6 MHz)
System crosstalk	<–56 dB to 30 MHz (<–80 dB to 4.4 MHz)

Source: Courtesy Dynair Electronics, Inc.

maximum signal bandwidth of 30 MHz. For comparison, the values for the 525-line bandwidth of 6 MHz are shown in parentheses.

The DA also provides switchable equalization for lengths of up to 590 feet of 75-ohm coaxial cable (values selected for Belden 8281, precision 75-ohm double-shielded video cable). An amplifier is available to extend the equalization range up to 1770 feet.

Switchers The characteristics of routing and production switchers are the most significant factors in the video signal path between picture signal sources and studio outputs. Switcher characteristics involve, in addition to video distribution amplifiers and mixing amplifiers, solid-state switching cross points. In the *off* condition a cross point is required to present and open circuit with no feed-through from the input bus. The lower impedance of the switching-point open-circuit capacitance introduces more crosstalk in wideband HDTV signals, compared to the narrower band-width of NTSC signals. Thus, the switcher requirements for HDTV are significantly more stringent in both bandpass and crosstalk.

The specifications of a 100 × 20 system presently being used by 1125 Productions in New York are listed in Table 11.3. Figure 11.1 is an illustration of the switcher with the control modules partially removed. Figure 11.2 is a schematic of the output module. Figure 11.3 is a simplified block diagram of the 10 × 10 video switching module.

11.3 Fiber-Optic Signal Transmission

11.3.1 Fundamentals

Transmission Principles A fiber-optic transmission system is based upon principles of operation similar to that of microwave and higher radio-frequency electromagnetic wave propagation systems. The information "carrier" is a light source in the infrared spectrum. The

Table 11.3 100 x 20 60-MHz Switching System Specifications

Frequency response (referred to 1 MHz)	100 kHz to 10 MHz ± 0.1 dB 20 MHz ± 0.25 dB 30 MHz ± 0.75 dB 60 MHz ± 1.5 dB
Group delay	100 kHz to 5 MHz, less than 5 ns
Crosstalk isolation[a]	0–5 MHz >55 dB 30 MHz >45 dB 30–60 MHz >25 dB
Transient response (10-MHz square wave 2 ns rise time)	8 ns rise and fall time 20% maximum overshoot and ringing
Slew rate	100 V/μs

[a]Same signal on all inputs except that under measurement.

Source: Courtesy Dynair Electronics, Inc.

Figure 11.1 Wideband routing switcher. (*Source:* Courtesy Dynair Electronics, Inc.)

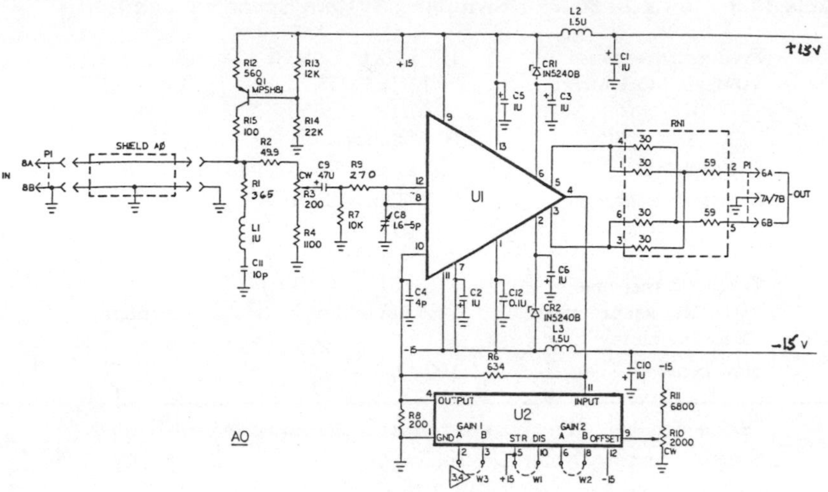

Figure 11.2 Routing switcher output module schematic. (*Source:* Courtesy Dynair Electronics, Inc.)

carrier usually is specified in wavelength, rather than frequency, and for convenience is measured in nanometers (10^{-9}), or billionths of a meter. Analog video and audio or digital data signals are processed in a variety of ways common to radio wave transmission to modulate the intensity of the transmission light source. The light source is either a light-emitting diode (LED) or a small solid-state laser not unlike those used in a compact-disk (CD) player. The transmisssion of information by fiber is accomplished by regulating the flow of light photons, rather than by regulating the amplitude and frequency of electrons in a conductor or radio waves in free space. The comparison of the three modes of transmission is shown in Figure 11.4.

The modulated light, at a power level of a milliwatt or less, is transferred via a lens element into the polished end of an optical fiber less than the diameter of a human hair. At the receiving end, ranging from a few feet to many kilometers away, the emerging light rays are coupled to a photodetector to provide an electrical signal corresponding in level to the intensity of the received light.

Light wavelengths used for communications systems are in three *windows* of 850, 1300, and 1550 nm. The attenuation is inversely proportional to the wavelength. The range of attenuation and trans-

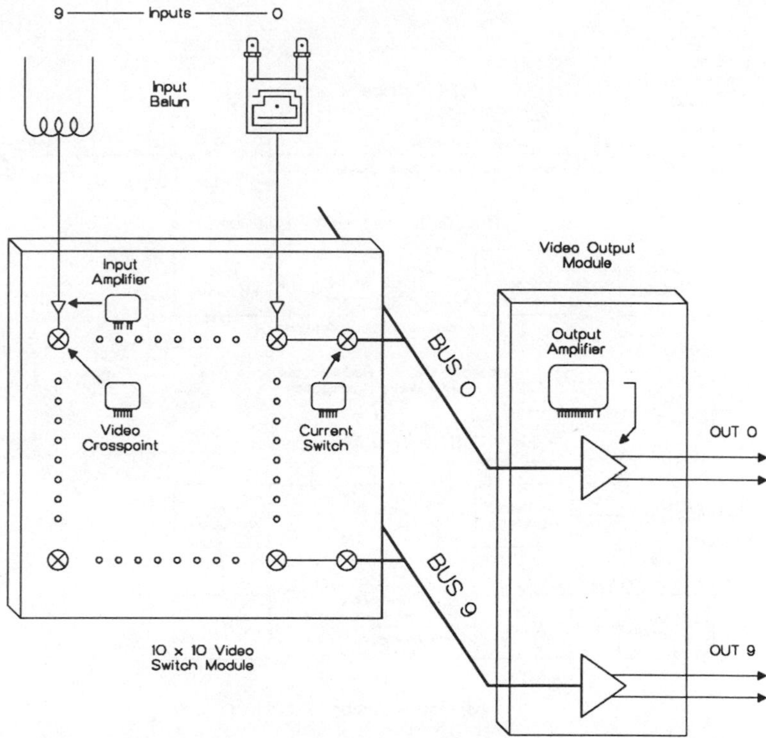

Figure 11.3 Simplified block diagram of routing switcher 10 x 10 switching module. (*Source:* Courtesy Dynair Electronics, Inc.)

mission for each of these windows and the various cable diameters is shown in Table 11.4.

Fiber-Cable Construction The center core of a fiber is drawn from glass of very high purity. A surrounding highly reflective glass cladding prevents the transmitted light rays from escaping and reduces attenuation to an extremely low value.

Two categories of fibers are used in signal transmission systems: "single mode" (SM) and "multimode" (MM) (see Figure 11.5). They are available in configurations ranging from a twin-fiber cable resembling an ac-power extension cord to a jacketed marine cable containing hundreds of individual fibers.

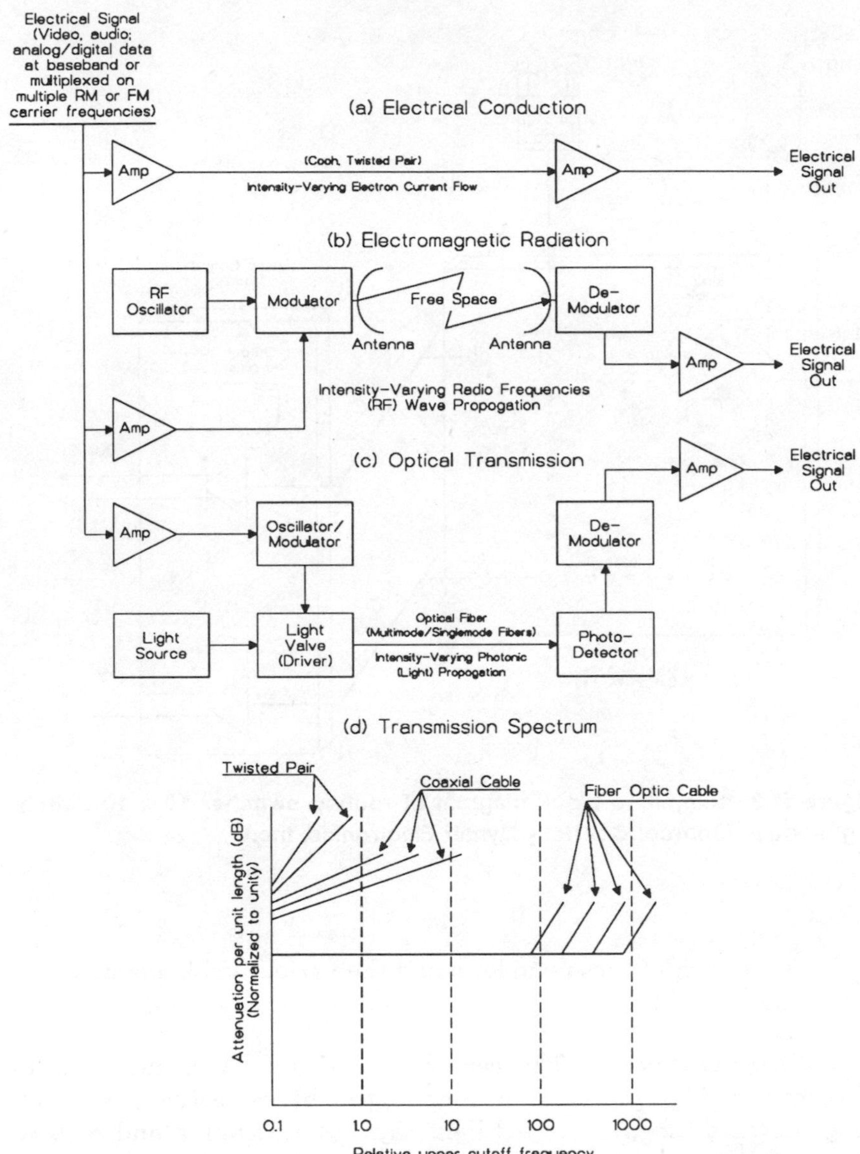

Figure 11.4 Systems for electrical transmission: (a) electrical conduction; (b) electromagnetic radiation; (c) optical transmission; (d) relative upper cutoff frequencies (MHz or Mb/s). (*Source:* Courtesy C. Robert Paulson, AVP Communications, Westborough, MA.)

Table 11.4 Optical Fiber Signal Attenuation Versus Light Carrier Wavelength.

Transmission window (wavelength – nanometer)	1 850 nm	2 1300 nm	3 1550 nm
Core/clad diameters (μm)	Typical attenuation range (dB/km) Typical transmission range (km)		
Multimode (MM) 100/140	4.5–6.0 0–2	3.5–5.0 Not available	Not available
Multimode (MM) 62.5/125	3.3–4.3 0–2	0.9–2.0 0–5	Not available
Multimode (MM) 50/125	2.7–4.0 0–2	0.8–3.0 0–10	Not available
Single mode (SM) 7–10/125			

Source: Courtesy C. Robert Paulson, AVP Communications, Westborough, MA.

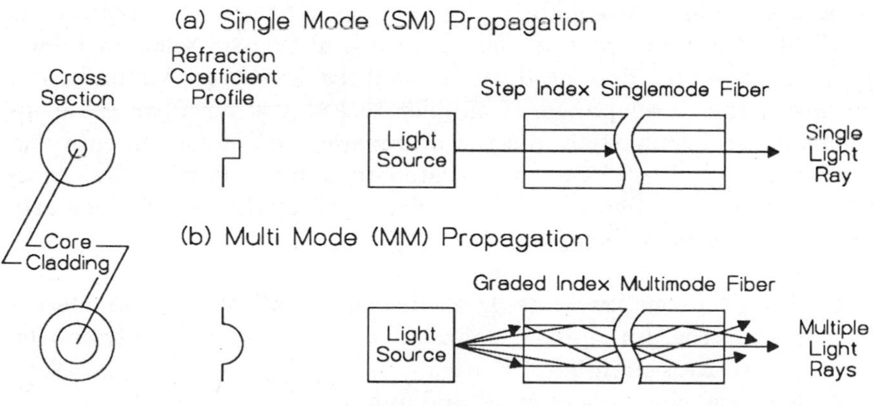

Figure 11.5 Fiber-optic cable construction: (a) single-mode (SM) propagation; (b) multimode (MM) propagation. (*Source:* Courtesy C. Robert Paulson, AVP Communications, Westborough, MA.)

11.3.2 525-Line Television Applications

Common-Carrier Service In the early 1980s, overly optimistic predictions of the demand for voice and data transmission services resulted in the construction of fiber-optic facilities in and between hundreds of metropolitan areas in excess of the demand.[9] Consequently, there is heavy competition among common carriers for customers to utilize the surplus of "dark" (unused) fibers. The active circuits in these cables carry digital bit streams at rates ranging from 100 Mb/s to over 1 Gb/s.

Some of the Regional Bell Operating Companies are offering switched transmission services at the DS-3 digital rate of 44.7 Mb/s. The D-3 terminal facilities are designed for 44.7 Mb/s transmission of 525-line compressed-bandwidth digital video and two audio channels. Thus, although the fiber network per se is capable of carrying signals of virtually unlimited bandwidth, it is limited to D-3 525-line service with its signal impairments.

Television Studio Use The advantages provided by fiber transmission in studios with the limited bandwidth of 525-line NTSC and 625-line PAL heretofore have not justified the higher cost of the necessary terminal equipment and the complexity of installation, compared to coaxial cable, albeit the cost of fiber is substantially lower than that of coaxial cable. Consequently, even though off-the-shelf hardware is available, the move to fiber for video signal transmission in television broadcasting and production facilities has been virtually nil. However, the development of simpler techniques for fiber cable installation and connection and the decreasing cost of components and terminal equipment have ameliorated to a large degree the objections to the use of fiber.[10–12] The major features that are particularly attractive are the following:

1. Simultaneous two-way transmission of all video and stereo-audio program signals, communication channels, cue channels, and control circuits.
2. Multiple channels of video and audio.
3. No need for equalization.
4. Immunity to external interference and crosstalk. Therefore, the use of fiber by industrial and educational organizations and cable systems is growing at a rapidly increasing rate, and the trend may be expected to spread to broadcasting, in new studio

installations where the multichannel and bidirectional capability further reduce the capital-cost differential between coaxial and fiber cable distributions systems.

11.3.3 High-Definition Television

Advantages of Fiber-Optic Cable The impracticability of equalizing coaxial cable for the 30 MHz bandpass required in HDTV studio facilities is the major factor dictating the use of fiber rather than coaxial cable for any distances other than a few feet. The advantages, in addition to virtually unlimited bandwidth capability of over 1 Gb/s, far in excess of the requirements for audio, video, and digital signals, are:

1. Immunity to electric and magnetic interference. Adjacent ac power wiring will not introduce any interference.
2. Grounding needed only for interconnected equipment carrying or processing analog signals.
3. No crosstalk among signals in adjacent fibers.
4. No pickup of radio, television, communication, or other broadcast signals.
5. Low cost for fiber cable, compared to coaxial copper cable.
6. Reduction in number of cables because of the capability of a fiber to carry a mix of signals, i.e., pictures, multichannel sound, intercom, control signals, and commands.

Fiber-optic transmission systems are suitable for analog AM or FM signal transmission and digital signal transmission. For analog transmission in television studio video and audio applications, most hardware currently available uses fm. Frequency modulation offers significant advantages for optical transmission of maximum signal-to-noise ratio and low differential gain. The characteristics of a typical system shown in Figure 11.4d indicate the excellent wideband transmission performance possible, compared to coaxial cable.

The signal voltage is used to control the intensity of the narrowband (0.1 nm wide) light output of a 1300- or 1550-nm solid-state laser, which is coupled to a glass fiber about 125 μm in diameter. The transmission loss in the fiber cable is less than 0.5 dB/km. At the receiving end, the very slightly attenuated light is coupled to a PIN-FET photo detector to produce an electrical output signal.

References

1. Proposed American Standard for Television. *SMPTE Journal* 96(11): 1150–1151 (November 1987).
2. Stumpf, R. J. A Film Studio Looks at HDTV. *SMPTE Journal* 96(3): 247–252 (March 1987).
3. Ono, Y. HDTV and Today's Broadcasting World. *SMPTE Journal* 99(1): 4–15 (January 1990).
4. Thorpe, L., K. Matsumoto, and T. Kubota. An HDTV Downconverter for Post-Production. *SMPTE Journal* 99(2): 124–135 (February 1990).
5. Gasper, J., H. Mahler, and G. Gabritsos. A Comparison of HDTV and Film—Overall Light Transfer Characteristics. *SMPTE Journal* 98(8): 556–562 (August 1989).
6. Thorpe, L., T. Yoshinaka, and K. Tsujikawa. HDTV Digital VTR. *SMPTE Journal* 98(10): 738–747 (October 1989).
7. The New Business of High Definition. Audio Visual Communications, December 1989, p. 31.
8. Benson, K. Blair. *Television Engineering Handbook*. McGraw-Hill, New York, 1986, p. 14.28.
9. Paulson, C. Robert. Television Signal Transmission: Another Technology in Transition. *SMPTE Journal* 98(5): 366–370 (May 1989).
10. Paulson, C. Robert. *Applications of Fiber Optics in Government and Military Facilities and Operations*. AVP Communications, Westborough, MA, 1990.
11. Paulson, C. Robert. Fiber Optic Technology. In *View*, Spring 1989, pp. 19–29.
12. Cable TV Industry Goes for Fiber Optics. Telecommunications Products Division, Corning Inc., Corning, NY, 1989.

Chapter

12

Picture Displays

Color video displays may be classified under the following categories:

1. Direct-view cathode-ray tube (CRT).
2. Large-screen display, optically projected from a CRT.
3. Large-screen display, projected from a modulated light beam.
4. Large-area display of individually driven light-emitting CRTs or incandescent picture elements.
5. Flat-panel matrix of transmissive or reflective picture elements.
6. Flat-panel matrix of light-emitting picture elements.

The CRT remains the dominant type of display for both consumer and professional 525/625-line television applications. The Eidophor and light-valve systems using a modulated light source have wide application for presentations to large audiences in theater environments, particularly where high screen brightness is required. Stadium displays, composed of discrete light-emitting picture elements, provide marginal resolution for 525/625-line reproduction where the high hardware and installation costs of several million dollars can be justified.

Matrix-driven flat-panel displays are used in increasing numbers for small-screen personal television receivers and for portable projector units. For the future, these types of displays may be expected to be developed as large-screen wall-mounted panels with a resolution capability suitable for home viewing of high-definition television transmissions.

12.1 Format Development

12.1.1 Screen Utilization

Visible Picture Area Prior to the development of cathode-ray tubes with rectangular faceplates for the visible phosphor screen, a variety of schemes were used to produce the largest image possible at the expense of a loss of picture information or geometric distortion of the image. Overscan, used initially to avoid the effect of low powerline voltage, was the normal practice of receiver designers until the advent of regulated power supplies and scanning circuits. It has been common practice to overscan by as much as 5–10%, and the accompanying loss of information in the cropped picture area has been accepted by viewers. The industry recognized the practice and forewarned television program producers by publishing a Society of Motion Picture and Television Engineers (SMPTE) Recommended Practice RP27.3 entitled "Safe Title and Picture Area."

Widescreen Film Transmission The procedure recommended by the SMPTE provides a satisfactory solution for the 3 × 4 aspect ratio of television studio productions and motion-picture films shot specifically for television distribution. However, in the staging of motion-picture films intended for theatrical distribution, generally no provision is made for the limitations of television displays. Instead, the full screen, in wider aspect ratios such as Cinemascope with accompanying stereophonic sound, is used by directors for a maximum dramatic and sensory impact.

Consequently, cropping of essential information may be encountered more often than not on the television screen. The problem is particularly acute in wide-screen features where cropping of the sides of the film frame is necessary in producing a print for television transmission. This is solved in one of three ways:

1. Letter-box transmission with blank areas above and below the widescreen frame. Audiences in North America and Japan have not accepted this presentation format, primarily because of the reduced size of the picture images and the aesthetic

distraction of the blank screen areas. Objections by European viewers are considerably less.

2. Print the full frame height and crop equal portions of the left and right sides to provide a 3 × 4 aspect ratio. This process frequently is less than ideal, as, for a example, a Western movie shootout between characters at the extreme left and right sides of the frame where both principals are cropped.
3. Program the horizontal placement of a 3 × 4 aperture to follow the essential picture information. Called *"pan-and-scan,"* the process is used in producing a print, or in making a film-to-tape transfer for television viewing. Editorial judgment is required to determine the *scanning cues* for horizontal positioning and, if panning is used, the rate of horizontal movement. This an expensive and laborious procedure and, at best, compromises the artistic judgments made by the director and cinematographer in staging and shooting and by the film editor in postproduction.

Aspect Ratio of Television Screens A research study of viewer reaction in the viewing of television pictures of moving images[1] indicated an overwhelming preference for a widescreen (16 × 9) aspect ratio over that of conventional systems (4 × 3). Widescreen pictures were compared with the those in the 4 × 3 format with the same height, same area, same diagonal, or same width.

Changing the image size, screen size, and viewing distance had little impact upon preference. The only variable of any significance was the relative size of the widescreen images. Equal-width images were less preferable than equal area, diagonal, or height images. The major finding of the study indicated a 70% majority favoring widescreen pictures, even though the height was reduced to achieve the wider aspect ratio. However, those preferring widescreen increased to 90% for pictures of equal height with increased width. In other words, the size of the image is the dominant parameter.

12.1.2 HDTV Screen Formats

Initial HDTV proposals increased the aspect ratio from the 525/625-line figure of 1.33 to 1.67. Subsequently this was opened up to a wider ratio of 1.78 (16:9). Motion-picture films, on the other hand, are produced in several formats. In addition to the basic 1.33 (4:3), the most common are 2.35 used for 35-mm anamorphic Cinemascope film, and 2.2 in a 70-mm film format. In addition, a 1.85 aspect ratio is provided on 35-mm film by means of narrower sprocket holes and a wider frame.

Therefore, for transmission on HDTV standards, widescreen film must be cropped or scanned as is done for 4:3 television transmission, albeit to a lesser degree. In addition, as described in Chapter 2 (see Figure 2.2), for FCC-required compatibility with current broadcasting standards, the side-cropping and horizontal positioning for an aspect ratio of 1.33 must be performed.

12.2 Cathode-Ray Tube Displays

12.2.1 Direct-View Screens

The direct-view cathode-ray tube (CRT) has been the dominant display device since the introduction of all-electronic television. This can be attributed to its high brightness and resolution, excellent gray-scale reproduction, and low cost compared to other types of displays. From the standpoint of television-receiver manufacturing simplicity and low cost, the packaging of the display device as a single component is extremely attractive. The tube per se is composed of only three basic parts: an electron gun, an envelope, and a shadow-mask phosphor screen.

The typical cost to a manufacturer of the tube assembly and the electronic components to accept a video-signal input is $200 or $300. The luminance efficiency of the electron optical system and the phosphor screen is very high. A peak beam current of under 1 μA in a 25-inch tube will produce a highlight brightness up to 100 ft-L. The major drawback is the power required to drive the horizontal sweep circuit and the high accelerating voltage for the electron beam. This is partially offset by the generation of the screen potential and other lower voltages by rectification of the scanning flyback voltage.

Screen Size As the consumer demands ever-larger picture sizes, the weight and depth of a CRT and the higher power and voltage requirements are serious limitations. These are reflected in sharply increasing receiver cabinet costs to accommodate larger tubes, a major share of the manufacturing cost, and in more complex circuitry for high-voltage generation.

To withstand the atmospheric pressure on the evacuated glass envelope, the CRT weight increases exponentially with the viewable

Table 12.1 Color CRT Diagonal Dimension Versus Weight

Diagonal, in(visible)	Weight, kg (lb)
19 V	12 (26)
25 V	23 (51)
30 V	40 (88)

diagonal. The figures for tubes designed to provide a 4:3-ratio display are shown in Table 12.1.

Nevertheless, manufacturers have continued to meet the demand for larger screen sizes with larger direct-view tubes. Examples are an in-line gun, 110° deflection tube with a 35-inch diagonal screen, initially designed for computer displays, but suitable for HDTV, produced by Mitsubishi and Matsushita. In the Trinitron configuration 37-, 38-, and 43-inch diagonal tubes have been demonstrated by Sony.

The weight of 35-inch and larger tubes and the depth of the receiver cabinet are of questionable practicability for home use. Consequently, a 27-inch tube is the largest size suitable for the majority of home viewing situations.

On the other hand, experimental tubes with rectangular screens, intended only for HDTV in the wider 16:9 aspect ratio, have been shown.[2,3]

High-Resolution Electron Guns Improved versions of both tri-dot delta and in-line guns used for computer-data displays are suitable for high-resolution television displays. The design of both types of guns can be adjusted for the lower beam current necessary to obtain a small spot size required for HDTV. The tri-dot gun provides small spot size at the expense of as many as 20 critical convergence adjustments for uniform resolution over the full tube faceplate. In return for slightly less resolution, this complexity is avoided by the use of in-line guns, which permit the design of a self-converging deflection yoke that will maintain dynamic horizontal convergence over the full face of the tube without the need for correction waveforms.

Figure 12.1 compares the spot size of 19 V tubes at beam currents of up to 1 μA for high-resolution and commercial receiver designs. The spot size at currents below 500 mA is improved substantially in the high-resolution tube. Thus, a maximum of color resolution is

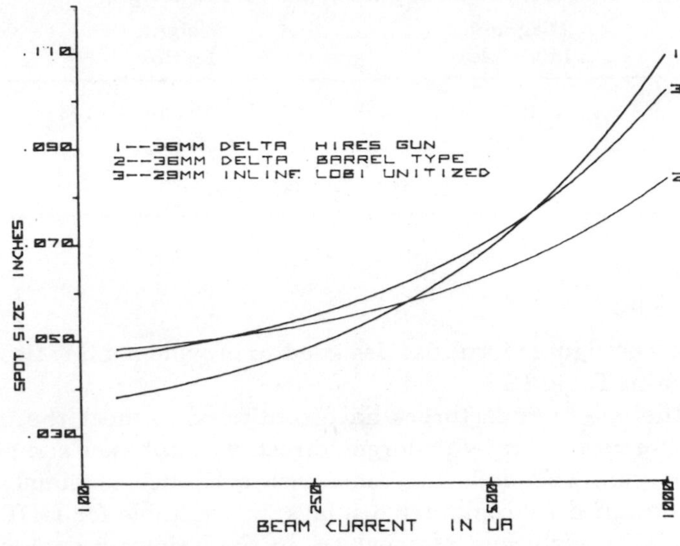

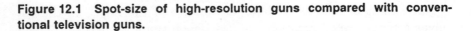

Figure 12.1 Spot-size of high-resolution guns compared with conventional television guns.

provided in the horizontal-scan dimension where the highest definition is required to reproduce the wide-bandwidth signals of HDTV. A delta-gun configuration provides slightly greater resolution at higher beam currents than an in-line gun, however.

The comparative resolution of three types of 19-inch (492-mm diagonal) tri-dot shadow-mask screens is shown in Table 12.2. The high-resolution version is capable of 1000-line definition, double that of the conventional tube. In addition, new electromagnetically focused guns, which provide an improvement in resolution, are suitable for HDTV CRT displays. Thus, it is readily apparent that the technology exists for manufacture of tubes with adequate resolution for HDTV standards, particularly in the more popular larger screen sizes.

High-Resolution Shadow-Mask CRT Direct-view 30- and 40-inch tubes with a wide-screen aspect ratio of 5:3 (1.67) have been developed by Matsushita, primarily for use as monitors in television studios for signals in either the 5:3 or 16:9 (1.78) format. However, they are also expected to find some use for home viewing of HDTV programming

Table 12.2 Comparative Resolution of 19-Inch Shadow-Mask Tri-Dot Screens

Tube type	Number of triads	TV lines
Conventional	400,000	500
Monitor	800,000	775
High resolution	1,400,000	1025

where the excessive weight and cabinet depth can be accommodated in return for very high brightness and contrast.

Construction of this large a glass tube presented several design and fabrication problems. The tensile stresses on the bulb must be minimized to avoid cracks resulting from temperature imbalance during and after manufacture and the danger of subsequent implosion. Study of the stresses in the design stages by computer simulation has reduced the maximum value to below 70 kg/cm^2, about the same level as in conventional tubes for a 4:3 aspect ratio.

The 90° deflection angle was chosen to provide accurate convergence of the three red-green-blue beams and uniform white balance over the full width of the screen, as well as the height. Second, the large neck diameter of 36.3 mm was chosen to permit the use of a large-diameter electron gun necessary to achieve a high degree of resolution at high values of beam current.

The screen is a configuration of phosphor-dot triads with a black matrix. The ideal pitch of the phosphor dots is determined by consideration of the resolution and moiré, and the shadow-mask mechanical strength. The degree of resolution of the screen, in turn, is determined by the pitch of the phosphor-dot triads (or the shadow-mask holes). Quantitatively, this is expressed as the modulation transfer function (MTF).

The relationship between the vertical pitch of the dot-triads and the scanning-line pitch can cause a disturbing moiré pattern in the picture. Reduction of moiré to an unnoticeable level requires an increase in the frequency of the pattern or a decrease in the amplitude. The pattern can be minimized by the use of certain specific values of triad pitch. However, selection of the pitch is limited by the thinner shadow mask for a finer pitch and by a sacrifice in resolution for a more coarse pitch. For the 40-inch tube, a pitch of 0.46 mm provides a resolution of 1000 television lines and satisfies the other

Table 12.3 Specification for Direct-View HDTV Tubes

Size	26 in[a]	30 in	40 in
Screen			
Horizontal	20.5	24.5	36.7
Vertical	15.4	14.6	19.3

[a]Masked for an HDTV aspect ratio of 5:3.

requirements of strength of the shadow mask and reduction in moiré pattern.

Specifications for shadow-mask 26-, 30-, and 40-inch tubes used in HDTV demonstrations and system investigations are listed in Table 12.3.

High-Resolution Trinitron A high-resolution 28V-inch tube was developed in 1985 by Sony for early demonstrations of HDTV by NHK in Japan. This led to the design in 1986 of a 37V-inch Trinitron with an aspect ratio of 5:3 (1.67) for HDTV demonstrations in Japan and the United States. In 1987 general agreement was reached by the industry on a wider aspect ratio of 16:9 (1.78). The following year this parameter was formalized by the Society of Motion Picture and Television Engineers (SMPTE) in their nine-page Proposed American National Standard SMPTE 240M. Accordingly, the design of the 37V-inch Trinitron was modified to meet the requirements of the SMPTE-proposed National Standard.

The design targets for the new tube are:

1. A flat rectangular screen with a square-cornered, 16:9-ratio viewing area.
2. Rectangular glass bulb exceeding the screen area only to the extent necessary for strength and safety.
3. Resolution of more than 1000 television lines.
4. Highlight luminous intensity of more than 95 cd/m^2.

The specifications are listed in Table 12.4. The outline and principal dimensions are shown in Figure 12.2.

The more stringent resolution requirements, particularly in the corners of the widescreen, are met by the use of an electron gun with a longer focus field than that of conventional tubes and a specially designed magnetic deflection yoke. In addition, fine spot size and

Table 12.4 Specifications for 38-Inch High-Resolution Trinitron

Maximum faceplate dimension (diagonal)	41.3 in (1048 mm)
Overall depth	31.4 in (798 mm)
Neck diameter	36.5 in (1.44 mm)
Weight	231 lb (105 kg)
Useful screen dimensions	
Horizontal	33.5 in (852 mm)
Vertical	18.8 in (477 mm)
Diagonal	38.5 in (977 mm)
Number of vertical color-stripe trios	1861
Resolution of monitor (TV lines)	
Horizontal	Over 1000
Vertical	Over 750
Deflection angle	90°
Anode voltage	32 kv
Luminance	95 cd/m^2

precise color tracking over the full range of luminance are realized at high beam currents. These characteristics are shown in Figures 12.3 and 12.4, respectively. The modulation transfer function (MTF) at 1100 lines is reduced only by slightly less than 6 dB at a beam current of 500 μA. At this current the spot size is 1.25 mm.

12.2.2 Optical Projection

Display Requirements To provide acceptable home viewing, a projection system must at least equal the performance of competing direct-view systems in terms of brightness, contrast, and resolution. The peak-white brightness has increased significantly by improvements in phosphors and signal processing. However, the CRT peak beam current can increase to as much as five times the average current. To avoid thermal overload of the phosphor screen, CRT projectors use beam-current limiters which operate when the average current exceeds the permissible maximum value. Thus, predominently white fields can be displayed with reduced average brightness, and some compromises in brightness and contrast must be accepted by viewers in exchange for a larger screen size. This problem is not encountered with light-valve projection systems.

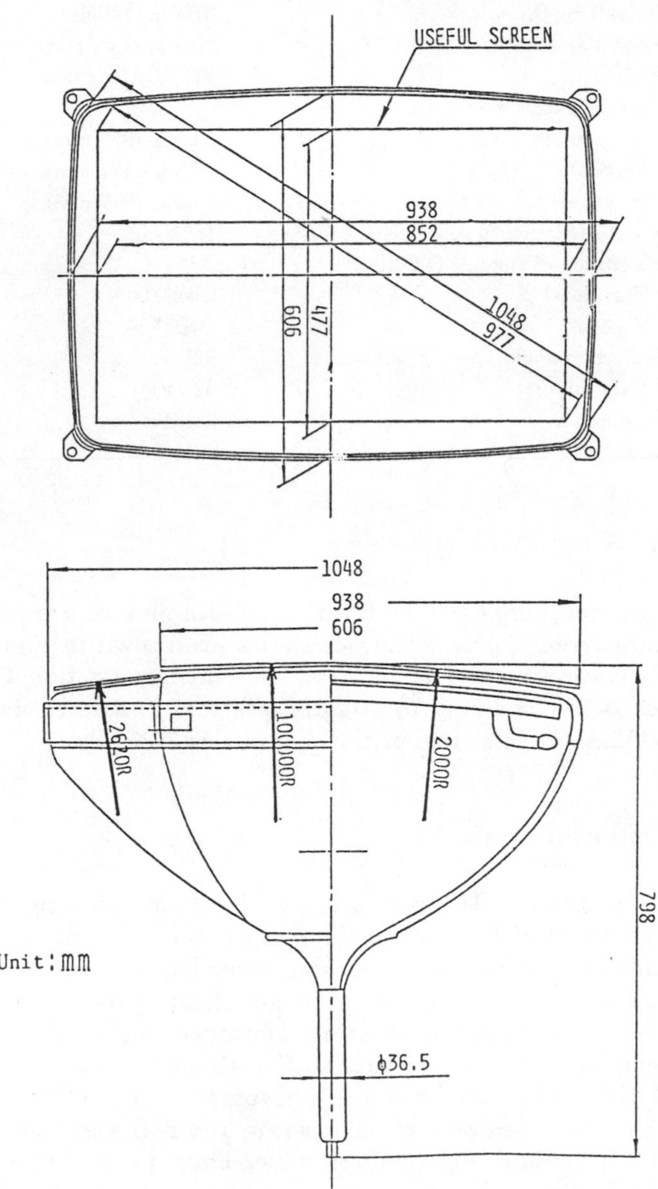

Figure 12.2 Principal dimensions of 38V-inch (965-mm) CRT.

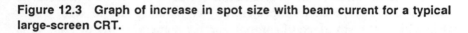

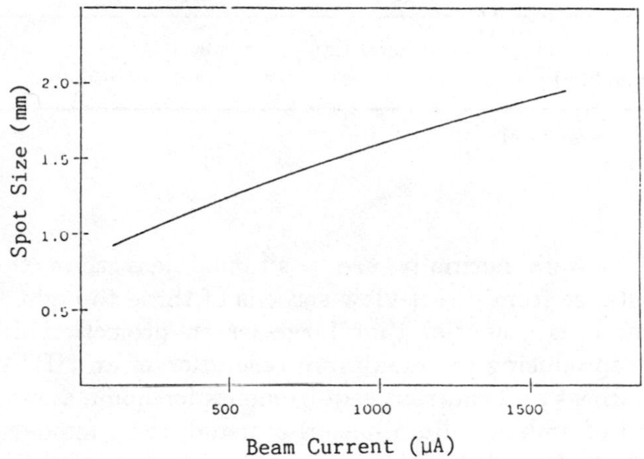

Figure 12.3 Graph of increase in spot size with beam current for a typical large-screen CRT.

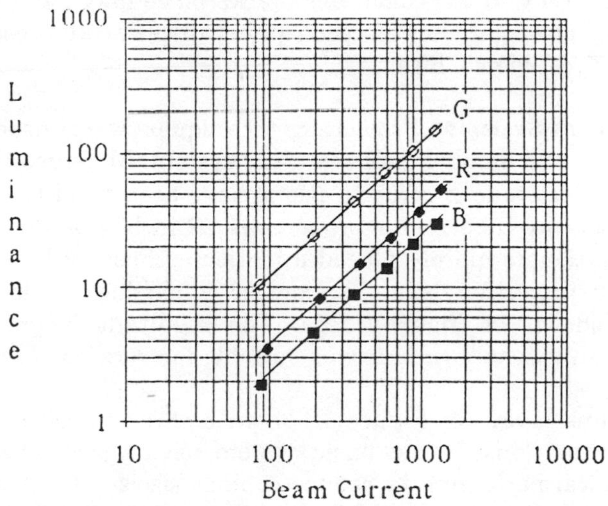

Figure 12.4 Beam current versus luminance characteristics.

Table 12.5 Brightness and Contrast Levels in Television Displays and Theaters

	Brightness (ft-L)[a]	Contrast ratio	Ambient fc
Television receiver	200–400(60-120)	30:1	5-15
Theater (35-mm film)	55(16)	100:1	0.1

[a]Ft-L × 3.426 = cd/m^2 = nit.

Because viewers normally are positioned less than the normal viewing distance from direct-view screens of three to eight times picture height, it is essential that large-screen projection displays be capable of reproducing the maximum resolution of an HDTV system.

The brightness and contrast requirements for home viewing, where a high level of ambient illumination is usual, are considerably more stringent than that of theater viewing of high-definition 35-mm film in relative darkness. The performance levels of direct-view television and motion-picture theater projection are listed in Table 12.5. Since this will be the standard of comparison for home viewing of HDTV, it is questionable whether large-screen CRT projection will be accepted by viewers as the answer to the need for a large-screen display, any more than it has for 525/625-line pictures. Consequently, the introduction of HDTV to the consumer marketplace may depend upon the availability of bright, large-screen panel displays at a cost competitive with direct-view CRTs.

Refractive Lens Systems Projection techniques were used with the first Nipkow mechanical disk-scanning system. More recently, refractive and reflective lens configurations have been used for the display of a cathode-ray tube raster on screens 20 inches or more in width. The first attempts merely placed a lenticular Fresnel lens, or an efficient *f*/1.6 projection lens, in front of a shadow-mask direct-view tube, as shown in Figure 12.5a. The resultant brightness of no greater than two or three FL was suitable for viewing only in a darkened room.4

The in-line layout in Figure 12.5b with three tubes, each with its own projection lens, is the basic system used for all multitube displays. Typical packaging to reduce cabinet size for front or rear projection is shown in Figure 12.5c and d, respectively. Because of the off-center positioning of the outboard color channels, the optical paths differ from the center channel, and keystone scanning-height

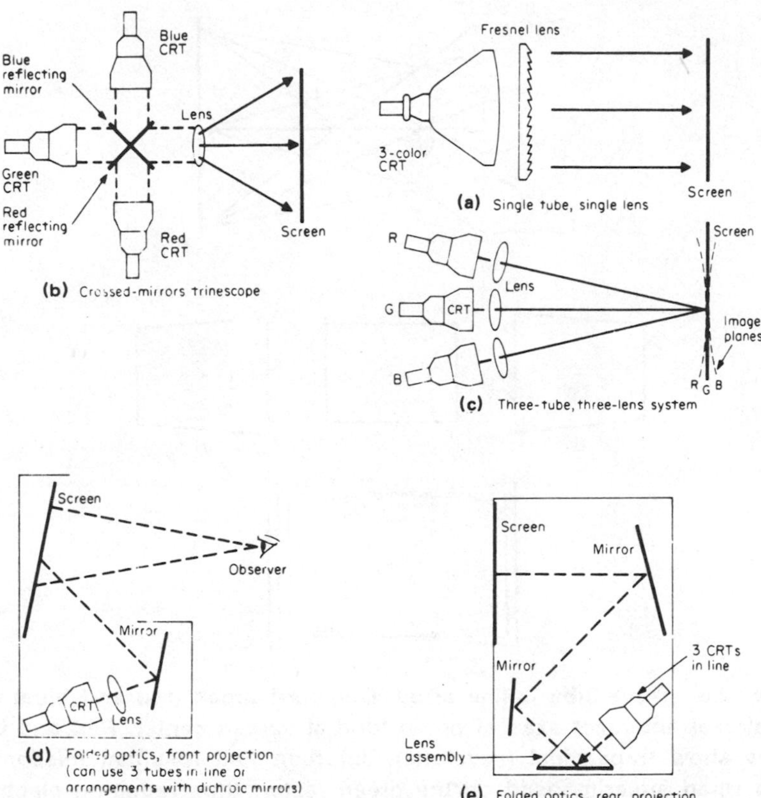

Figure 12.5 Projection optical configurations: (a) single-tube, single-lens rear-projection system; (b) three-tube, three-lens rear-projection system; (c) folded optics, front projection with three tubes in line and a dichroic mirror; (d) folded optics, rear projection.

modulation is necessary to correct differences in optical throw from left to right. The problem, shown diagramatically in Figure 12.6, is more severe for wide-screen formats.

Reflective Lens Systems One form of the in-line array benefits from the relatively large aperture and the light-transmission efficiency of Schmidt reflective optics by combining the electron optics and phosphor screen with the projection optics in a single tube. The principal components of an integral Schmidt system are shown in Figure 12.7.

Electrons emitted from the electron gun pass through the center opening in the spherical mirror of the reflective optical system to

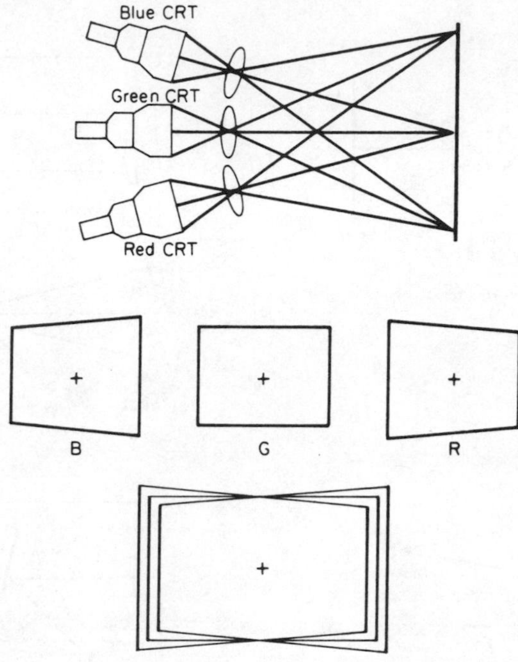

Figure 12.6 Three-tube in-line array. Outboard tubes (red and blue) and optical axes intersect axes of green tube at screen center. Red and blue rasters show trapezoidal (keystone) distortion and resultant misconvergence when superimposed on the green raster, thus requiring electrical correction of scanning geometry and focus.

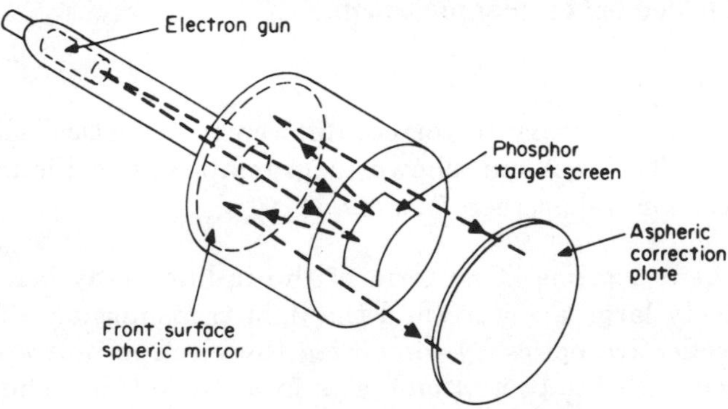

Figure 12.7 Projection CRT with integral Schmidt optics.

Table 12.6 Projection-Lens Performance

Lens	Aperture	Image diagonal (in)	Focal length	Magnification	Response at 300TVL
Fresnel	f/1.7	12.0	11.8	5	6%
Refractive					
Glass	f/1.6	7.7	6.7	8	33%
Acrylic	f/1.0	5.0	5.0	10	13%
Schmidt	f/0.7	30.0	3.4	30	15%

scan a metal-backed phosphor screen. Light from the phosphor (red, green, or blue, depending upon the color channel) is reflected from the spherical mirror through an aspheric corrector lens, which is the face of the projection tube.

Schmidt reflective optical systems are significantly more efficient than refractive systems, because of the lower *f*-stop characteristic and the lesser amount of attenuation by glass in the optical path. In addition, greater magification is possible with cost-effective designs. On the other hand, limiting resolution is lower than that of high-quality, expensive glass projection lenses. A comparison of the characteristics is shown in Table 12.6.

Projection Screens As with direct-view CRT screens, resolution of projection screens can be affected by the choice of material and construction. This is a lesser problem with front-projection screens, although granularity or lenticular patterns of highly directional surfaces can be a limiting factor, with either front and rear projection of HDTV.

In general, any screen element, such as a matrix arrangement of lenticular segments and black stripes to broaden the horizontal viewing angle by quantizing the image, can limit image detail. For 525/625-line television, this does not provide a limiting aperture. The picture quality for high-resolution applications may be degraded not only by a loss in resolution but also by the creation of artifacts from the greater number of scanning lines and smaller picture elements.

The brightness of the image in both front and rear projection screens is determined by the directional characteristic of the screen material. In either type, the brightness for a given projector luminance output (lumens) varies in proportion to the reciprocal of the

Table 12.7 Screen Gains

Type	Gain (cd/m^2)
Flat white paint (magnesium oxide)	0.85–0.90
Semigloss white paint	1.5
Aluminized	1–12
Lenticular	1.5–2
Beaded	1.5–3
Ektalite (Kodak)	10–15
Scotch-light (3M)	Up to 200

square of any linear dimension (width, height, or diagonal) of the screen as follows:

$$B = \frac{L}{A}$$

where

B = perceived brightness, $\frac{cd}{m^2}$ (nit)

L = projector light output, *lumen*

A = screen viewing area ($H \times W$), m^2

Thus, an increase in screen width from the conventional aspect ratio of 4:3 (1.33) to an HDTV ratio of 16:9 (1.78) requires an increase in projector light output of 33% for the same screen brightness.

To improve the apparent brightness, screens can be designed with directional characteristics. This characteristic, called *screen gain G*, changes the brightness equation as follows:

$$B = G \times \frac{L}{A}$$

Table 12.7 lists some typical front-projection screens and the gains.

Contrast, the ratio of brightness at maximum and minimum video levels, is affected by the quality of the optical components and ambient-light conditions. Flare and stray light in the optical system will raise black level and reduce contrast. Ambient light will produce a similar degradation, particularly on highly directive front-projection screens. On the other hand, rear-projection screens can be designed with low reflectance to reduce the effect of ambient light. This can be

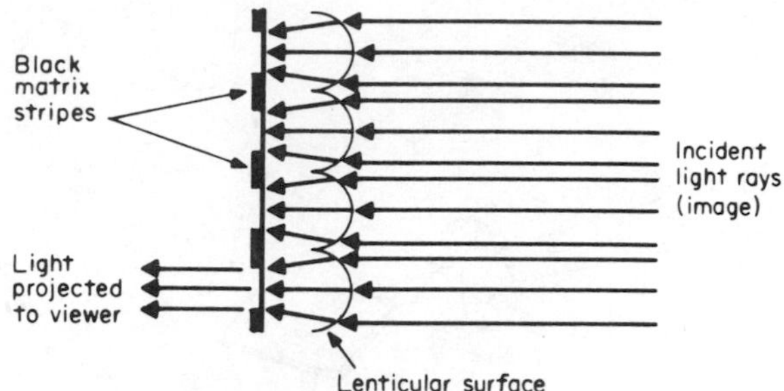

Figure 12.8 High-contrast projection.

accomplished by the use of a ribbed lenticular-lens configuration on the projector side, and light-absorbing black striping on the viewing surface as shown in Figure 12.8.

12.3 Light-Valve Projection Displays

12.3.1 Eidophor Reflective Optical System

Light-valve systems are capable of producing images of substantially higher resolution than that required for 525/625-line systems. Thus, they are ideally suited to large-screen theater displays of HDTV. The Eidophor, named from a Greek term meaning "image bearer," manufactured by Gretag, was developed by Dr. Fischer in Zurich, Switzerland.

Schlieren Optics In a manner similar to film projectors, a fixed light source is modulated by an optical valve system called *Schlieren optics* located between the light source and the projection optics. In the basic Eidophor system, collimated light typically from a 2-kw xenon source is directed by a mirror to a viscous oil surface in a vacuum by a grill of mirrored slits as shown in Figure 12.9.

The slits are positioned relative to the oil-coated reflective surface so that when the surface is flat no light is reflected back through the slits. An electron beam scanning the surface of the oil with a television picture raster deforms the surface in varying amounts, depending upon the video modulation of the scanning beam.

Where the oil is deformed by the modulated electron scanning beam, light rays from the mirrored slits are reflected at an angle which permits them to pass through the slits to the projection lens.

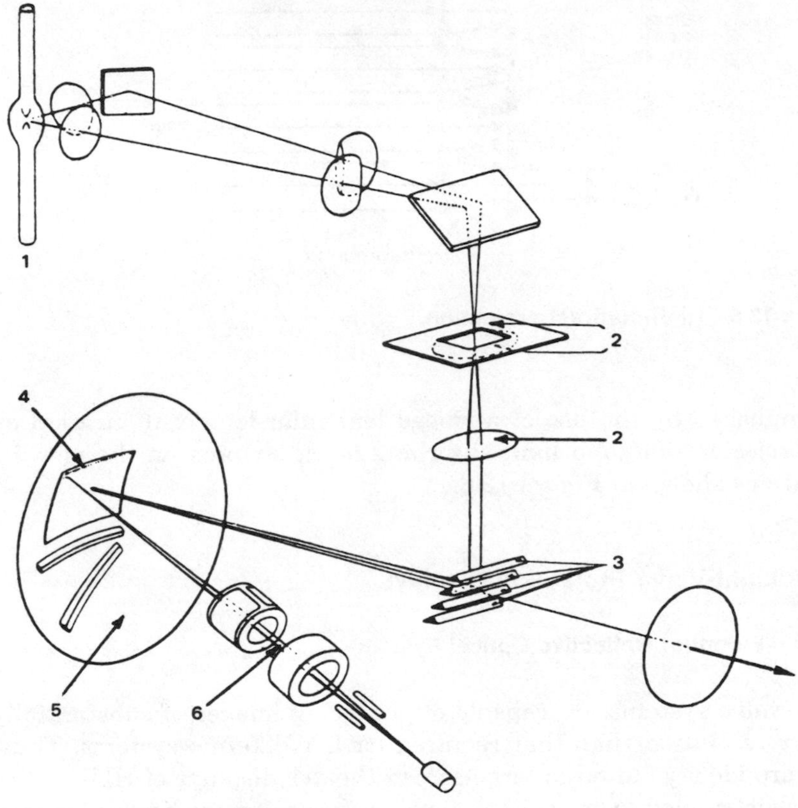

Figure 12.9 Diagram of Eidophor projector optical system.

The viscosity of the liquid is high enough to retain the deformation over a period slightly greater than a television field.

Color Display System Projection of color signals is accomplished by the use of three units, one for each of the red, green, and blue primaries converged on a screen. Brightness levels of 600 cd/m^2 (175 ft-L) are achievable on a 4-foot-wide screen.

12.3.2 Talaria Transmissive Color System

A later design, the Talaria, developed by a team headed by Dr. William E. Glenn, then at the General Electric Company, also uses the

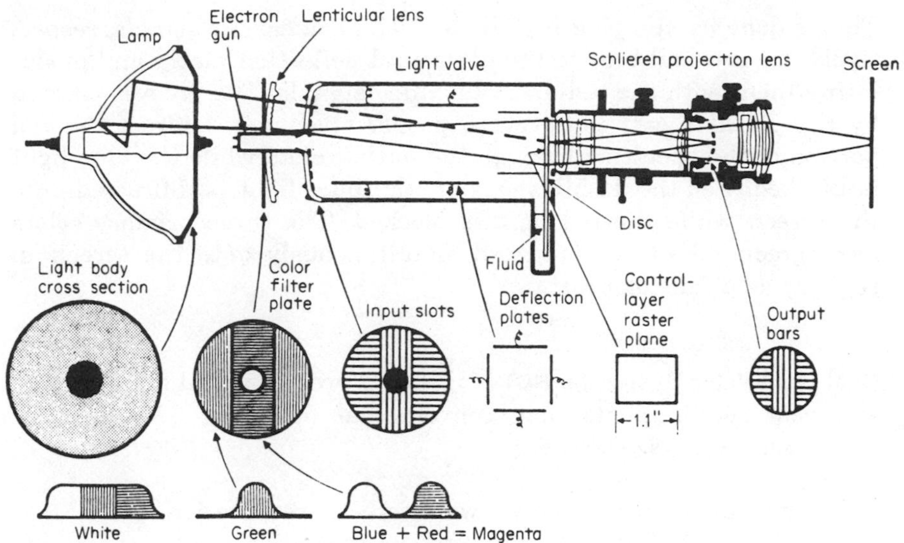

Figure 12.10 Diagram of the basic General Electric single-gun Light-Valve system.

principle of deformation of an oil film to modulate light rays with video information. However, the oil film is *transmissive* rather than *reflective*. In addition, for full color displays, only one gun is used to produce red, green, and blue colors. This is accomplished in a single light valve by the more complex Schlieren optical system shown in Figure 12.10.

Color Separation Colors are created by writing diffraction grating, or grooves, for each pixel on the fluid by modulating the electron beam with video information. These gratings break up the transmitted light into its spectral colors, which appear at the output bars where they are spatially filtered to let only the desired color be projected onto the screen (see Figure 12.10).

Green light is passed through the horizontal slots and is controlled by modulating the width of the raster scan lines. This is done by means of a high-frequency carrier, modulated by the green information, applied to the vertical deflection plates.

Magenta light, composed of red and blue primaries, is passed through the vertical slots and is modulated by diffraction gratings created at right angles (orthogonal diffraction) to the raster lines by velocity modulating the electron beam in the horizontal direction.

This is done by applying 16-MHz and 12-MHz carrier signals, respectively, for red and blue to the horizontal deflection plates and modulating them with the red and blue video signals. The grooves created by the 16-MHz carrier have the proper spacing to diffract the red portion of the spectrum through the output slots while the blue light is blocked. For the 12-MHz carrier, the blue light is diffracted onto the screen while the red light is blocked. The three primary colors (red, green, blue) are projected simultaneously onto the screen in registry as a full-color picture.

Dual Light-Valve System for HDTV* HDTV presents additional, more stringent requirements than conventional 525- and 625-line color television displays. These are:

1. Increased light output for wide-screen presentation.
2. Increase in horizontal and vertical resolution.
3. Broader gamut of color response to meet proposed system specifications.
4. Projection image aspect ratio of 16:9, rather than 4:3.

To meet these requirements, the General Electric designers chose to use a system with two light valves (TALARIA MLV-HDTV)[5] shown in Figure 12.11.

One monochromatic unit with green dichroic filters produces the green spectrum. Because of the high scan rate for HDTV (33.75 kHz), the green video is modulated onto a 30-MHz carrier instead of the 12 or 16 MHz used for 525- or 625-line displays.

Adequate brightness levels are produced using a 700-watt xenon lamp for green light valve and a 1300-watt lamp for magenta (red and blue) light valve.

A second light valve with red and blue dichroic filters produces the red and blue primary colors. This high-resolution magenta light valve is a new design, intended specifically for HDTV systems. The red and blue colors are separated by the use of orthogonal diffraction axes. In other words, red is produced when the writing surface diffracts light vertically. This is accomplished by negative-amplitude modulation of a 120-MHz carrier, which is applied to the vertical diffraction plates of the light valve. Blue is produced when the writing surface diffracts light horizontally. This is accomplished by mod-

*Contributed by Campolo F. Recuay, General Electric Co.[5]

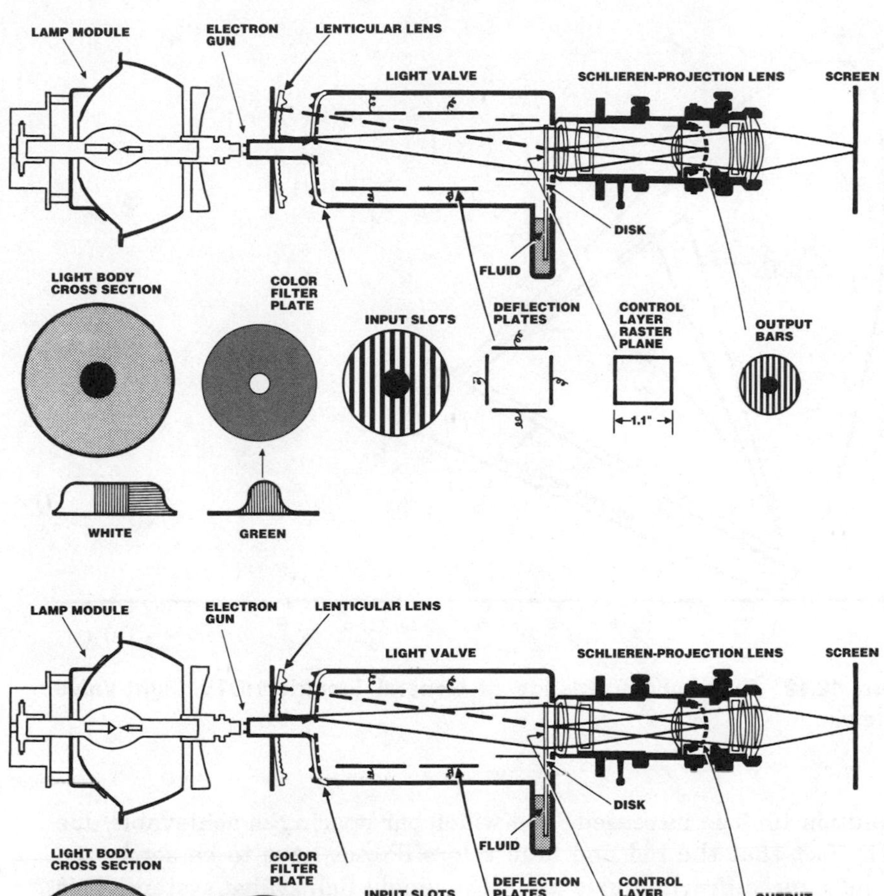

Figure 12.11 Diagram of the two-channel HDTV Light-Valve system.

ulating a 30-MHz carrier with the blue video signal and applying it to the horizontal plates as is done in the green light valve.

The input slots and the output-bar system of the conventional light valve are used, but with wider spacing of the bars. Therefore, the

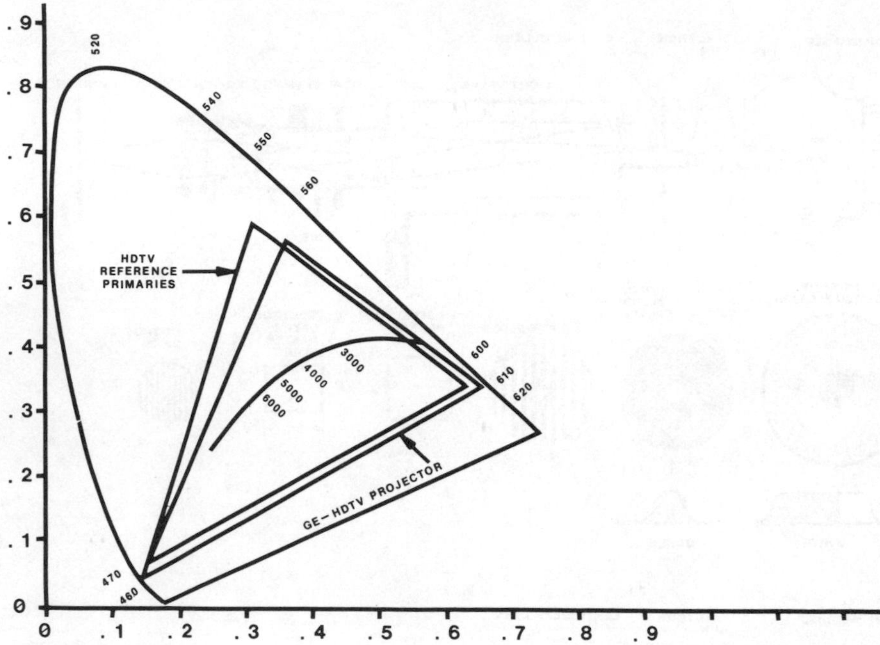

Figure 12.12 Color characteristics of General Electric HDTV Light-Valve system.

resolution limit is increased. The wider bar-spacing is achievable due to the fact that the red and blue colors do not have to be separated on the same diffraction axis as in the single light-valve system. This arrangement eliminates the cross-color artifact present with the single light-valve system, and therefore improves the overall colorimetric characteristic (see Figure 12.12).

High-resolution electron guns contribute in providing the required resolution and modulation efficiency for HDTV systems up to 1250 lines. The video carriers have been optimized to increase the signal-bandwidth capability to 30 MHz.

The writing surface in the General Electric system has been kept at the conventional 4:3 ratio in order to permit use of the basic hardware, and the required 16:9 widescreen presentation is obtained by means of a cylindrical anamorphic lens. Future designs will provide the wider aspect ratio by means of an appropriate change in the light-valve scanning configuration.

The performance specifications for the General Electric HDTV display system are listed in Table 12.8.

Table 12.8 HDTV Light-Valve Display Specifications

Brightness	2500 lumens
Resolution	
Horizontal	800 television lines per picture height
Vertical	700 lines
Contrast	200:1
Aspect ratio	16:9
Projection lens	Anamorphic 3.3x

12.3.3 Laser-Beam Projection Scanning System

Two approaches to laser-projection displays have been implemented on a limited scale, but since have been supplanted by the Eidophor or light valve for commercial applications. The most successful system employed three optical-laser light sources whose coherent beams are modulated electro-optically and deflected by electro-mechanical means to project a raster display on a screen.[6] The vertical scanning function is provided by a rotating polygon mirror, and the higher-rate horizontal scan by a vibrating mirror. A block diagram of the system is shown in Figure 12.13.

The other system employs an electron-beam pumped monocrystalline screen in a CRT to produce a 1-inch (25-mm) raster.[7] The laser-screen image is projected by conventional optics as shown in Figure 12.14.

12.4 Flat-Panel Displays

Flat-panel displays have become increasingly common as an alternative to the cathode-ray tube in highly portable test equipment, laptop computers, and small-screen television receivers. The basic technique used in addressing and driving flat-panel displays differs from the deflection systems used for CRT displays. Instead of the electron beams in a CRT that may be deflected by means of electromagnetic or electrostatic fields, the flat panel is composed of an assembly of discrete light-emitting elements that must be selected and driven by electrical signals corresponding to the intensity of television picture elements.

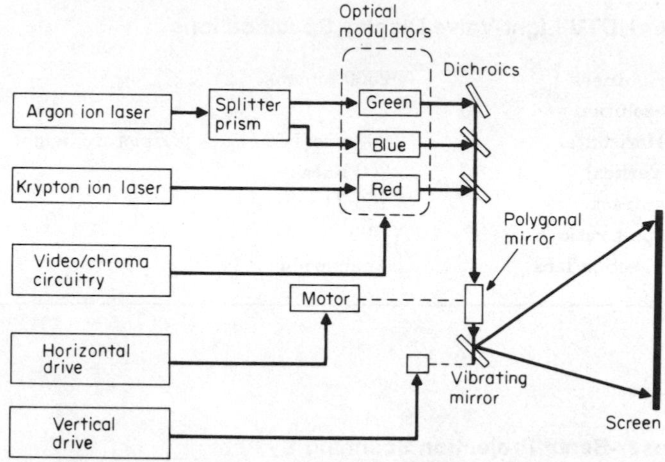

Figure 12.13 Block diagram of laser-scanner projection display.

12.4.1 Display Technologies

Current panel designs are based upon one of three technologies:

1. Gas-discharge or plasma display panels.
2. Thin-film (TF) electroluminescence (EL) panels.
3. Liquid-crystal displays (LCDs).

Plasma and Electroluminescent Panels The major drawback of plasma displays has been the high power drain, although significant

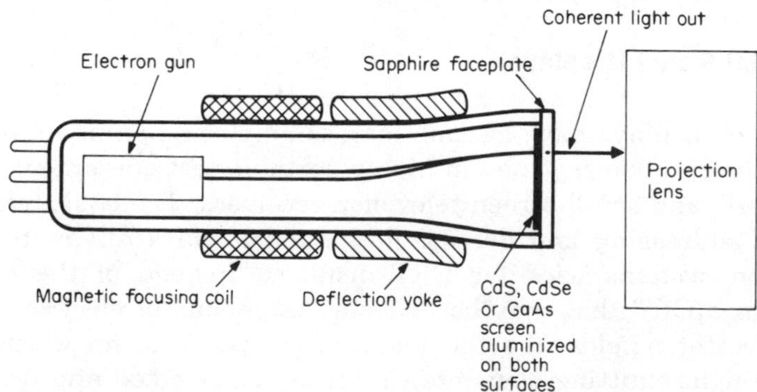

Figure 12.14 Laser-screen projection CRT.

progress has been made in recent years in surmounting this obstacle to consumer-product applications. In addition, both plasma and TF-EL displays have a limited gray scale in the order of under 20:1. This is sufficient for most computer and data displays, but is far below that necessary for acceptable television displays and to compete with motion-picture film.

Liquid-Crystal Displays Twisted-nematic LCD displays have a somewhat greater gray scale of up to 32:1 in computer applications, with a practical limit of 100:1 in direct-drive circuits. Unfortunately, for television displays, the already low contrast in such direct-drive circuits decreases with an increase in the number of television lines and the number of picture elements per line. This is a serious shortcoming for 525/625-line applications and a prohibitive limitation at the higher number of lines and pixels required for HDTV.

The solution for television applications has been to use a matrix of active transistor elements to drive each picture element (pixel). This permits each picture element to be switched on by the time-multiplexed drive signal and to stay on at the appropriate level until the next drive signal. In effect, it is a form of *sample and hold* for each pixel at a television frame rate. This technique, known as *matrix addressing,* produces high contrast for displays of 1000 or more rows.

Thin-Film Transistor LCD Whatever the approach, HDTV has practical limitations because of the number of transistors required. For example, over a million transistor drivers would be required for each of the red, green, and blue primary colors. The solution has been to fabricate thin films of transistors (TFT)[8] that can be coated over an entire substrate and selectively etched away. The TFTs serve as drivers for the panel of LCDs.

Design Limitations At this time, production of TFT-LCD color television displays are limited in screen size to 13-inch diagonal. Until the limited viewing angle of the twisted-thread LCD is solved, the market will be limited to small-screen "personal" viewing. Furthermore, the high production-rejection rate has resulted in a cost higher than that which can be supported in the consumer marketplace.

12.4.2 Matrix Addressing

Matrix addressing requires that the picture elements of a panel be arranged in rows and columns as illustrated in Figure 12.15 for a

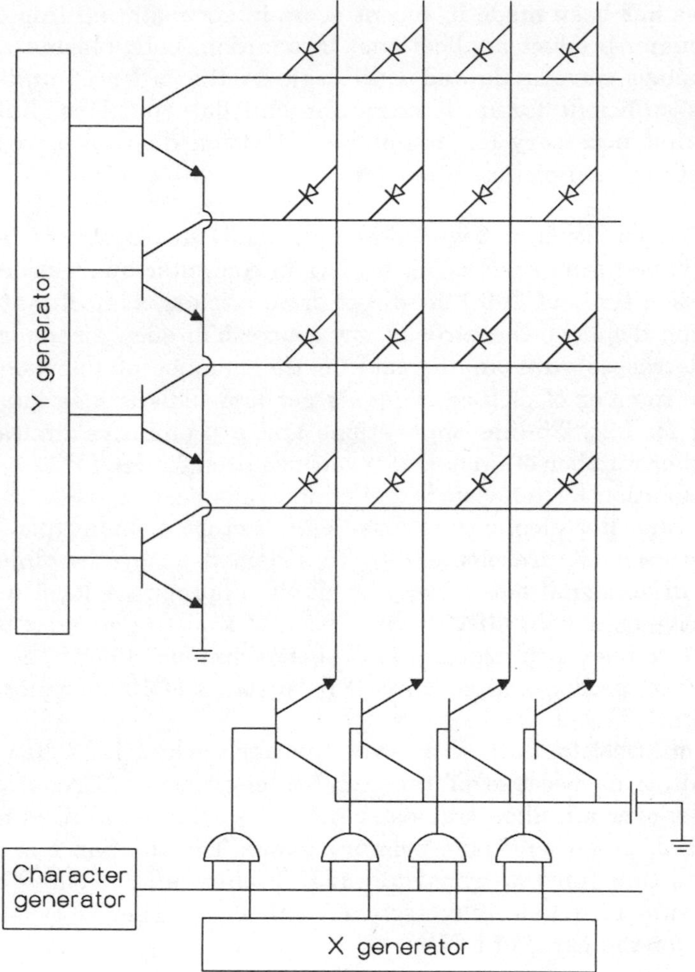

Figure 12.15 Matrix addressing for flat-panel displays.

simplified 4 × 4 matrix. Each element in a row or column is connected to all other elements in that row or column by active transistors. Thus, it is possible to select the drive signals so that when only a row or column is driven, no elements in that row or column will be activated. However, when both the row and column containing a selected element are driven, the sum of the two signals is sufficient to activate that element.

The reduction in the number of connections required, compared to the discrete-element drive approach, is very large, since the number of lines is reduced from one-per-element to the sum of the number of rows and columns in the panel. For example, in the small 4 × 4 array shown in Figure 12.14, the reduction is only 16 to 8, but for a 500 × 500 array, the reduction from 250,000 to 1000 is significant.

References

1. Pitts, K. and N. Hurst, Sarnoff Research Center, How Much Do People Prefer Widescreen (16 × 9) to Standard NTSC (4 × 3)? *IEEE Transactions on Consumer Electronics,* 35(3): 160–169 (August 1989).
2. Ashizaki, S., Y. Suzuki, K. Mitsuda, and H. Omae, Matsushita Electronics Corp. Direct-View and Projection CRTs for HDTV. *IEEE Transactions on Consumer Electronics,* 34(1): 91–98 (February 1988).
3. Uba, T., K. Omae, R. Ashiya, and K. Saita, Sony Corp. 16:9 Aspect Ratio 38V-High Resolution Trinitron for HDTV. *IEEE Transactions on Consumer Electronics* 34(1): 85–89 (February 1988).
4. Yamamoto, Y., Y. Nagaoka, Y. Nakajima, and T. Murao. Supercompact Projection Lenses for Projection Television. *IEEE Transactions on Consumer Electronics,* August 1986, p. 202.
5. Recuay, Campolo F., General Electric Co. Large Screen Projection Displays for HDTV. Presented in Milan, Italy, May 1989.
6. Benson, K. B., (Ed.) *Television Engineering Handbook.* McGraw-Hill, New York, 1986, p. 12.31.
7. Benson, K. B., (Ed.). *Television Engineering Handbook.* McGraw-Hill, New York, 1986, p. 12.13.
8. Credelle, T. L. TFT-LCD for Video Applications. *Society for Information Display* 5(3): 8–11 (March 1989).

Chapter

13

Program Production

13.1 Motion Picture Film Origination

13.1.1 Film Characteristics

Resolution The limiting image resolution on a print made by the normal 35-mm photographic process used for motion picture production is a cumulation of losses in the camera optics and in the photographic development and printing process. Table 13.1 lists the typical losses for the components of the filming process.

Microscopic examination of test pattern photographs has shown that no discernible modulation transfer frequency (MTF) fluctuations are present in a print at approximately 1100 television lines (50 cycles/mm).[1]

Image Steadiness Frame unsteadiness in motion-picture film results from frame-to-frame registration errors. For the most part these errors are caused by the mechanical characteristics of the perforations of the negative and positive film stocks, film transports in cameras, printers, projectors, and shrinkage of film stock with time. The frame-positioning errors at each of these stages occur at random intervals. Thus, the overall effect on the projected image is the root-mean-squared (rms) sum of all of the contributing errors.

Analysis of test pattern photographs[1] indicated that observed resolution fluctuated rapidly between 600 and 800 lines, for a visual

Table 13.1 Response at 1100 Television Lines (TVL) for Components of 35-mm Motion-Picture Film Process

Components	Response, 1100 TVL (%)
Camera lens, Panavision	30.0
Negative film, Kodak 5247	30.0
Positive film, Kodak 5381	60.8
System	5.4

average of 700 lines. Therefore, it can be concluded that the visual resolution in the average motion picture theater will be less than 700 TVL/PH. For the viewing audience seated further than 3 × PH, resolution of 700 lines or less may be expected.[1]

HDTV Resolution Requirements Thus, for HDTV to equal the projected resolution capability of good quality 35-mm film, at least 700 TVL are necessary. Taking into account the Kell factor for interlaced scanning (see Section 3.4), this indicates that 35-mm film is a suitable program-origination medium for an HDTV system of at least 1000 interlaced scanning lines with a maximum horizontal bandwidth of over 20 MHz.[1]

Film Formats In addition to the higher level of technical performance required of an HDTV film-to-video transfer system, a variety of film formats must be accommodated. They are:

1. Conventional television aspect ratio of 1.37.
2. Wide-screen nonanamorphic aspect ratio of 1.85.
3. Wide-screen anamorphic aspect ratio of 2.35.
4. Positive and negative films.
5. Pan and scan for the 1.37 picture aspect ratio required for simulcast or augmentation channels.
6. 16-mm films (optional).

The major source of 35-mm program material consists of films produced for television broadcast and prints of features made for theatrical release. In anticipation of approval of standards for HDTV, some filmed television programs are being produced in widescreen formats. On the other hand, a majority of theatrical productions are in widescreen formats—either horizontally compressed anamorphic Cinemascope (2.35:1) or nonanamorphic wide screen (1.85:1).[2]

For presentation of wide-screen films on conventional receivers by means of *simulcast* or *augmentation-channel* transmission, it is necessary that a *"pan-and-scan"* operation be performed to provide a picture framed in an aspect ratio window of 1.37 to cover significant action or information. The scanned copies of widescreen films for television presentation normally are produced in a programmed film printer at a film-processing laboratory or in a video postproduction facility during a film-to-tape operation.

Furthermore, in order to obtain the highest level of quality it is advisable that the original negative or, as a less than optimum alternate, a dupe negative be used for a transfer.

13.1.2 Film-to-Tape Transfer Systems

Three types of film-to-video transfer systems presently are used for 525- and 625-line television service. All are adaptable to operation on the proposed HDTV standards. However, one is dependent upon the availability of high-resolution CCD sensors on a production rather than experimental basis.

1. Telecine with intermittent pull-down film projector and a three-channel photoconductive camera.
2. Continuous-motion film transport with CRT flying-spot light source and three photoelectric transducers.
3. Continuous-motion film transport with three channels of CCD line sensors.

Projector-Camera Telecine A system employing a 35-mm intermittent pull-down projector and a three-channel photoconductive color camera has been used for the film-to-video HDTV transfers produced since 1986 by Japan Broadcasting Corporation (NHK) and since 1988 by Sony in their film-to-HDTV transfer service.[3] The system was developed in Japan by Ikegami Electronics with the technical assistance of the NHK for the anticipated HDTV television broadcasting service and video-theater markets.

A single 35-mm projector, an optical multiplexer to permit the addition of a second film projector and a slide projector, and the camera are shown in Figure 13.1. The multiplexed arrangement with the addition of a second projector permits the uninterrupted transmission of two or more reels. The camera and film projector are shown

Figure 13.1 Telecine camera and control unit with multiplexed 35-mm intermittent pull-down projector. (Courtesy Ikegami Electronics.)

in outline form in Figure 13.2. The projector, made by Seiko, provides exceptional image stability by means of double register pins in the intermittent claw 2-3 pull-down assembly.[4] In addition, a liquid gate reduces the visibility of any scratches that unavoidably may be present on the base side of a film, ensures a smooth movement of spliced film through the gate, and reduces the possibility of damage to valuable negatives. The latter is an important consideration in high-resolution systems.

Projection speed is continuously variable from still frame to 40 frames/s. This feature is useful for scene-by-scene color correction and in the pan-and-scan preparation of a transfer to the conventional television aspect ratio of 1.37. In the event transfers are required of films made specifically for television at 30 frames/s, either of two

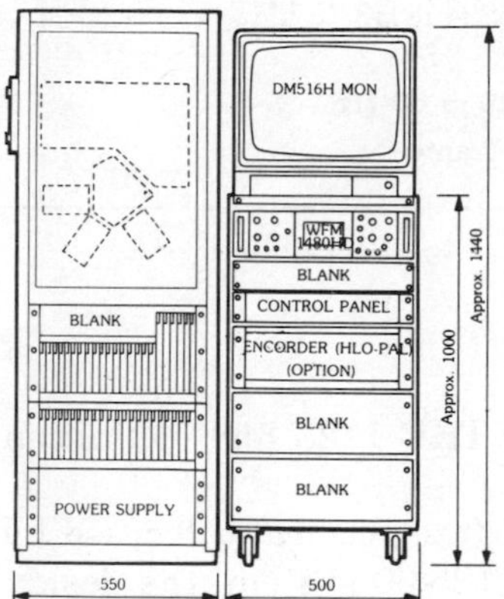

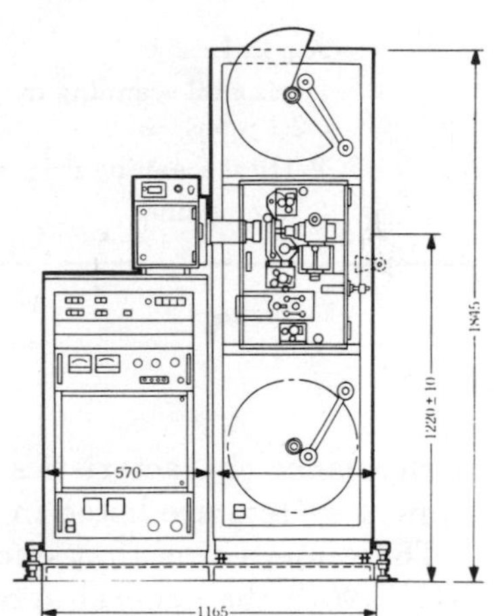

Figure 13.2 Outline drawing of camera and 35-mm projector assembly. (Courtesy Ikegami Electronics.)

television-signal synchronized speeds of 24 or 30 frames/s may be selected.

One-inch (25-mm) MS (magnetic focus, electrostatic deflection) Saticon photoconductive pickup tubes are used in the Ikegami TKC-1125 camera. The unity transfer characteristic of the Saticon tube provides a stable black level and uniform gray scale (see Section 10.2.1). Preset color matrixing and scene-by-scene adjustment of color correction permit optimum color balance and saturation. The

Table 13.2 Performance Specifications of Ikegami TKC-1125 Telecine Camera

Amplitude-frequency response (100 kHz reference)	60–30 MHz: ±0.5 dB
Modulation depth at 800 TVL (green channel)	35% at center 25% at corners
S/N ratio (Y signal)	44 dB peak/rms
Registration and geometric errors	0.05% (0.9 H & V) 0.1% (other areas)
Aspect ratio	16:9 (option 5:3)
Horizontal scanning frequency 2:1 interlace	33.76284 or 33.75 kHz
Vertical scanning frequency	59.940 or 60 Hz
Scanning lines	1125/frame

performance characteristics of the Ikegami TKC-1125 high-definition camera system are listed in Table 13.2.

The camera was first offered with the 5:3 aspect ratio for use by NHK. With the general agreement on an 1125/60 system, the design was revised to provide the wider aspect ratio of 16:9.

Continuous-Motion Flying-Spot Scanner System[5,6] In April 1989 Rank Cintel introduced their MkIII CRT flying-spot transfer system. A new design of the cathode-ray scanner tube and the electron-optics system provided a higher resolution capability and signal-to-noise ratio adequate to meet high-definition system performance requirements. The scanner is capable of handling all film formats and performing programmed pan-and-scan transfer operations. Unlike previous designs of film scanners, the flying-spot CRT and its deflection system are sealed in a factory-tested magnetically shielded assembly. This eliminates the need for operating and maintenance adjustments and thus assures the maximum resolution capability required for HDTV.

An 1125/60 version, the MkIII-HD telecine, has been field tested by a broadcaster and cable-system operators in anticipation of HDTV standards for the NTSC and PAL markets being approved. A limited number of systems have been sold for 1125-line transfer service in production of videotape for theater video systems.

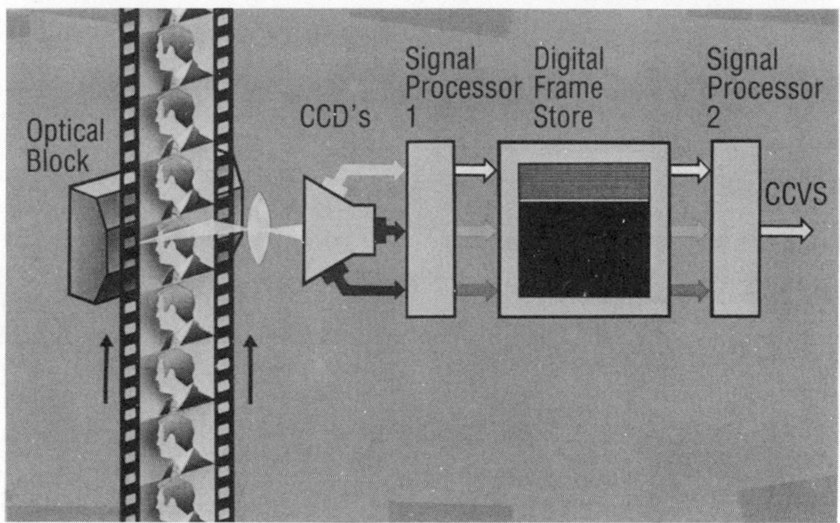

Figure 13.3 CCD film-scanning and signal-processing sequence. (Courtesy BTS Broadcast Television Systems.)

However, in view of the fact that broadcast standards are not yet agreed upon by regulatory bodies and equipment users, Rank Cintel has stated publicly that the company will manufacture equipment to meet the requirements of whatever systems are adopted as national standards. In other words, all of the functions and features provided in 525- and 625-line versions will be available in all high-definition models.

Continuous-Motion CCD-Sensor Systems[7] The film transport for CCD systems is similar to that of flying-spot scanners. However, rather than the line scan of a continuously moving film provided by a *flying-spot* of light, a light source projects the image of a continuously moving film on a horizontal row of red, green, and blue light sensors, each trio corresponding to one pixel of a line. The line scan is accomplished by sampling the voltages sequentially. The vertical-scan function is provided by the continuous motion of the film. The output signals from the line scan are stored as frames in a digital memory. The digitally stored frames can be read out and processed to provide an analog output color signal in synchronism with a reference video signal (see Figure 13.3).

13.2 Videotape Origination

13.2.1 Tape-to-Film Laser-Beam Transfer

Limitations of CRT Transfer The major restriction to the widespread use of electronic cameras for the production of programs for theater distribution has been the quality limitations in the production of release prints on motion-picture film, ideally in a 35-mm format. Kinescope film recording of images from a cathode-ray tube, at best, is capable of providing marginal resolution and contrast range for 525- and 625-line television viewing.

The technical design problems are concerned primarily with the difficulty in producing the high brightness needed to properly expose suitable recording film and to match the spectral characteristics of color picture tubes and color film. For theater viewing the requirements can be met only by the use of a high-definition television system and a film recording system that do not utilize an intermediate cathode-ray tube imaging step.

Film Exposure by Laser Beams This can be accomplished by the use of converged high-resolution laser beams to expose color film directly (see Section 7.3). The vertical scan is provided by a moving mirror and the line scan by a mirrored rotating polygon. This was demonstrated in 1971 by CBS Laboratories in the United States.[8] Subsequently NHK, and other companies under their guidance in Japan, developed prototype 16-mm recorders using a similar system. This led to the design of an improved system for 35-mm film recording of 1125-line high-definition television signals.[9]

The red (He-Ne), green (A_r^+), and blue (He-Cd) laser beams are varied in intensity in accordance with the corresponding video signals by acoustic optical modulators (AOM) with a center frequency of 200 MHz and a modulation bandwidth of about 30 MHz. The three color signals, converged by dichroic mirrors, are deflected horizontally by a 25-sided, 40-mm diameter polygon mirror rotating at 81,000 r/m.

Recording Film Experiments in direct recording on 35-mm color print film, such as EK 5383 have shown that the performance characteristics of a direct laser-beam recording can meet the resolution and noise level requirements of high-definition television and is superior to a conventional 35-mm print. A comparison of the modulation transfer functions (MTF) for various color films and a scanned laser beam are shown in Figure 13.4.

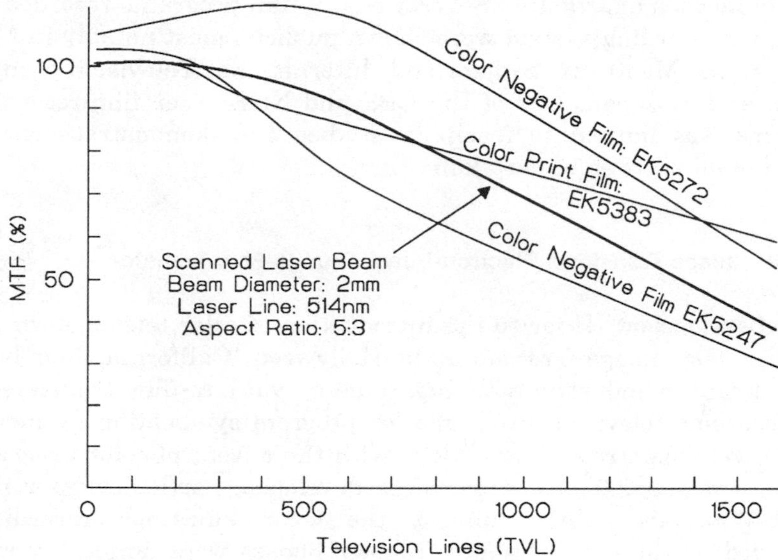

Figure 13.4 Modulation-transfer function (MTF) of typical films and scanned laser beam.

Higher resolution and less granular noise, in exchange for a 100:1 lower exposure speed, can be obtained by the use of color internegative film. This slower speed, in the order of ANSI 1 to 5, is within the light-output capability of a laser-beam recorder. Furthermore, sharpness can be improved significantly in laser recording by the use of electronic contour correction, a technique not possible in conventional negative-positive film photography.

Recording Camera The recording camera in the NHK system used a continuously moving, rather than an intermittent, film-transport mechanism. This required a scanning-standards converter to translate the video signal from interlaced to a noninterlaced sequential scan at a frequency of 33.75 kHz. The aspect ratio of the high-definition system was the standard of 5:3 proposed at the time of the system development by NHK.

The sound recording was accomplished also by laser-beam means to produce a stereo sound quality superior to conventional 35-mm optical-print techniques.

NHK Production Experience Several HDTV film programs recorded on the laser recording system were shown publicly, most notably in May 1983 at the Montreux, Switzerland, International Television Symposium, and in Japan. Use of the CBS and NHK laser film-recording systems was limited to feasiblity studies and demonstrations. No commercial use was undertaken.

13.2.2 Image Transform Electron-Beam Tape-to-Film Transfer

Early Development Prior to the introduction of color television in the early 1950s, Image Transform in Hollywood, California, had been providing the industry with high-quality video-to-film transfers of monochrome television programs for program syndication by means of electron-beam recording (EBR). With the advent of color programming, a demand for color transfers developed, particularly for network "specials." Unfortunately, the color kinescope recordings provided by the networks and transfer houses were noticeably inferior to a live release.

Color Program Transfer To fill the need, using their basic monochome EBR equipment, Image Transform developed an electron-beam recording system for color television programs on 35-mm film as black-and-white color-separation masters from which a color internegative and color release prints could be produced. When processing 30 frames per second videotape playback signals, the black-and-white master is exposed at three times "video real time." This exposes designated separate red, green, and blue frames for each frame of the final color film (Figure 13.5). The black-and-white master exposure rate is 72 frames per second to produce 24 frames per second color-film release prints. A commercial service was inaugurated in 1972.[10,11]

The system has been used also for recording high-definition wideband NTSC signals, for special nonstandard-signal applications requiring higher resolution, and for 24-frame video[10] out of real time by the use of a computer to control the VTR playback and the frame rate of the system read-in/read-out.

13.2.3 Sony Electron-Beam Tape-to-Film Transfer

The introduction of 1125/60 HDTV (SMPTE 240M) as a practical signal format for program production and presentation in theaters, as

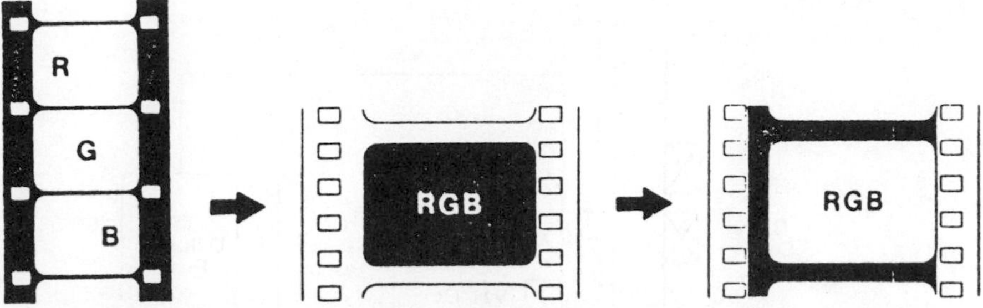

Figure 13.5 Steps for production of successive-exposure master and color internegative in the Image Transform tape-to-film transfer system.

well as for broadcast-television viewing, prompted further investigation of EBR technology to meet the system specifications of HDTV.[12]

The recording system operates out of real time to expose a black-and-white film master. For this purpose, a Sony Model HDVS VTR was modified to provide slow-speed red, green, and blue component-color playback capability at one frame per second. A block diagram of the recording system is shown in Figure 13.6.

The high-definition red, green, and blue analog video signals are converted to digital signals for: (a) subsequent gamma correction to compensate for the characteristics of the master recording film and color negative, and (b) conversion from 30-frames per second interlaced television scan to 24-frames per second progressive scan in a red, green, and blue sequence. The progressive-scan mode was chosen for a maximum of resolution and a minimum of artifacts. The sequential R, G, B digital signal is converted to analog to feed the EBR (see Figure 13.7).

The sharply focused electron beam scans the film surface in a vacuum, producing a latent image in the emulsion. The conductivity of raw film stock is lowered substantially by evaporation of moisture from the film in the vacuum chamber necessary for the electron-beam scanning process. A reduction in conductivity can result in the buildup of a static charge and accompanying distortion of the beam scanning beam. Therefore, the scanning operation is limited to one scan per frame.

The film is moved through the aperture gate in the electron-beam recorder chamber by an intermittent claw pull-down mechanism. Monitoring of the scanning beam for maintenance, and adjustment of scanning parameters such as aspect ratio and size, when not

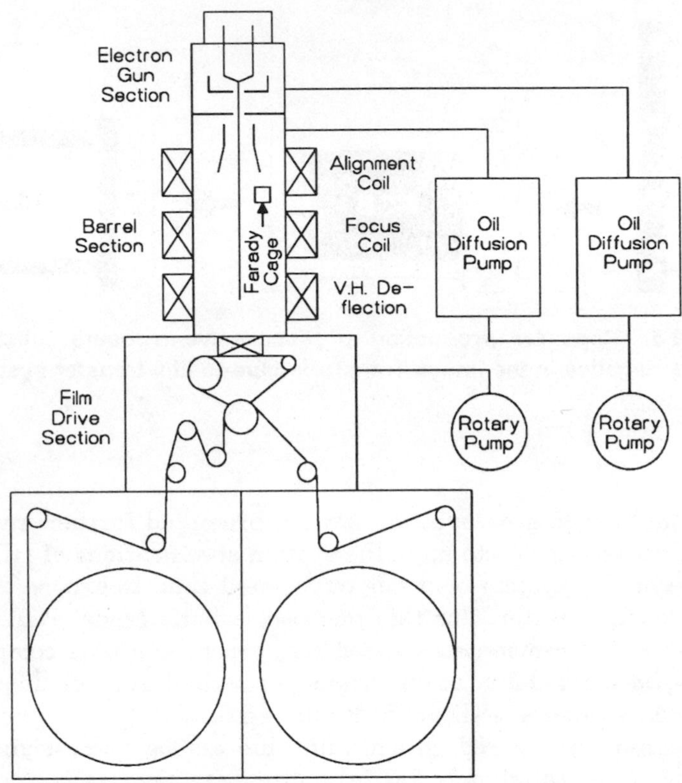

Figure 13.6 Outline drawing of Sony tape-to-film electron-beam recording unit.

recording is provided by a fluorescent screen which can be set in place of the film pressure plate.

EBR Writing Speed The virtually still-frame film speed of one frame/s permits considerable latitude in the choice of black-and-white film for the EBR master recording. After extensive experiments, a progressive scan with a 20-nA beam at a raster scan rate 2.5 times the original video signal rates of 33.75 kHz/horizontal line and 30 Hz vertical/frame was found to be optimum. Recording of a 46-MHz burst fed into the EBR chamber by the EBR electron gun has indicated an excellent modulation depth.[12] The film stock used typically in the Sony system is Fuji 71337 fine-grain release positive. This provides a density of 2.0 with an EBR current of 30 nA.

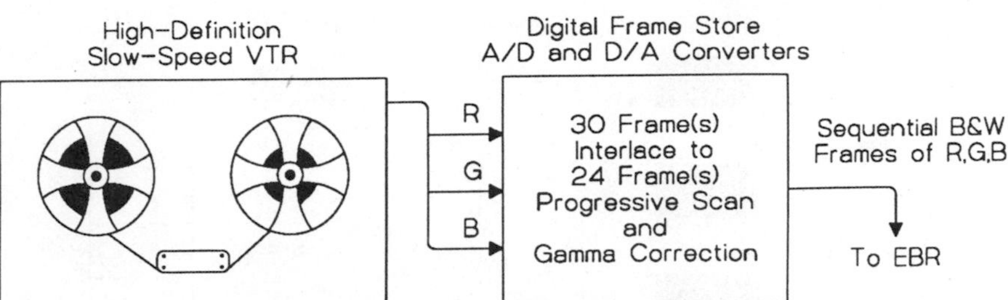

Figure 13.7 Sony tape-to-film electron-beam recording system. (Courtesy Sony Advanced Systems.)

Color Film Production The sequential black-and-white exposures of red, green, and blue video signal produced by the EBR process are transferred to a color internegative in an optical step printer, as shown in Figure 13.8. Two shutters are used: one to blank the pull-down of the master projector and, at one-third the speed, a second to blank the pull-down of color camera.

Each frame of the color negative film is exposed sequentially to the black-and-white master frames corresponding to the red, green, and blue video signals through appropriate red, green, and blue color filters mounted on a rotating disk, thus providing a color composite. Typically, using EK-5272 color internegative, the color filters employed are Wratten numbers 29 for red, 99 for green, and 98 for blue.

13.3 HDTV Videotape Systems

13.3.1 Analog 1-Inch Systems

Early Developments In the early 1980s an HDTV system based on 1125 horizontal lines at a field rate of 60 Hz was proposed by segments of the television industry. The bandwidth capabilities were 20 MHz for luminance and a maximum of 7 MHz for each of the two color difference signals. These specifications were further refined by the SMPTE in their proposal 240M presented to the American National Standards Institute (ANSI) in 1987. The luminance bandwidth and chrominance bandwidths were increased to 30 MHz and 15 MHz, respectively, in the SMPTE 240M document.

Videotape recording and reproducing equipment for the 1125/60 HDTV system was first available in the mid-1980s.[13] FM signal

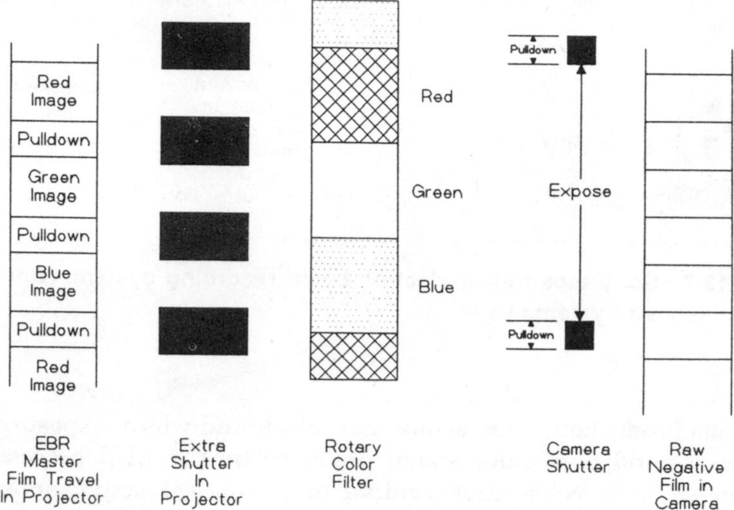

Figure 13.8 Relative timing of components in EBR step-printing process. (Courtesy Sony Advanced Systems.)

processing, similar to conventional NTSC and PAL systems, was employed in the prototype equipment. By 1987 over 100 analog recorders were in use for the production of programs in the HDTV format and for further engineering investigation and developoment.

Limitations The most significant shortcoming highlighted by the production experience with the prototype recorders was the progressive degradation of picture quality by the increase in video noise and transients introduced with each additional generation. This limitation of analog recording is of major concern in conventional 525- and 625-line systems and has led to the development of the D2 digital recording system. The limitations are much more evident in HDTV presentations where the viewing distance relative to picture size may be as close as three times picture height. It is a serious consideration in HDTV program production and presentation on a large screen to theater audiences. For example, a noticeable mismatch will result when a fourth generation or higher is intercut with first generation material.

The Need for Multiple Generations Whether a program is shot in multiple takes from a single camera or simultaneously on several cameras, assembly into a contiguous artistic production dictates that the

Table 13.3 Analog and Digital Parameters of SMPTE 240M 1125/60

Parameter	Analog bandwidth (MHz)	Digital Sampling (MHz)
Luminance (Y)	30	74.25
Color (R-Y)	15	37.125
Color (B-Y)	15	37.125

director and editor be given complete flexibility in postproduction procedures. Consequently, the use of matting, blowups and zooms, and other special effects more often than not results in multiple generations to achieve the desired creative result. Furthermore, substantial savings in postproduction costs can be realized if up to eight or more generations are of acceptable quality, suitable for matting, matched cuts, and other manipulations of images and frames.

This requirement of no degradation of quality in up to 10 or more generations has prompted the move to digital signal processing in all stages of production up to the final distribution copy.

13.3.2 Digital 1-Inch Systems

SMPTE 240M Digital Format The basic analog and digital parameters of the 1125/60 SMPTE 240M Production System are listed in Table 13.3.

Each of the luminance and color components are linearly quantized at 8 bits per sample to produce a theoretical p-p/rms video signal-to-noise ratio of 56 dB. As shown in the following equation, this results in a data rate of 1.188 Gbits/s.

$$8\ (74.25 + 2 \times 37.125) = 1.188\ \text{Gb/s}$$

The sampling frequencies are 5.5 times those selected in CCIR Recommendation 601.[8] These are in a video coding ratio of 22:11:11. The corresponding ratio of the components in a 625-line system is 4:2:2. This 5.5 relationship between the two systems results in a simple conversion between component-video analog standards and the digital standard.

Recording Tape The choice of recording tape is limited because of the high resolution capability required to record the short wavelengths of the digitized HDTV signal. Metal-particle tape was selected by Sony over evaporated because of the extensive and satisfactory experience with it in development of the D-2 format. The formulation selected has a coercivity (H_c) of 1450 Oe and a retentivity (B_r) of 2450 G.

The tape coating, base thickness, and particle orientation were chosen to accommodate the mechanical configuration of the Type-C 1-inch equipment (Sony BVH-3000).

Video Heads Performance data on D-2 525- and 625-line recorders indicated that a 30-dB high-frequency signal-to-noise ratio for recording and playback will result in a sample error rate, with error correction, greater than 1×10^{-6}. This is adequate for good quality HDTV television picture reproduction and can be obtained with a writing speed twice that of a Type C VTR. Using the same drum diameter as the Type C with a rotational speed of 7200 rpm, the writing speed is 51.5 m/s. The minimum recorded wavelength is 0.69 μm. The principal parameters of the 1125/60 HDTV and 525/59.94 systems are compared in Table 13.4.

The very high data rate of 1.188 Gbits/s is not amenable to sequential recording by a single head. A practical solution was to apportion the data into parallel recording channels. Accordingly, approximately 150 Mbits are assigned to each of eight channels. Sixteen video heads, in groups of four each, are mounted in quadrature on the rotating head drum as shown in Figure 13.9. Two additional heads on opposite sides of the drum between the video-head assemblies provide the flying-erase function. The record/playback amplifiers are mounted on the head drum. Connection to the record and playback circuits is accomplished by means of a wide-band 18-channel rotary transformer.

Recorded Tracks The layout of the recorded tracks is shown in Figure 13.10a. Figure 13.10b is an enlarged view showing the 16 diagonally recorded tracks comprising a single television field. The HDTV video tracks are a quarter the width of Type C, thus permitting a single field to be subdivided into 16 parallel tracks.

Stationary Heads Longitudinal tracks recorded by stationary heads are: time code, two cue channels, and eight digital-audio signals. The latter are recorded in the DASH (digital sudio, stationary head) format.

Five of the eight audio heads are intended to permit complete flexibility in insert and assemble audio editing. The functions of each are

Table 13.4 Principal Parameters of 1125/60 HDTV and 525/59.94 Type-C Videotape Recording Systems

Parameter	1125/60 HDTV	525/59.94 Type C
Tape width	25.4 mm (1 in)	25.4 mm (1 in)
Tape length (63 min)	3080 m (1010.5 ft)	4320 m (3026 ft)
Tape speed	805.2 mm/s(31.7 in/s)	244 mm/s(9.606 in/s)
Head-to-tape speed	51.5 m/s(2027 in/s)	25.59 m/s(1007.5 in/s)

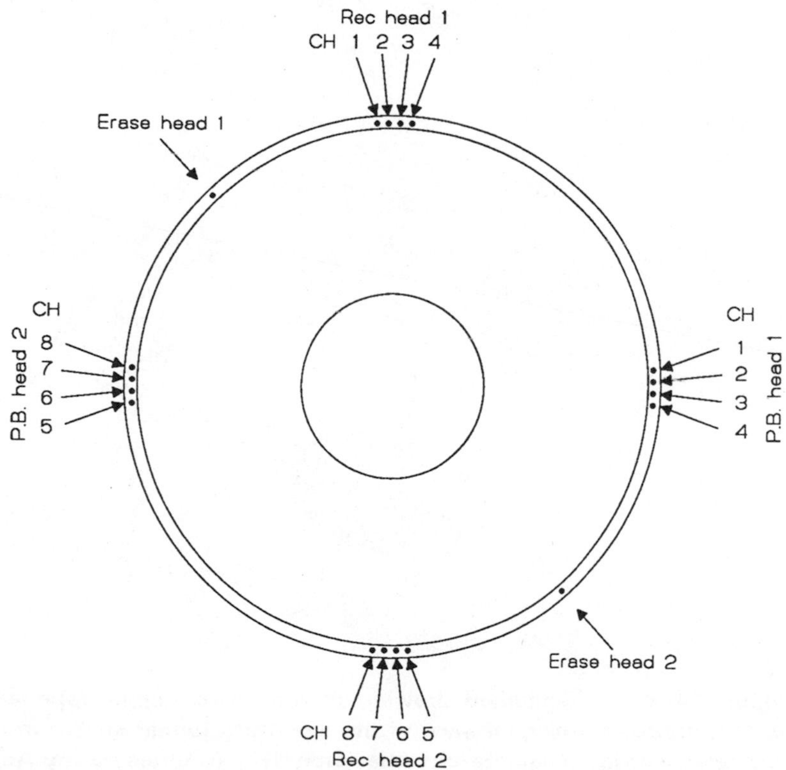

Figure 13.9 Placement of rotating heads on Sony HHD-scanner. (Courtesy Sony Advanced Systems.)

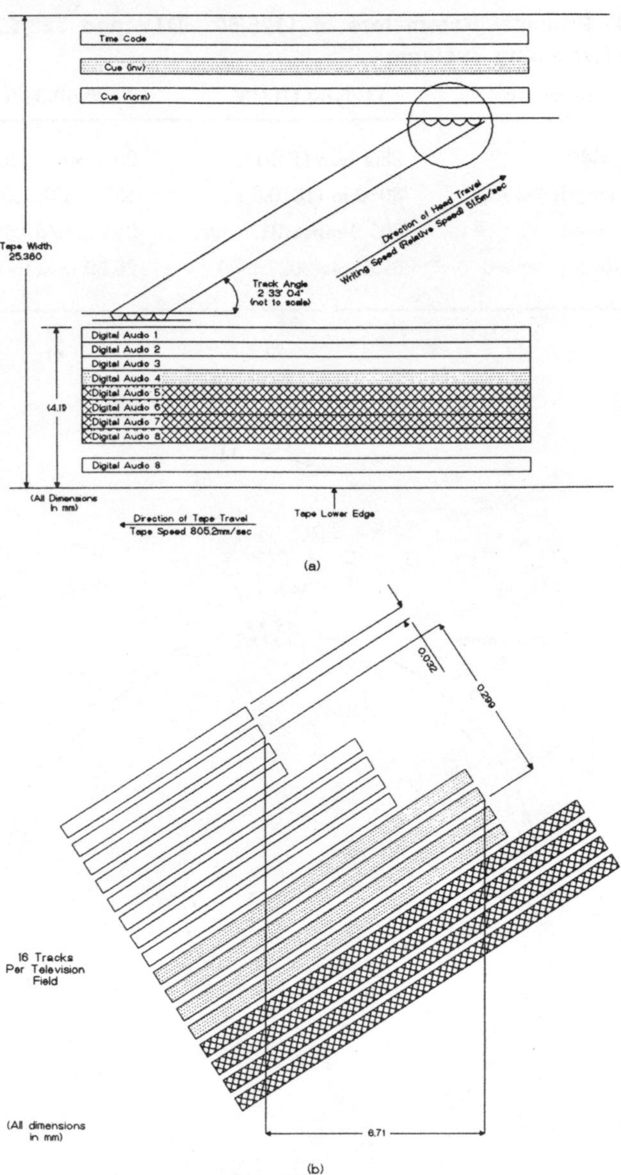

Figure 13.10 (a) Simplified drawing of Sony HDD digital tape-recording track format. (b) Enlarged view of the tape-track format shown in (a). The 16 tracks shown constitute one television field. (Courtesy Sony Advanced Systems.)

the following: digital recording, digital playback, digital confidence, analog record/playback, and analog erase.

13.3.3 Analog 1/2-Inch Cassette System*

Applications The open-reel one-inch HDTV videotape systems are suitable for program production where equipment size and the need for highly qualified technicians as operators are not serious restrictions. On the other hand, when broadcast and cable networks begin HDTV program transmissions, the large industrial marketplace will be anxious to follow suit. However, the equipment must have the operational simplicity and compact packaging approaching, if not equalling, that of current 525- and 625-line video-cassette recording (VCR) equipment. Furthermore, with the introduction of large-screen projection HDTV displays for home viewing, the consumer market for the capability of viewing cassette movies and to time-shift recording of broadcast programs in HDTV will develop rapidly.

In anticipation of this evolution, Toshiba, in conjuction with NHK Engineering Service has developed an HDTV VCR which, at this stage of development, promises to meet the industrial need and can serve as the basis for a subsequent consumer design.[14,15]

Cassette and Tape The cassette is slightly larger than the standard VHS design. At the tape speed of 119.71 mm/s (4.713 in/s) the maximum recording time is 63 minutes. This is the same recording time as the 1-inch HDTV format described in Section 4.3.2. A metal particle tape formulation with a coercivity of 1500 Oe is the recording medium.

Design Specifications The basic specifications developed in conjunction with NHK are listed in Tables 13.5, 13.6, and 13.7.

Recorded Helical Tracks Figure 13.11 is a simplified drawing of the recorded tape-track pattern. The tape wrap around the drum is approximately 204°. Allocation of the signals on each track is as follows:

FM video (azimuth angle of adjacent tracks, ±15°)	180.0°
Gap for editing and head switching	6.5
PCM audio	17.5

*Adapted from A 1/2-Inch Cassette HDTV VTR, Kizu, S., N. Endo, and K. Ogi, Reference 15. (Courtesy Toshiba Corp., Kawasaki, Japan.)

Table 13.5 Mechanical Specifications of 1/2-Inch VCR System

Recording time	63 min
Cassette dimensions	205(w) × 121.5(d) × 121.5(d) × 25(h) mm
Tape length and width	453 m, 12.65 μm
Tape thickness	13.5 μm
Magnetic material	Metal particle, coercivity 1500 Oe
Drum diameter	75.0 mm
Drum rotational speed	90 r/s
Writing speed	119.71 mm/s (4.713 in/s)
Helical angle	21.4° (stationary) 4.27° (relative)
Lap angle	204°
Azimuth angle	± 15°
Track pitch	24.8 μm with 4.8-μm guard band

Source: Courtesy Toshiba Corp.

Control, time code, and an analog audio track are recorded longitudinally by stationary heads.

The two tracks labeled A and B constitute a segment; three segments make up one television field. Because the drum rotational speed is 90 r/s, head-switching occurs in picture and differences among heads will appear as "banding" and "skew." These degradations are avoided by the use of field memories to reconstruct the TCI lines into such a form that the upper portion of the picture is recorded onto the beginning portion of the tracks (segments) and the lower portion of the picture is recorded onto the end portion of the tracks (segments). In other words, the subsequent lines are distributed into six tracks. This method of "shuffling" practically eliminates the visiblity of banding and skew.

Table 13.6 Video Recording Signal Specifications of 1/2-Inch VCR System

Video-signal bandwidth	Y = 20 MHz Pr, Pb = 7 MHz (line sequential)
Recording format	Y, C time-compressed integration, 2 channels, 3 segments, FM recording
Sampling frequency	F_y = 48.6 MHz, F_c = 16.2 MHz
TCI sampling frequency	tci = 28.998 NHz
FM deviation	Sync-tip to reference white 7.26 MHz

Source: Courtesy Toshiba Corp.

Table 13.7 Audio Recording Signal Specifications of 1/2-Inch VCR Systems

Program channels	PCM 4 channels
Sampling frequency	48 kHz
Quantizing level	16 bits
Recording data rate	48.6 Mb/s
Modulation system	8–14
Error correction	Reed-Solomon product code
Interface	Digital and analog
Auxiliary audio channel	Analog 1 channel
Recording type	Bias recording
Time code	Saturation recording
Control signal	Saturation recording

Source: Courtesy Toshiba Corp.

Recording Signal Processing Digital techniques are used to process the video signal prior to recording and after playback. The luminance signal is time-expanded and the two color signals are time-compressed to produce signals with equal bandwidths of 12 MHz for

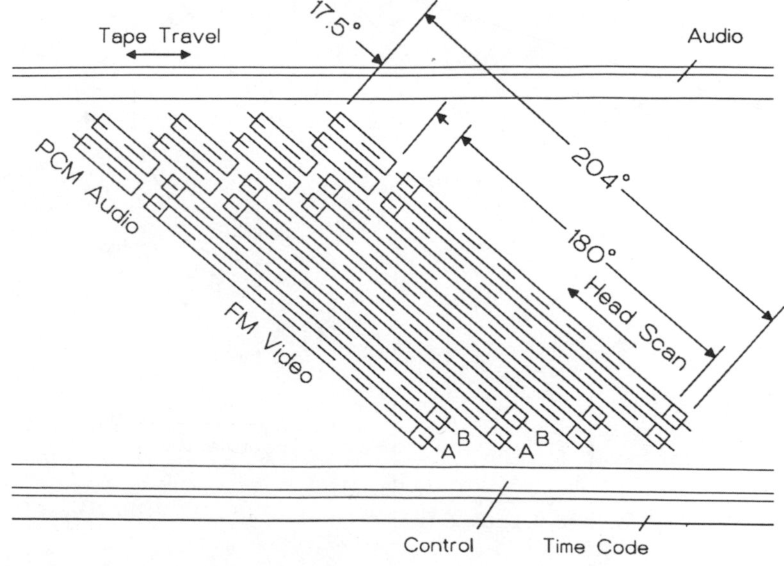

Figure 13.11 Simplified drawing of Toshiba 1/2-inch VCR tape-track pattern. (Courtesy Toshiba Corp.)

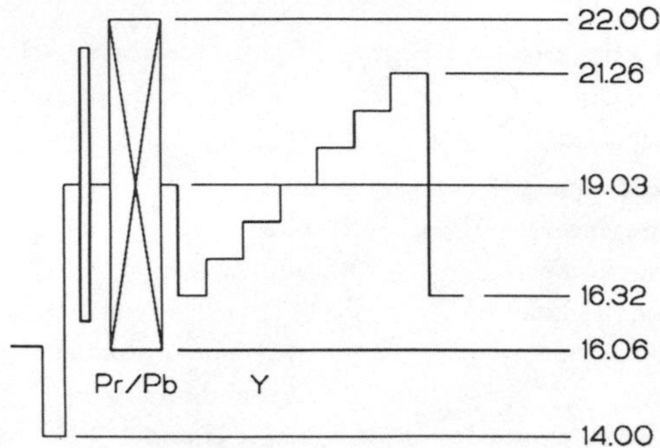

Figure 13.12 Deviation frequencies of FM video-signal carrier.

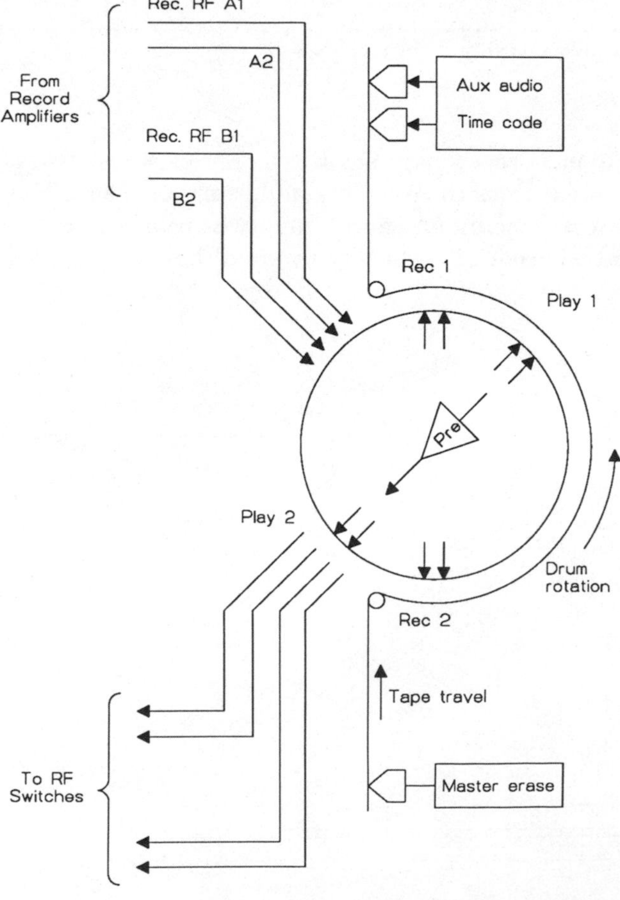

Figure 13.13 Relative angular placement of heads on 1/2-inch VCR drum. (Courtesy Toshiba Corp.)

each. This digital processing is termed "time-compressed integration" (TCI). The purpose is to reduce the overall bandwidth of the three signals.

After D/A conversion, the analog signal is applied to the recording heads as a frequency-modulated carrier with sync tips and peak white at 14 MHz and 21.26 MHz, respectively. The deviation frequencies for a stair-step video signal are shown in the waveform drawing, Figure 13.12.

Head Configuration Two pairs of recording heads, two pairs of playback heads, and two pairs of advance playback heads are mounted within the rotating drum. The relative angular placement is shown in Figure 13.13. The playback heads are located in a delayed staggered position at 45° relative to the recording heads. While recording, the playback heads act as confidence heads by picking up the recorded signal just after recording. The recording and playback timings between the video and audio signals are adjusted by means of a digital memory.

Eight preamplifiers for the playback heads are mounted within the rotating drum. The RF bandwidth necessary for video recording is 4–34 MHz. However, this has been extended to a low limit of 200 kHz to cover the spectrum of digital audio data.

References

1. Kaiser, A., H. W. Mahler, and R. H. McMann. Resolution Requirements for HDTV Based Upon the Performance of 35mm Motion-Picture Films for Theatrical Viewing. *SMPTE Journal* 94(6): 654 (June 1985).
2. ANSI Standard PH22.195 1977.
3. Tanaka, H. and L. J. Thorpe. Experiences of an HDTV Production Facility, Television—Merging Multiple Technologies. SMPTE Television Conference, January 26–27, 1990.
4. Benson, K. B. *Television Engineering Handbook.* McGraw-Hill, New York, 1986, pp. 17.4, 17.5.
5. Millward, J. D. Flying-spot Scanner on 525-lines NTSC Standards. *SMPTE Journal* 90(9): 786, 1981.
6. Benson, K. B. *Television Engineering Handbook.* McGraw-Hill, New York, 1986, p. 17.8.
7. Benson, K. B. *Television Engineering Handbook.* McGraw-Hill, New York, 1986, p. 17.9.

8. Beiser, L. et al. Laser-Beam Recorder for Color Television Film Transfer. *SMPTE Journal* 80(9): 699–703 (September 1971).
9. Sugiura, Y., Y. Noriri, and K. Okada. HDTV Laser-Beam Recording on 35mm Color Film and Its Application to Electro-Cinematography. *SMPTE Journal* 93(7): 642–651 (July 1984).
10. Comandini, P. Signal Processing in the Image Transform System. *SMPTE Journal* 86(8): 547–549 (August 1977).
11. Comandini, P. and T. Roth. Film Recording in the Image Transform System. *SMPTE Journal* 87(2): 82–84 (February 1978).
12. Thorpe, L. J. and Y. Ozaki. HDTV Electron Beam Recording. *SMPTE Journal* 97(10): 833–843 (October 1988).
13. Thorpe, L. J., T. Yoshinaka, and K. Tsujikawa. HDTV Digital VTR. *SMPTE Journal,* 98(10): 738–747 (October 1989).
14. Tanimura, H., Y. Hashimoto, and T. Yoshinaka. HDTV Digital Tape Recording. Third International Colloquium on Advanced Television Systems, Ottawa, Canada, October 4–8, 1987.
15. Kizu, S., N. Endo, and K. Ogi. A 1/2-Inch Cassette HDTV VTR. SMPTE 24th Annual Television Conference, January 26–27, 1990.

Chapter

14

Future HDTV Development and Utilization

14.1 Introduction

Since the early 1980s high-definition television has been heralded as the next major development in television home entertainment. The public interest has built up to a level where hardly a day goes by without mention in the daily press of HDTV developments, plans by industry for equipment manufacture, and anticipated adoption of system standards in the United States and overseas. On the production front, currently in the United States and Canada HDTV programs are being produced in accordance with the SMPTE 240M 1125/60 Production Standard for theatrical presentations and in anticipation of approval of an HDTV system for broadcast. Overseas, Japan is inaugurating an HDTV broadcast service using the 1125/60 system.

This chapter reviews HDTV developments over the last decade and recent progress toward standardization in the United States by industry and the FCC. The implication of possible adoption of simulcast transmission standards and their implementation is discussed

---

Contributions by C. Robert Paulson, AVP Communications, Westborough, MA.

with emphasis on the problems presented to the local television broadcaster, national and regional networks, and cable systems.

14.2 System Development

14.2.1 Compatible or Simulcast Transmission

Although HDTV equipment has been available since 1984, standardization for broadcasting service has been slowed by lack of agreement on how the public can best be served. The primary consideration is whether to adopt a system fully compatible with NTSC or a simulcast system requiring additional transmission spectrum and equipment.

The principal proponents of the two approaches in the United States are the Advanced Compatible Television (ACTV) Research Consortium and The 1125/60 Group. The 1125/60 system, formalized by the SMPTE in its 240M Production Standard (see Chapter 7), is not compatible with 525-line NTSC standards. Viewers with NTSC receivers would continue to be served on existing channels with NTSC down-conversions of the HDTV signals.

14.2.2 Advanced Compatible Television (ACTV)

ACTV, developed by the Sarnoff Research Center (see Section 8.2), was demonstrated at the NAB Convention in April 1989. On April 20, prior to NAB, a short segment was transmitted from the Sarnoff Research Center in New Jersey to New York for broadcast in New York over a WNBC-TV evening news program to demonstrate the NTSC compatibility of ACTV. Consisting of two companion systems, the design was developed to comply with the FCC's "Tentative Decision" of September 1988, which required an HDTV broadcast standard to be compatible with NTSC receivers (see Section 8.2.8).

The basic signal, ACTV-I, was intended to provide a widescreen picture of improved picture and sound quality on new HDTV receivers, while being compatible with NTSC receivers on a single 6-MHz channel. A second signal, ACTV-II, would have provided full HDTV service on a second "augmentation" channel when such additional spectrum might be available.

The FCC, in March 1990, informally announced that it intends to select a simulcast standard for HDTV broadcasting in the United States and would not consider any augmentation-channel proposals, thus strengthening the position of the 1125/60 system supporters.

Nevertheless, the Advanced Television Research Consortium, composed of NBC, Sarnoff Research Center, and Philips' and Thomson Consumer Products, is continuing their development of the ACTV system in parallel with their embarking on the development of a simulcast system of their own design.

14.2.3 1125/60 System

Inception of Broadcast Service In the early 1980s Japan Broadcasting Company (NHK) initiated a development program directed toward providing a high-definition television broadcasting service. The video-signal format was based upon a proposed 1125/60 standard published in 1980 by NHK (see Section 7.2 and Table 7.1). Cameras, developed in cooperation with NHK, were provided by Sony. Experimental broadcasts were transmitted to prototype receivers in Japan over the MUSE satellite using this format.[1,2]

The 1984 Olympic Games in the United States, televised by NHK for viewers in Japan, was the first event of worldwide interest covered using high-definition television.[3] The HDTV signals were "pan-scanned" to a 1.33 aspect ratio and transcoded to 525 lines for terrestrial transmission and reception over regular television channels.

On June 3, 1989, NHK started regular HDTV program transmissions for about an hour each day using their MS-2 satellite. When a new satellite is launched in 1990, the program service will be increased to eight hours a day in the first stage of commercial application as the standard for broadcasting HDTV to the home via direct-broadcast satellite (DBS).

After considerable study by the cognizant engineering committees, a production specification (SMPTE 240M) describing the 1125/60 format was proposed for adoption as an American National Standard and published in April 1987 by the Society of Motion Picture and Television Engineers.[4]

Program Production Currently in North America, several teleproduction companies in major cities are using 1125/60 equipment for program production and postproduction editing, sometimes in place of 35-mm film or in combination with 35-mm film. For example, the 1125/60 system was used by the CBS television network to produce a "made-for-TV" feature, "Littlest Victims." The program was down converted to 525-line standards and aired over the network on Ap 23, 1989. In addition, some edited program masters have been c verted to film for theatrical distribution.

Another use for HDTV has been the conversion of videotape master material, composited from 35-mm film original negatives, to film internegatives for subsequent intercutting into film productions in the "film-editing" manner (see Section 13.2).

14.2.4 PAL and SECAM HDTV Systems

European Systems Evaluation Under the Eureka EU95 research program, eleven countries are cooperating to develop and evaluate a 1250-line HDTV system. Participants are representatives from Philips N.V. (Netherlands), Thomson S.A. (France), Robert Bosch GmbH (West Germany), Nokia Group (Finland), and Sviluppo della Televisione ad Alta Definizione Europea (Italy). The 1250/50 "Eureka" HDTV format is being evaluated as the possible future DBS transmission to the entire Western European Community.

On the other hand, because of more pressing economic and political problems, the Eastern Bloc has shown little interest in pursuing the development and implementation of a higher resolution system than their current 625-line SECAM standard.

Nevertheless, at a meeting in Rome in June 1990, government ministers agreed to embark upon the second phase of the Eureka EU95 project. This will consist of an evaluation and test of equipment for pilot HDTV services comparable to that planned by the ATSC in the United States (see Section 14.2.1). Among the leading supporters of that effort will be Philips N.V. and Thomson S.A., who signed an accord in May 1990 to launch a European HDTV service by 1995.

14.2.5 Eastman Kodak Electronic-Intermediate System

HDTV Video Postproduction In a further development of electronic techniques in film production, Eastman Kodak announced in 1989 a long-range program to develop an "Electronic-Intermediate (EI)" digital-video postproduction system. At the 1989 SMPTE Conference in Los Angeles, they introduced their concept of an HDTV system intended primarily for use by large-budget feature film producers to provide new creative dimensions for special effects without incurring the quality compromises of normal edited film masters.

Original camera negative 35-mm film will be the input to the Elec-onic-Intermediate system. Transfer of widescreen frames to a digi- frame storage will occur at a rate substantially slower than real time, probably about one frame per second. Sequences can be displayed a frame at a time for unfettered image manipulation and compositing. This is an electronic implementation of the time-standing-still milieu in which film directors and editors have been trained to exercise their creativity.

Technically, by selecting a nominal single-frame-per-second transfer rate from color negative to digital-video recording, the EI frame can have perhaps as many image samples per line as a real-time electronic postproduction system. The serial digital-transfer rate out of the film scanner for any given resolution specification is reduced by 24:1. It is of interest to note that the image-storage device in the prototype system was a Sony HDD 1125/60 digital component recorder.

Images are digitized and stored with a resolution of 10 bits per R, G, and B color, with another 10 bits assigned as the keying signal for a total of 40 bits per image sample. However, because the transfer and processing occurs in less than real time, the highest digital throughput rate in the EI system is less than that required to operate in real time in the SMPTE 240M 1125/60 Production Standard 8-bit subsampled digital component system. The processed images will be outputted at a rate up to as much as 4 frames per second. Electron-Beam Recorders (EBRs) are available to provide a 35-mm negative film without discernible artifacts.

Production Implementation Kodak is creating a consortium of manufacturers and software developers to design and produce all of the elements of the EI system. The announced application is limited strictly to the creation of artistic high-resolution special effects on film. On the other hand, the EI system has the long-range potential of developing into a means for electronic real-time distribution of theatrical films.

Acceptability of the EI system as a postproduction format will be determined by a measure of the electronic intermediate-negative film quality relative to a camera-original negative.

14.3 Industry Standardization

14.3.1 Advanced Television Systems Committee

Recommendations In the meantime, standardization activity by industry has gained momentum. The Advanced Television Systems Committee (ATSC), the U.S. body that makes recommendations to the State Department and proposes industry standards, withdrew its previous support of the SMPTE 240M 1125/60 production standard and decided instead to favor a worldwide HDTV standard based on the common image format (CIF) advocated by the CCIR.

Laboratory Tests Concurrent with the study of the various system proposals, the ATSC is equipping their Advanced Television Test Center in Alexandria, VA, to begin in late 1990 evaluating means for transmission of seven proposed formats for their suitability as the U.S. standard for VHF and UHF terrestrial broadcasting to homes. The tests are scheduled for completion by September 30, 1992. At that time an advisory committee will recommend a specific system proposal, or a combination of features of several systems, as the most desirable system. The final decision is expected to be announced by the FCC in the spring of 1993.

The formats variously are confusingly categorized by their proponents as IDTV (Improved Definition Television), ACTV (Advanced Compatible Television), EDTV (Enhanced Definition Television). Unlike the 1125/60 system, they all share a field-rate compatibility of 59.94 Hz with NTSC. All HDTV formats proposed specify a 16:9 aspect ratio.

14.3.2 Federal Communications Commission Actions

Systems Selection On March 21, 1990, the FCC announced through its chairman that it favored a technical approach in which high-definition programs are broadcast on existing 6-MHz VHF and UHF channels separate from the 6-MHz channels used for conventional (NTSC) program transmission. Although the statement does not refer to the bandwidth requirements for HDTV, the implication is that only a single channel will be allocated for transmission of an HDTV signal. It follows that this limitation to a 6-MHz channel will require the use of video-compression techniques.

In addition, it was stated that no authorization would be given (for now) for any enhanced television system so as not to detract from development of full high-definition television. A date of the spring of 1993 was suggested by the FCC as the time for a final decision on the selection of an HDTV broadcasting system. At that time, presumably a schedule will be set up for the eventual phasing out of 525-line NTSC broadcasting.

Simulcast Policy Under the simulcast policy, broadcasters would be required to transmit NTSC simultaneously in one channel of the

VHF and UHF spectra and the chosen HDTV standard in another 6-MHz television broadcast channel.

This approach is similar to that followed by Britain in their introduction of color television, which required monochrome programming to continue on VHF for about 20 years after 625/50 PAL color broadcasting on UHF was introduced. Standards converters working between 625-line PAL color and 405-line monochrome provided the program input for the *simulcast* black-and-white network transmitters. The British policy obviously benefited "old standard" receiver owners who did not wish to invest in new color receivers, and permitted program production and receiver sales for the "new standard" to develop at a rate compatible with industry capabilities. All television transmission now is on UHF with the VHF channels being reassigned to other radio services.

For the development of HDTV, the advantage of simulcasting to viewers with NTSC receivers is that they may continue to receive all television broadcasts in either the current 525-line standard or the new HDTV standard, albeit the latter without the benefit of widescreen and double resolution, but without the expense of purchasing a dual-standard receiver or a new HDTV receiver.

Presumably, although it has not been defined by the FCC, the HDTV channels would also carry the programs available only on NTSC standards. Ideally, for the viewer these programs would be converted to HDTV tranmission standards for reproduction in the narrower 1.33 aspect ratio and at the lower resolution of 525-line standards.

A less desirable solution would be to carry programs available only in NTSC standards without conversion to HDTV and require HDTV receivers to be capable of switching automatically between standards. A third choice would be not to carry NTSC-only programs on the HDTV channel and require HDTV receivers to be both HDTV/NTSC channel and format switchable. These questions must be resolved before broadcasters embark upon an HDTV equipment procurement program and the design and manufacture of receivers are undertaken.

14.4 Simulcast Transmission

14.4.1 Station and Cable System Requirements

Standards Conversion The simulcast policy would, at least for a number of years, require the operation of two transmitters by broadcasters and two station channels on cable systems: one for viewers with 525-line NTSC receivers and the other for transmissions to viewers with HDTV receivers. Assuming that home-viewer demand for HDTV programming develops, to retain a competitive position in the local market and to support local advertisement sponsors, broadcasters and cable operators will find it necessary to originate HDTV program material in addition to network program feeds. This will dictate that a standards converter will be available to down-convert the local HDTV program material for the NTSC transmitter.

Initially, at least, not all local and network programs will be on HDTV standards. If HDTV receiver designs are required by the FCC to have dual-standard capability, hopefully switched automatically, the viewers with HDTV receivers will be able to view both program formats. Although this approach by legislation is somewhat akin to the mandate that all tuners be capable of UHF reception when VHF channels were reassigned and UHF service was inaugurated, the cost of a dual-standard receiver will be higher than one capable only of HDTV operation.

On the other hand, the incremental cost for dual-standard flexibility is overshadowed by the cost of larger direct-view and projection displays,[5] which will be demanded by viewers for maximum enjoyment of high-resolution images.

Thus, if the unlikely situation exists where many receivers will have only HDTV capability, up-conversion by the broadcaster of NTSC signals to HDTV standards by broadcasters will be necessary to provide a contiguous HDTV signal transmission format with a mix of NTSC and HDTV source signals.

Station Terminal Equipment The interstudio and master control upgrading are not simple equipment replacements. Interconnection among facilities and components must be capable of handling a video bandwidth of over 30 MHz. This will necessitate in most cases replacement of existing coaxial-cable circuits with fiber optics (see Section 11.3). New dual-standard or additional HDTV picture and waveform monitoring equipment will be required.

In many instances larger console and equipment-rack space will be needed. Expansion of maintenance space and additional test equipment designed to accommodate wideband signals will be needed. Viewing rooms may require redesign to cope with the wider aspect ratio and the large-screen projection monitors more suitable for HDTV than smaller direct-view CRTs.

Cable System Terminal Equipment The basic cable service, relaying off-air broadcasting channels, is not affected by the use of a second channel for companion HDTV transmission other than the need for HDTV monitoring and test equipment. Local origination of programs and announcements may be carried on a channel authorized for NTSC transmissions and, dependent upon a favorable FCC ruling, on an HDTV channel. If the cable operator elects to originate programs on HDTV standards, the studio and head-end equipment requirements are comparable to that of a broadcasting station.

An up-converter will be required if all transmissions on HDTV channels must be on the high-definition standards. If local originations are produced on HDTV, a down-converter will be required to accommodate viewers of the NTSC channels. Simultaneous local originations on several HDTV channels would require down-converters for each. The cost may be prohibitive and thus may limit most local originations to the NTSC channels and limit the local HDTV progamming to live and taped features and news.

Local Program Production In addition to the new equipment requirements, of no minor significance is the complexity and cost of HDTV program production to accommodate the conventional 4:3 screen and the widescreen. This dictates either staging all local originations to place the significant action in the narrower area of the NTSC 4:3 aspect ratio or providing a costly pan-and-scan editing operation in the conversion from the widescreen 16:9 HDTV format to NTSC 4:3. If neither of these expedients is employed, the NTSC pictures will suffer from loss of what may be important information in the widescreen side panels. With the inevitable decline of the audience share for NTSC programming as the HDTV viewing audience increases, any limitations placed on creative staging to benefit the narrower NTSC aspect ratio will be highly undesirable, and any additional postproduction costs for a scanned NTSC transfer will be difficult to justify. VHF and UHF broadcasters appear to be faced with a Hobson's choice as 1993, the FCC's year of decision, approaches.

Network Programming The high cost of a second network feed to provide an input for the NTSC broadcast station transmitter, as well as the lack of adequate common carrier facilities, will favor the NTSC conversion being provided at local stations. Similarly, to avoid the need for a second tape of syndicated programs, the NTSC conversion can be handled most economically at the station level.

Another important consideration will be the insertion of local spots and identifications on both HDTV and NTSC channels. With NTSC conversion equipment at the station, these insertions can originate from the local program control room carrying the network feed, and there is no need for parallel NTSC sources and program assembly.

14.4.2 Broadcast Network Requirements

Facilities The additional complexity of providing an NTSC signal feed presents problems to network and cable-system operators similar to those encountered by local broadcast stations. It is unlikely that a feed of NTSC signals will be required because of the cost of a second common-carrier network transmission circuit. Instead, as noted in Section 14.4.1, an NTSC conversion probably will be performed at time-zone delay distribution points and at local stations. Nevertheless, down-conversion equipment will be necessary at network centers for a variety of viewing purposes and in "pan-and-scan" operations. In the event that NTSC-only originations are required to be carried on the HDTV channel, up-conversion facilities will be required as well.

Program Assembly Assembly of the various elements to feed a broadcast network is a complex operation involving a large number of video and audio sources. These range from live and recorded programs to commercials and spot announcements, all of which are integrated in a production control (PC) room. This will require switching facilities to handle signals in both NTSC and HDTV formats from in-house videotape equipment and off-premise sources. Commercials received in NTSC shortly before airing will require up-conversion to HDTV.

To avoid the need for a second PC room assigned to NTSC sources, NTSC recordings may have to be integrated into HDTV programs and the down-conversion for NTSC program feeds provided on each feed.

The switching of signal sources and outputs is considerably more complex when a network is split as, for example, regional networks for winter sporting events with commercials for snow tires in the north and suntan lotion in the south.

14.5 Recorded Program Distribution

14.5.1 Video Cassettes and Disks

The introduction of HDTV programming and purchase of receivers capable of either high-definition or dual-standard reproduction will create a small but immediate demand for programs recorded on HDTV videotape cassettes and disks, with initially only a slight effect on rental of NTSC recordings. Sales of recordings undoubtedly will fall off with viewers' purchase or anticipation of future purchase of HDTV receivers. Nevertheless, the inventory burden on the retail suppliers of recordings will be increased materially over the next decade as NTSC receivers are phased out and HDTV takes over.

In other words, the demand for NTSC recordings will exist over a period approaching less than that of simulcast broadcast transmission. However, as in the case of Regular-8 and Super-8 amateur motion-picture film and some of the video-cassette formats, the market will diminish to a level where the product may be discontinued.

14.6 Schedule for Initiation of United States Broadcast Service

14.6.1 Federal Communications Commission Test Schedule

The commission has announced a test schedule composed of several phases. Tests will begin in the fall of 1990 and continue for a year into 1991. Eight systems are being considered for participation in the tests. In addition to laboratory tests, field tests will be made under all types of transmission and reception conditions. The field tests will include effects between NTSC signals, an HDTV signal on an NTSC signal, and an NTSC signal on an HDTV signal. This will require a major effort on the part of industry, not unlike the NTSC color television tests, to supply equipment and engineers.

All tests are to be completed by September 30, 1992, after which an advisory committee can select either a specific system or a combination of system features for adoption as a United States broadcast standard. The final decision will be announced by the FCC in the spring of 1993.

14.6.2 Broadcasting and Cable System Preparation for HDTV

Broadcast Stations and Networks Since the final decision on selection of a transmission-system standard will not be announced until 1993, broadcasters' preparation will be limited to planning in broad terms what facilities will be required and the budget requirements. Signal generation and terminal equipment undoubtedly will have to provide baseband capabilities of at least 30 MHz bandwidth. In advance of the decision, any plant updating or expanding should provide for this increased bandwidth.

Some limited facilities will be needed for client viewing of tapes provided by production companies equipped with 1125/60 facilities. This requirement may become more urgent if HDTV receivers, VCRs, and cassettes are available in advance of adoption of a broadcast standard in the United States.

Cable Systems The FCC Rules and Regulations at present do not apply to nonbroadcast services. The possible availability of HDTV receivers, VCRs, and cassettes may create a demand for HDTV programming to be provided in a simulcast two-channel format in advance of an authorization for broadcasting. In addition, system planning is dictated for the preparations necessary to accommodate HDTV when broadcast standards are adopted.

14.7 New Markets for HDTV

The major impetus for HDTV development is to provide the home viewer with a sharper widescreen image comparable to theatrical presentation of motion pictures. Therefore, the requirements for broadcasting and recorded video have been first priority. On the other hand, the future of HDTV will depend to a large degree on the participation of many industries not concerned with entertainment. This sharing of the development and manufacturing costs among several industries unquestionably will result in a reduction in HDTV start-up costs for broadcasters and cable system operators.

Some of the closed-circuit industries having a significant need for television imaging systems with higher definition than the conventional 525- or 625-line systems are listed below.

Printing and publishing
Computer graphics
Medical profession
Biochemistry
Image and document storage and retrieval
Teleconferencing
Monitoring of manufacturing operations
Sales promotion
Flight simulation and training

Thus, the future applications of high-definition television will extend far beyond entertainment programming. A wide variety of industrial and educational fields will find HDTV a necessary technology to meet the continuing increase in the level of imaging requirements.

References

1. Rzeszewski, T. S. A Technical Assessment of Advanced Television. *Proc. IEEE* 789–803 (May 1990).
2. Jurgen, R. K. Chasing Japan in the HDTV Race. *IEEE Spectrum* 26–30 (October 1989).
3. Thorpe, L. J. The HDC-300—A Second Generation HDTV Camera. *SMPTE Journal* 364–375 (May 1990).
4. Engineering Report. Report on SMPTE Standard for Signal Parameters—1125/60 High-Definition Production System, SMPTE 240M. *SMPTE Journal* 401–402 (May 1990).
5. Mackell, M. J. Likely Costs of Consumer Advanced Television (ATV) Technology. *IEEE Trans. Consumer Electronics* 63–71 (May 1989).

Index